Neuroethics

Basic Bioethics

Arthur Caplan, series editor

A complete list of books in the Basic Bioethics series appears at the back of this book.

Neuroethics

An Introduction with Readings

edited by Martha J. Farah

The MIT Press
Cambridge, Massachusetts
London, England

For information about special quantity discounts, please email special_sales@mitpress.mit.edu

This book was set in Sabon on 3B2 by Asco Typesetters, Hong Kong.
Printed and bound in the United States of America.

Library of Congress Cataloging-in-Publication Data

Neuroethics : an introduction with readings / edited by Martha J. Farah.
p. ; cm. — (Basic bioethics)
Includes biographical references and index.
ISBN 978-0-262-06269-5 (hardcover : alk. paper) — ISBN 978-0-262-51460-6 (pbk. : alk. paper) 1. Neurosciences—Moral and ethical aspects. 2. Neurology—Moral and ethical aspects. I. Farah, Martha J. II. Series: Basic bioethics.
[DNLM: 1. Neurosciences—ethics. 2. Bioethics. 3. Brain—physiology.
WL 21 N4935 2010]
RC343.N42 2010
174′.96168—dc22

2009052778

10 9 8 7 6 5 4 3 2 1

To the communities that have stimulated and supported me:
Penn's Center for Neuroscience & Society and Center for Cognitive Neuroscience, and the Neuroethics Society

Contents

Series Foreword

I am pleased to present the twenty-eighth book in the series Basic Bioethics. The series presents innovative works in bioethics to a broad audience and introduces seminal scholarly manuscripts, state-of-the-art reference works, and textbooks. Such broad areas as the philosophy of medicine, advancing genetics and biotechnology, end-of-life care, health and social policy, and the empirical study of biomedical life are engaged.

Arthur Caplan

Preface

One morning in 2002, as I was scanning the newspaper one more time in search of a good excuse to delay getting down to work, a word popped out at me. Ironically, it was a word of great relevance to my work, lurking in the unlikeliest of places, William Safire's *New York Times* column. *Neuroethics*. I was intrigued, and so I Googled it. "Do you mean *Euroethics*?" Google helpfully prompted, detecting a possible typo. Fast forward to this morning, when Google reports 168,000 hits for neuroethics.

It is not hard to understand the explosive growth of neuroethics as a field of study. Spurred by the advent of functional magnetic resonance imaging (fMRI) in the 1990s, cognitive and affective neuroscience have finally come of age. We have learned enough about the neural bases of human thought and feeling to explain, predict, and even control some aspects of human behavior. This knowledge brings risks as well as benefits, and much of neuroethics is concerned with understanding the impact of neuroscience on society and assessing the inevitable trade-offs between risk and benefit. What makes this intellectually interesting, in addition to pragmatically important, is that the brain is the organ of the human mind. Neuroscience, more than any other branch of the life sciences, intersects with the fascinating realms of human identity, autonomy, and agency.

Since my first encounter with the word and the field, I have planned—and scrapped plans for—books of various kinds on neuroethics: a single-author monograph, a trade book, a handbook, a reader. In the end I found my way to writing this book, a kind of hybrid between a single-author book and a reader. I chose this format because it allowed me to accomplish three goals while avoiding the wasted effort of repeating in different words what other authors had already so clearly and beautifully expressed.

The first of my goals was to highlight what I consider to be the key issues of neuroethics and the inter-relations among them. Chapter 1 describes my "parse" of the subject matter of neuroethics, elaborated in the five chapters and accompanying readings that follow. There are sure to be interesting topics not included here. Furthermore, I hope that in 10 years the book's organization will be outdated, as progress leads to consensus and closure on some issues and raises new matters for debate. For the time being, though, I believe the readings of this book cover the most socially relevant and intellectually challenging issues of neuroethics. In the chapters that precede each set of readings, I have tried to clarify the relations among the issues discussed in each set of readings and across the different sets.

The second of my goals was to facilitate the teaching of neuroethics. As a new and interdisciplinary field, with neither a textbook nor a recognized canon, neuroethics may seem a daunting subject to teach. My hope is that, by organizing the material into a small number of key issues, by including reader-friendly briefings on the associated neuroscience, by highlighting the cross-cutting ethical issues in each set of readings, and by suggesting questions for discussion, I will induce at least a few instructors to try teaching a course on neuroethics. The book can also help instructors incorporate neuroethics into established courses in neuroscience, psychology, bioethics, or philosophy.

The third of my goals was to impart to readers a basic understanding of neuroscience, at least those aspects of it that relate to the neuroethical issues discussed here. "Is" and "ought" are of course distinct concerns, but they both belong in discussions of neuroethics. Ignorance of the empirical facts of neuroscience tends either to limit neuroethics to vague generalizations or to focus it on unrealistic scenarios. I hope this book will find its way into the hands of both scientists and humanists. For the former, the brief summaries of selected topics in neuroscience can serve to jog memory and to fill in gaps in knowledge of the fast-developing fields of cognitive and affective neuroscience. For the latter readers, this material can help them appreciate the scientific capabilities and constraints that shape the neuroethical issues discussed here. I also hope it will serve as a foundation for further learning about neuroscience.

In addition to filling the reader in on the relevant neuroscience, I have tried to provide some context for understanding the social and ethical issues discussed in the readings. However, I saw no point in trying to substitute my own words for those of the brilliant writers whose work is included here. I have therefore left it to them to supply the depth and nuance, as well as the varied and sometimes conflicting perspectives, that bring neuroethics to life.

My list of people to thank therefore begins with the authors of the readings included here. Their writing is instructive, insightful, and a pleasure to read. In many cases the authors allowed me to abridge their articles for the sake of focus and consistency. To the authors who were so good natured about my slicing and dicing of their prose, heartfelt thanks!

I also thank the two wonderful editors with whom I worked at MIT Press, Clay Morgan and Barbara Murphy. Clay, who stepped into the project after Barbara left MIT, was a tremendous source of information, encouragement, and gentle reminders about the march of time! Three other people who played important roles in the creation of this book are Laura Betancourt, Irena Ilieva, and Joe Boland, at the time research assistants in my lab, who obtained permissions, tracked down elusive references, and cheerfully provided all manner of clerical support for which they were sorely overqualified. Finally, I thank my colleagues, students, and staff, who make it a pleasure to come to work each day.

Neuroethics

1

Neuroethics: An Overview

Chemists can tell us how molecules interact and change according to general principles rooted in physics. No surprise there—the relation between chemistry and physics is a textbook example of intertheoretic reduction in the philosophy of science. Beginning in the mid-twentieth century, biologists began to explain the functions of cells in terms of the molecules that make them up. This has been worked out in detail for many cellular functions and in gist for the rest. Even those special cells called neurons, with their special tricks of signaling and changing connections to one another, are being explained in terms of more fundamental physical and chemical processes.

While cellular neuroscientists are steadily filling in our understanding of what neurons do and the molecular machinery by which they do it, systems neuroscientists armed with computational models are showing us how groups of these cells in combinations can do even more tricks. The behavior of large ensembles of neurons can, in turn, be studied by neuroscientists and psychologists by putting people in scanners, stimulating specific brain areas, or observing the effects of brain lesions. Perception, memory, decision making, and many other mental functions have been associated with the activity of specific sets of localized populations of neurons. At this relatively molar level of description, the brain's operations can be linked upwards to psychology as well as downwards to biology.

It is here, at this juncture between psychology and the natural sciences, that neuroethics comes in. In principle, and increasingly in practice, we can understand the human mind as part of the material world. This has profound implications for how we regard and treat ourselves and each other. It gives us powerful new ways to predict and control human behavior and a jarringly material view of ourselves. Neuroethics is the field that grapples with these developments.

1.1 A New Name for a New Field

Does the field of neuroethics really need its own name, distinct from philosophy of mind or bioethics? Newly named fields evoke skepticism and even disdain in the academic world, and some authors have questioned whether there is anything fundamentally new in neuroethics besides the name (Schick, 2005; Wilfond & Ravitsky, 2005). As academics we are shocked, shocked by attempts to market academic work, and nothing seems more like marketing than a brand name. But I have come to believe that the field is distinct enough from other established disciplines that a distinct label is warranted.

To be sure, virtually all bioethical issues concerning any organ system or medical specialty have counterparts involving the brain, neurology, and psychiatry. These issues, some of which will be discussed later in this chapter, make up part of neuroethics and could easily enough retain the label "bioethics" rather than form part of a newly designated field. They will have a familiar ring to bioethicists, and the principles and precedents of bioethics have an important contribution to make toward understanding these cases. But there is more to neuroethics than classic bioethics applied to neuroscience. New ethical issues are arising as neuroscience gives us unprecedented ways to understand the human mind and to predict, influence, and even control it. These issues lead us beyond the boundaries of bioethics into the philosophy of mind, psychology, theology, law, and neuroscience itself. It is this larger set of issues that has attracted so many new and established scholars to the area and earned it a name of its own: neuroethics.

This book is an introduction to the field of neuroethics, with an emphasis on the second type of neuroethical issue just described. Although the more familiar bioethical issues are important and invariably acquire interesting new twists when manifest in the context of neuroscience, it is the relatively newer neuroethical issues that are most in need of explication.

1.2 Understanding Neuroethics

What, specifically, do people need to know to understand these issues? Based on my experience teaching neuroethics to undergraduates and graduate students, as well as talking to people about it everywhere from professional meetings to the local dog run, I believe that one important body of knowledge is neuroscience itself. In each of the following five chapters, I have therefore tried to summarize the key ideas and findings from neuroscience that are relevant to the neuroethical issues discussed in this book. These include very brief overviews of neurotransmission and psychopharmacology, the neural bases of emotional memory and personality, principles of brain imaging, the neuropsychology of responsible behavior, and recent work on imaging consciousness in the damaged brain. I have tried to identify the most relevant parts of neuroscience for understanding the neuroethical issues of each section and the specific readings in particular. My hope is that this information will ground the reader's understanding of the neuroethical issues in real science (as opposed to vague abstractions about smart pills and science fiction scenarios about cyborgs) and might even inspire and embolden non-neuroscientist readers to learn more.

If a grasp of neuroscience is one essential component of understanding neuroethics, then an appreciation of the ethical issues is the other. By "ethical" issues I mean to include the full range of concerns regarding the impact of neuroscience on the individual human person and on society as a whole, including the moral, legal, and policy implications of that impact. My goal is not to deliver a comprehensive review of this subject matter but to offer readers a representative sample of the most interesting and well-articulated ethical issues and to give them a sense of the diversity and nuance of different perspectives on those issues.

Whereas neuroscience is largely a matter of fact, the ethical implications of neuroscience can be seen very differently by different people. For this reason, the bulk of this book is made up of the writings of others, in some cases abridged to highlight a specific neuroethical theme within the author's broader original topic. The field of neuroethics has some singular voices, and I wanted to let them speak for themselves here. There are nevertheless commonalities and unifying themes among the most opposed viewpoints presented here, and these are highlighted in the chapters that precede each set of readings.

1.3 Classic Bioethical Issues in Neuroethics

The remainder of this chapter is an overview of the many and varied issues of neuroethics, beginning with the relatively familiar or "classic" bioethical issues of neuroethics and concluding with the newer ethical challenges posed by contemporary neuroscience. I characterize some issues as classic bioethical issues because, although they involve neuroscience, the ethical issues are not fundamentally different from those arising in other branches of life science. That is, although the brain is central to these issues, from an ethical perspective its role is not substantially different from that played by other organ systems in analogous situations. These issues are no less interesting and important for having underlying commonalities with other issues in bioethics, as the examples reviewed here will show.

The development of predictive tests for incurable neurodegenerative diseases raises a variety of ethical concerns. For example, brain imaging has enabled researchers to better understand vulnerability to Alzheimer's disease, mechanisms of disease onset, and treatment response. Positron emission tomography (PET) scanning (see chapter 4), in particular, measures relevant brain function more sensitively than conventional behavioral tests of clinical dementia research. PET research has revealed neuroimaging correlates of incipient Alzheimer's disease, which in some cases may herald the clinical onset several years in advance (Scheltens & Korf, 2000). With the enthusiastic backing of PET scanner manufacturers, the medical community has been encouraged to consider using this method as a diagnostic test in the differential diagnosis of patients already showing signs of cognitive decline. In 2004, the U.S. government agreed to provide Medicare reimbursement for such scans under specific circumstances.

No one has yet proposed scanning asymptomatic elderly individuals to predict future disease or mental status, but one can imagine numerous motivations for doing so. For insurance companies, personnel departments, and even the individual himself or herself, prediction of Alzheimer's disease would allow for more rational planning for the future. The ethical question, of course, is what price this added planning capability. The knowledge that one is bound to develop Alzheimer's disease is a terrible burden, particularly as there is no cure. Although this dilemma results from recent advances in neuroscience, relevant ethical analyses have been developed by bioethicists working on the implications of genetic testing (Bell, 1998). The main ethical concerns are privacy rights (should your insurance company or boss know the test results?) and quality of life (what are the effects on patient well-being of knowing versus not knowing?). These are common to genetic and neuroimaging-based prediction.

Another important ethical issue raised by neuroscience is the safety of some of its newly developed research methods. One such method is transcranial magnetic stimulation (TMS), which alters brain function using powerful magnetic fields. It is noninvasive in the sense that the magnet remains outside the head, but the magnetic fields pass through the skull and other tissue and induce electrical currents in cortical tissue. For some applications, a single pulse (onset followed by offset of magnetic field) is used, but more commonly repetitive pulses are used (rTMS). The effects of TMS vary according to where the field is focused, its strength, and its pulse frequency and can either increase or decrease cortical activity near the stimulation site as well as in other brain regions to which the stimulated area projects.

The ability to target specific brain areas for temporary activation or deactivation makes TMS a valuable research tool, and cognitive neuroscientists have embraced it (Sack & Linden, 2003). The impressive ability of TMS to bring about scientifically informative brain changes raises the question: What other kinds of brain changes does it cause? Concern about the side effects of TMS, especially rTMS, has accompanied its use from the start. We now know that high-frequency, high-intensity rTMS can provoke seizures, even in people with no seizure history, although guidelines developed in the 1990s have succeeded in eliminating this phenomenon (Wasserman, 1997).

TMS also shows promise as a treatment modality for a variety of neuropsychiatric illnesses (Loo & Mitchell, 2005) and was approved in 2008 by the U.S. Food and Drug Administration (FDA) for the treatment of depression in specific kinds of patients. FDA regulation of medical devices is generally less stringent than regulation of drugs. This was all too apparent, in the view of many, when the FDA in 2005 approved vagal nerve stimulation as a treatment for depression based on extremely weak evidence of effectiveness. Brain stimulation with TMS and with implanted devices are among the most promising new therapeutic modalities, which lends urgency to questions of clinical trial design and the approval process for devices. Safety, efficacy, and regulatory controls on brain stimulation are neuroethical issues, as they concern the way in which society manages advances in clinical neuroscience, but their ethical, legal, and social dimensions do not differ substantially from those in the evaluation and regulation of other biotechnologies.

A more widely used application of magnetism in neuroscience is functional magnetic resonance imaging (fMRI). As will be discussed in chapter 4, this has been the workhorse of cognitive neuroscience research since the 1990s, thanks to its ability to measure brain activity with a useful degree of spatial and temporal resolution, without the need for radioactive tracers or injected contrast media. Current research involves placing the human subject in a magnetic field of strength 1.5 or 3 tesla, and all indications are that this is safe. Until recently, technical limitations prevented the use of stronger fields; they could be created only across spaces too small to accommodate a human head. However, it is now possible to scan humans at 7 tesla and higher.

Strong static magnetic fields can affect blood pressure, cardiac function, and neural activity. In addition to static fields, image acquisition with MRI involves exposure to varying magnetic fields and radiofrequency fields, which pose risks that range from activation of nerves and muscles to heating of tissue. Subjects in high-field scanners sometimes report seeing lights as a result of induced currents in their retinas and/or optic nerves. Although safety studies have suggested that such effects are benign, little is known about the long-term effects of these newer and more powerful scanning protocols.

As with TMS, high-field MRI raises important questions about the risks to which we put human research subjects. How thoroughly should such methods be tested for safety before they are used in research with humans? Who should decide? These are important ethical questions that must be addressed as researchers push the envelope of brain fMRI. However, they are not substantially different from questions regarding the safety of new methods for studying any other part of the body. Although high-field scanning is mainly of interest in the study of brain function, the ethical issues it poses are not fundamentally different from those surrounding any new scientific method that has potential risks and benefits and that is used in the study of any organ system.

Another bioethical issue that arises in connection with fMRI concerns brain abnormalities found by chance in the course of research scanning. fMRI studies generally include a nonfunctional scan of brain structure to enable localization of the brain activity revealed by fMRI relative to the anatomy of each research subject. The structural scans are of sufficient sensitivity and resolution that anatomic abnormalities and signs of disease will often show up. This raises the question of what researchers should do with these incidental findings. There is currently no universally accepted procedure for dealing with incidental findings from research scans (Illes et al., 2004). Of course, the ethical issues raised by incidental findings from brain scans are not fundamentally different from those that would be raised by imaging other organ systems. Indeed, one of the most relevant legal precedents does not come from imaging at all but from testing of blood lead levels. In 2001, a Maryland state appeals court decided that researchers studying the effects of lead abatement should have notified families of children with dangerously high levels of lead in their blood.

The issues just reviewed are the most commonly discussed "classic" bioethical issues of neuroethics, but they are not the only ones. Most bioethical issues have some intersection with neuroscience. For example, stem cell therapy has been the focus of much discussion in bioethics, and therapeutic targets include neurologic diseases such as Alzheimer's and Parkinson's diseases (Goldman, 2005). Future genetic technologies for selecting or altering the traits of a child are likely to include mental traits such as intelligence and personality, which are functions of the brain, as well as other physical traits (Chapman & Frankel, 2003). Issues of drug industry marketing, regulation, and safety are nowhere more relevant than with drugs for neuropsychiatric illness, as the chronic nature of such illnesses make treatments more profitable and questions of long-term safety more pressing (Antonuccio, Danton, & McClanahan, 2003).

1.4 New Ethical Challenges from Neuroscience

In contrast with the issues just reviewed, some neuroethical issues arise specifically because the brain is the organ of the mind. Neuroscience is giving us new, and in some instances very powerful, ways to understand people and to control their behavior. Of course, nothing is entirely without precedent if one describes it in abstract enough terms. My point here is simply that some neuroethical issues are *relatively* novel and emerge *primarily* because of the very special status of the brain in human life. These issues are the focus of this book.

One set of such issues emerges from recently developed technologies for monitoring and manipulating the brain. It remains to be seen how these developments will intersect with our strongly held beliefs about the value of privacy, freedom, fairness, and responsibility. One of the main tasks of neuroethics is to assess the likely impact of neuroscience on these and other moral and cultural ideals. This requires a realistic understanding of the capabilities of neuroscience as well as an awareness of the ways in which society already compromises one ideal for the sake of another (e.g., trading freedom for safety).

The use of psychopharmacology to change or enhance normal brain function raises a host of neuroethical issues, discussed in chapters 2 and 3 and the readings that accompany them. These issues are not hypothetical; use of prescription neuropsychiatric medications by healthy persons is at an all-time high. In addition to concerns about safety and distributive justice, which might belong in the "classic bioethical issues" category, neuropsychological enhancement raises profound questions about human effort and just deserts (did I earn my A if used Ritalin?) and personal identity (am I the same person off Prozac as on?).

Other new ethical issues arising from the application of neurotechnology include those posed by fMRI and other brain imaging methods. The main concern in these cases is not with safety or incidental findings but with privacy of thought. Unlike imaging other bodily organs, imaging the brain reveals information about the mind. Researchers have found imaging correlates of individual differences in personality and intelligence, which can be applied outside the research laboratory; for example, by employers and marketers. fMRI and other methods are being adapted for lie detection and behavior prediction, which has attracted attention from the intelligence and criminal justice communities. These trends raise new questions about whether, when, and how to ensure the privacy of one's own mind.

Of course, to the extent that functional neuroimaging is not up to the task of reliably delivering such information—and at present it is not—another problem arises: The high-tech aura of brain images leads many people to accept them uncritically. The danger is that people will be judged based on wrong information about their personalities, abilities, truthfulness, or behavioral dispositions. The neuroethics of brain imaging is the focus of chapter 4 and its accompanying readings.

Some of the most profound ethical challenges from neuroscience come not from new technologies but from new understandings. Neuroscience is calling our age-old understanding of the human person into question. In place of the folk psychology with which we have traditionally understood ourselves and each other, neuroscience is offering us increasingly detailed physical mechanisms. Personality, self-control, responsibility, consciousness, and even states of transcendent spiritual experience have become subjects of study in cognitive neuroscience. Much as the natural sciences became the dominant way of understanding the world in the eighteenth century, so neuroscience may be responsible for a kind of second enlightenment in the twenty-first century, naturalizing our understanding of humanity and transforming the way we think about ourselves. Such a transformation could help bring about a more understanding and humane society, as people's behavior is seen as part of the larger picture of causal forces surrounding them and acting through them. But it could also reduce us to machines in each other's eyes, mere clockwork devoid of moral agency and moral value.

Although many people believe that, in principle, human behavior is the physical result of a causally determined chain of biophysical events, most of us also put that aside when making moral judgments. We do not say, "But he had no choice—the laws of physics made him do it!" However, as the neuroscience of decision making and impulse control begins to offer a more detailed and specific account of the physical processes leading to irresponsible or criminal behavior, the amoral deterministic viewpoint will probably gain a stronger hold on our intuitions. Whereas the laws of physics are a little too vague and general to displace the concept of personal responsibility in our minds, our moral judgments might well be moved by a demonstration of subtle damage to prefrontal inhibitory mechanisms wrought by, for example, past drug abuse or childhood neglect. This has already happened to an extent with the *disease model* of drug abuse. The implications of neuroscience for morality in general and the law in particular are discussed in chapter 5 and the readings that follow.

Our intuitive understanding of persons includes the idea that they have an essence that persists over time. The changes wrought by normal development and life experience are understood as elaborations on a foundational personal identity that is constant throughout life. We also have the intuition that persons are categorically either alive or dead. Furthermore, most people also believe that persons have a nonmaterial component such as a spirit or soul. Yet none of these beliefs fit with the idea that a person is his or her brain. As physical objects, brains can and do change in countless ways in response to injury, disease, drugs, and, less commonly but no less realistically, implants, grafts, and other surgical interventions. There is no principled limit to the ways in which a brain can physically change and thus no immutable core to the neural substrates of a person. How can this fact be squared with the notion of an enduring personal identity or essence? As for life and death, there exists a continuum of levels of function linking the brains of fully living beings like you and me, and those of indisputably cold, dead corpses. Legal systems and religions have both grappled with the question of where to draw the line between us and those corpses, in part because any particular place is somewhat arbitrary. The standard medicolegal definition of death, which can apply to a warm, breathing body, seems counterintuitive to most. Finally, as neuroscience reveals progressively more about the physical mechanisms of personality, character, and even sense of spirituality, there is little about a human being left to attribute to an immaterial soul. The incommensurate realms of personhood and brain function, which figure indirectly in all of the neuroethical issues discussed in this book, are the focus of chapter 6 and the readings that accompany it.

References

Antonuccio, D. O., Danton, W. G., & McClanahan, T. M. (2003). Psychology in the prescription era: Building a firewall between marketing and science. *American Psychologist, 58*(12), 1028–1043.

Bell, J. (1998). The new genetics in clinical practice. *British Medical Journal, 316*, 618–620.

Chapman, A. R., & Frankel, M. S. (2003). *Designing our descendants: The promises and perils of genetic modifications*. Baltimore, MD: Johns Hopkins University Press.

Goldman, S. (2005). Stem and progenitor cell–based therapy of the human central nervous system. *Nature Biotechnology, 23*, 862–871.

Illes, J., Kirschen, M. P., Edwards, E., Stanford, L. R., Bandettini, P., Cho, M. K., et al. (2004). Incidental findings in brain imaging research. *Science, 311*(5762), 783–784.

Loo, C., & Mitchell, P. (2005). A review of the efficacy of transcranial magnetic stimulation (TMS) treatment for depression, and current and future strategies to optimize efficacy. *Journal of Affective Disorders, 88*(3), 255–267.

Sack, A. T., & Linden, D. E. (2003). Combining transcranial magnetic stimulation and functional imaging in cognitive brain research: Possibilities and limitations. *Brain Research Reviews, 43*, 41–56.

Scheltens, P., & Korf, E. S. C. (2000). Contribution of neuroimaging in the diagnosis of Alzheimer's disease and other dementias. *Current Opinion in Neurology, 13*(4), 391–396.

Schick, A. (2005). Neuro exceptionalism? *American Journal of Bioethics, 5*(2), 36–38.

Wassermann, E. M. (1997). Risk and safety of repetitive transcranial magnetic stimulation: Report and suggested guidelines from the International Workshop on the Safety of Repetitive Transcranial Magnetic Stimulation, June 5–7, 1996. *Electroencephalography and Clinical Neurophysiology, 108*, 1–16.

Wilfond, B. S., & Ravitsky, V. (2005). On the proliferation of bioethics subdisciplines: Do we really need "genethics" and "neuroethics"? *American Journal of Bioethics, 5*(2), 20.

2

Better Brains

Articles about smart pills and mood brighteners have appeared in periodicals ranging from the *New York Times* to *Seventeen* magazine. Millions of television viewers watched "desperate housewife" Lynette get hooked on her children's Adderall, and millions more saw the ambitious Jenna of *30 Rock* work around the clock with the help of an experimental drug, stopping only after the performance-enhanced research mice were found feet up in their lab cages.

Why is the public so fascinated by brain enhancement? One reason is that this neuroethical issue touches many people in their everyday lives in a way that other neuroethical issues such as brain privacy or free will do not. Anyone with a child preparing to take the SAT or a job that requires occasional all-nighters will become aware that others in the same position appear to be benefitting from prescription stimulants. Perhaps the analogy with doping in sports also intrigues us. Our feelings about brain enhancement are almost guaranteed to be both strong and ambivalent, given how unreservedly we celebrate self-improvement and individual success yet also value fairness, hard work, and "clean" (drug-free) living.

The prevalence of brain enhancement depends in part on one's definition. For example, should we count students who fail to meet diagnostic criteria for attention deficit–hyperactivity disorder (ADHD) but who have a few symptoms for which their doctor has prescribed Adderall? Even excluding such cases, epidemiologic studies suggest that use of prescription stimulants such as Adderall as a study aid is commonplace on American college campuses, with 7% of students reporting so-called nonmedical use (McCabe, Knight, Teter, & Wechsler, 2005). Less is known about other segments of the population and other drugs, but judging from reports in the popular press, it seems clear that a variety of neuropsychiatric medications are being used to improve attention, alertness, and mood by healthy people in many different walks of life

(Madrigal, 2008; Maher, 2008). Drugs under development for the treatment of elderly patients with dementia or mild cognitive impairment are expected to find an even larger market in healthy older adults (Duncan, 2008).

The readings in this section explore the ethics of brain enhancement and focus on the intersecting spheres of human values, societal pressures, and policy alternatives. We begin with a briefing on the science behind brain enhancement, aimed at demystifying some of the drug effects discussed in the readings.

Psychopharmacology Basics

The science behind current and near-term brain enhancement is psychopharmacology. In the coming decades, we may see nonpharmacologic methods of influencing brain function used for enhancing normal brains. This category includes transcranial magnetic stimulation and transcranial direct current stimulation (TMS and tDCS, Fregni et al., 2005; Reis et al., 2008), deep brain stimulation (e.g., Hamani et al., 2008), and neural prostheses (Hochberg et al., 2006; see also Perkowitz, this volume). However, for the time being these methods are too new, and in most cases too invasive, to justify nontherapeutic use.

Psychopharmacology is a science for which the "applications" often come before the basic science. There is much that we still do not know about why some currently used medications work as they do. Nevertheless, certain basic principles relating drug action and psychological effect are well established. The most basic of these principles is the role of neurotransmitters in brain function and hence psychological function.

Most people are familiar with the idea that neurons perform their functions as parts of an enormous and highly interconnected network. What earned Ramón y Cajal the 1906 Nobel Prize in Physiology or Medicine was the discovery that this network is not physically continuous. The communicating parts of every neuron are separated from those of other neurons by gaps called synapses. How does one neuron influence another without being connected? The answer is a series of chemical reactions and interactions known as neurotransmission. In the simplest case the first, or presynaptic, neuron releases neurotransmitter molecules into the synapse, where they simply diffuse around until they encounter and interact with "receptor" molecules on the surface of the second, or postsynaptic, neuron, in turn triggering other chemical processes within that neuron. The computational power of the brain is expanded by having more than one type of signal crossing the synapses: There are many different neurotransmitters, receptors, and other specific molecular actors that influence neurotransmission. Drugs affect brain function by adding to or decreasing the effects of these molecules.

The drugs discussed in this section and the next tend to affect, directly or indirectly, one or more of the following neurotransmitters: dopamine, norepinephrine, acetylcholine, and serotonin. The first two are involved in attention, the third in memory, and the fourth is known primarily for its role in mood and personality. This is an extreme oversimplification in that each is essential for multiple different psychological abilities and states. Nevertheless, it will do for the current purpose, which is to sketch the mechanisms of action for some well-known neuropsychiatric drugs that have crossed over into enhancement use.

Dopamine and norepinephrine neurons are targets of stimulant drugs such as Adderall (mixed amphetamine salts) and Ritalin (methylphenidate) and play a role in attention and related psychological factors such as subjective energy, wakefulness, and motivation. Adderall improves these functions by increasing the rate at which dopamine and norepinephrine are released from neurons. Ritalin intervenes in the brain's neurotransmitter "recycling" system, decreasing the rate at which the neurotransmitter molecules are taken out of the synapse and returned to the presynaptic neuron, a mechanism known as *reuptake inhibition*. In either case, the result is more neurotransmitter available in the synapses between neurons.

Both Adderall and Ritalin are classified as stimulant drugs in that their effects on dopamine and other neurotransmitters have the effect of increasing people's arousal, energy level, and attentional focus, the latter due in part to their effects on prefrontal cortical function. They are widely used for the treatment of ADHD. More relevant to the readings that follow, they are also increasingly used to enhance schoolwork by healthy normal college students, especially in the United States (McCabe et al., 2005).

Acetylcholine plays a crucial role in learning and memory, and the degeneration of acetylcholine-producing neurons is responsible for the memory loss that accompanies Alzheimer's disease. Aricept (donepezil) is a drug that increases acetylcholine levels by interfering with an enzyme that normally breaks down acetylcholine and is thus modestly helpful in staving off the cognitive decline of early Alzheimer's disease. It has also been explored as a cognitive enhancer in research with healthy young and middle-aged research subjects, who show subtly but reliably enhanced learning with the drug (Grön, Kirstein, Thielscher, Riepe, & Spitzer, 2005; Yesavage et al., 2002).

Serotonin is implicated in mood, temperament, and personality, and drugs that influence serotonin levels include antidepressants. Like Ritalin, Prozac (fluoxetine) and other antidepressants of the selective serotonin reuptake inhibitor (SSRI) class increase neurotransmitter levels by decreasing reuptake. However, whereas attention improves measurably in the hours after a single oral dose of one of the stimulants, mood improves only after weeks of continual SSRI administration, despite the immediate effect that a dose of SSRI has on serotonin activity. The reason for this is a question for research at present, a reminder that scientific understanding often lags behind therapeutic success in psychopharmacology.

Of course, the neurotransmitters mentioned above are not only influenced by prescription drugs. A wide variety of naturally occurring substances are enjoyed for their effects on brain chemistry. For example, the nicotine in tobacco directly boosts acetylcholine activity. The caffeine in coffee and tea acts on adenosine receptors, which in turn affect dopamine activity. Physiologically, there is no categorical difference between such natural stimulants and prescription stimulants, although there are certainly differences in potency. There are also numerous differences in society's attitudes toward, and legal restrictions on, the use of these substances. This brings us to the social and ethical context of brain enhancement and the issues addressed by the readings.

Neuroethics of Brain Enhancement

The Readings

The neuroethics of brain enhancement encompasses many different issues, perhaps because brain enhancement touches so many different spheres of society, from physicians to schoolchildren to pharmaceutical companies. Each of the readings that follow engages some subset of these issues.

Flower provides an overview of the social and ethical context of enhancement in general, including brain enhancement. Although his tone is clearly wary of this growing trend, he reminds us that the ethical terrain is complex, crisscrossed by distinctions among medical, dietary, and street drugs, legal and illegal uses, and the evolving roles of physician and patient, making it unrealistic to attempt any blanket recommendations for or against enhancement. *Farah et al.* review some of the evidence that cognition can be enhanced by drugs and then survey the main social and ethical problems raised by this practice, classifying them into four categories of concern: safety, coercion, distributive justice, and "personhood and intangible values." *Diller* takes a historical and cultural approach to the emergence of attention enhancement in the 1990s. He describes various societal factors that set the stage for skyrocketing rates of stimulant use, especially by children, including changing expectations of children, new pressures on doctors and their patients, and the growing acceptance of lifestyle pharmacology. The *President's Council on Bioethics* (under the leadership of Leon Kass) points out that, even if concerns about safety, freedom, and fairness could be resolved, most of us would still feel uncomfortable with brain enhancement. The question they address is why. This is an important question because, as they point out, such discomfort is not necessarily an indication that brain enhancement is wrong. We must attempt to analyze the reaction and discover whether any important values or principles are violated by the practice of enhancement. They identify four ways in which brain enhancement clashes with certain beliefs about humanity and its relation to the world. Finally, *Greely et al.* take us back to the realm of pragmatic considerations, specifically how society can and should respond to the growing trend toward brain enhancement. In this article, the authors (including myself) accept the premise that brain enhancement is inevitable and, at least in some ways, beneficial, and we sketch out a number of different policy mechanisms that could be used to maximize the benefit to humanity while minimizing the harm.

Selected Cross-cutting Issues

A number of important themes recur within this group of readings. One is the blurry line between illness and health and therefore between treatments that are therapeutic and treatments that are enhancements. Flower cites and then dismisses several criteria for illness and health. He concludes that the current "forensic" approach to the distinction between therapeutic and nontherapeutic drug use is misguided because there is nothing intrinsic to a drug or drug effect that determines, independent of cultural context, its therapeutic or nontherapeutic nature. Diller notes that the evolving diagnostic criteria for ADHD have led to increases in the number of individuals counted as having this disorder, providing a very concrete illustration of Flower's point that society plays an important role in drawing the line between illness and health or therapy and enhancement.

Pondering the moral status of brain enhancement also leads to more general questions about the nature and purpose of medicine and its relation to other aspects of human life. The concept of *medicalization* is almost inescapable in discussions of brain enhancement and is discussed at length by Flower. Although reclassification of a problem from nonmedical to medical is not an intrinsically bad thing (as in, for example, the recognition that epilepsy is not a form of possession but a neurologic condition), in most discussions of contemporary phenomena, medicalization is taken to be problematic. It suggests that the "patient" has relinquished his or her own role in solving the problem and turned it over to the doctor, becoming the passive recipient of a "treatment." To the extent that the problem to be solved, or wish to be fulfilled, is the sort of thing that humans could be expected to address with their own agency, as the President's Council on Bioethics would say, this diminishes that agency and that humanity. The dehumanizing and disempowering potential of brain enhancement is also noted in Farah and colleagues' discussion of "personhood and intangible values" and implicitly, but powerfully, in Flowers' closing reflections on *Brave New World*.

One response to the concern about medicalization is to reject the disease model of enhancement altogether. After all, we do not convince ourselves that we have a caffeine deficiency when we order coffee; we just want a lift and know that we can get it at Starbucks. Perhaps the alternative to medicalizing our lifestyle problems is to "barista-ize" our doctors. The idea that physicians might become lifestyle enhancement specialists (and from there, suggests Flower, social engineers) may not sit well with all of them or with the insurers who pay for their services but seems inevitable as long as the trend toward brain enhancement continues and involves prescription drugs (see readings by Farah et al. and Greely et al.).

Economics figures in the enhancement debate in several ways. As Flower points out, economic factors alone are likely to position insurers and national health systems against brain enhancement. In contrast, drug companies will generally favor this trend, which so greatly expands their potential market. Greely et al. note that employers will be motivated to encourage brain enhancements that improve productivity. Similarly, Diller points out that economically driven shifts toward families in which both parents work, classrooms with more children and fewer teachers, and physicians who cannot take the time to explore psychosocial influences and treatments all contribute to the appeal of pharmacologic approaches to the challenges of childhood. Diller also points out that the question of long-term efficacy and safety for stimulants remains unanswered because of the high cost of long-term studies.

Questions for Discussion

1. Each of the authors in this section mentions or discusses the safety of brain enhancement. What are some of the distinctive problems associated with evaluating and regulating the safety of enhancement compared with evaluating and regulating the safety of therapy? Consider how the safety of therapies are evaluated and regulated and adapt these methods to enhancements. Broaden the category of safety concerns to include all manner of negative effects on individuals and society, not just health side effects, and again consider the question of how one could realistically assess and manage the safety of brain enhancement.
2. The caffeine in coffee and the nicotine in cigarettes enhance mood and cognition. To what extent do the concerns expressed by the President's Council on Bioethics apply to the use of these substances? What does your answer imply for the validity of the arguments of the President's Council on Bioethics? Are there morally relevant differences between dietary, over-the-counter and prescription enhancers, which make some permissible and others not? What are they?
3. Explore different perspectives on the practice of brain enhancement by role playing. Groups of three can enact the following situations: (a) An executive is exasperated with her secretary's inattention to details and disorganization and as a last resort asks the secretary to try a cognitive enhancer. The secretary is reluctant and talks it over with her husband and then meets again with her boss. (b) A middle-school student is performing below his potential and finds school boring, although he does not have a learning disability or medical condition. His teacher meets with the mother to recommend trying a stimulant-like nutritional supplement, making a case that the supplement will help her son. The mother then explains the idea to her son, asks him how he feels about it, and encourages him to try the supplement. (c) A military pilot is embarking on a long and dangerous mission and is ordered by his superior to use a powerful new cognitive enhancer, the long-term effects of which are unknown. He considers refusing and discusses what to do with his navigator, who has been given the option of using the enhancer but is not required to.

References

Duncan, D.E. (2008). The ultimate cure. *Portfolio*. May 12 2008, http://cms.preview.portfolio.com/news-markets/national-news/portfolio/2008/05/12/Analysis-of-Neurotech-Industry/.

Fregni, F., Boggio, P., Mansur, C., Wagner, T., Ferreira, M., Lima, M.C., et al. Transcranial direct current stimulation of the unaffected hemisphere in stroke patients. *Neuroreport*, *16*, 1551–1555.

Grön, G., Kirstein, M., Thielscher, A., Riepe, M.W., & Spitzer, M. (2005). Cholinergic enhancement of episodic memory in healthy young adults. *Psychopharmacology 182*, 170–179.

Hamani, C., McAndrews, M.P., Cohn, M., Oh, M., Zumsteg, D., Shapiro, C.M., et al. (2008). Memory enhancement induced by hypothalamic/fornix deep brain stimulation. *Annals of Neurology*, *63*, 119–123.

Hochberg, L.R., Serruya, M.D., Friehs, G.M., Mukand, J.A., Salen, M., Caplan, R., et al. (2006). Neuronal ensemble control of prosthetic devices by a human with tetraplegia. *Nature*, *442*, 164–171.

Madrigal, A. (2008). Wired readers' brain-enhancing drug regimens. *Wired*. April 24, 2008, http://www.wired.com/medtech/drugs/news/2008/04/smart_drugs.

Maher, B. (2008). Poll results: Look who's doping. *Nature*, *452*, 674–675.

McCabe, S.E., Knight, J.R., Teter, C.J., & Wechsler, H. (2005). Non-medical use of prescription stimulants among US college students: Prevalence and correlates from a national survey. *Addiction*, *100*, 96–106.

Reis, J., Robertson, E.M., Krakauer, J.W., Rothwell, J., Marshall, L., Gerloff, C., et al. (2008). Consensus: Can transcranial direct current stimulation and transcranial magnetic stimulation enhance motor learning and memory formation? *Brain Stimulation*, *1*(4), 363–369.

Yesavage, J.A., Mumenthaler, M.S., Taylor, J.L., Friedman, L., O'Hara, R., Sheikh, J., et al. (2002). Donepezil and flight simulator performance: Effects on retention of complex skills. *Neurology*, *59*, 123–125.

Reading 2.1

Lifestyle Drugs: Pharmacology and the Social Agenda[1]

Rod Flower

In everyday usage, the noun *lifestyle* is a word that everyone recognizes and understands, but in the grammar of pharmacology it has acquired a new, adjectival role to describe the properties of an eclectic group of drugs. What, then, are these substances? It is not as easy as you might think to answer this question, but let us start our quest for information in the conventional places. Your trip to the library will probably be fruitless because the term is absent from most reference works; use the term as a keyword to search PubMed, and you will retrieve few publications before the late 1990s. Even today, there is not a huge academic literature on the subject. However, this contrasts sharply with the profuse and often shrill discussion of lifestyle drugs by other communities. No doubt stimulated by the fact that the worldwide lifestyle drug market is currently worth US$20 billion and rising to an estimated US$29 billion by 2007 (Atkinson, 2002), financial analysts have produced a series of weighty (and often expensive) reports on the topic (Bogdanovic & Langlands, 2001; Research and Markets Ltd., 2001). But, above all, it is the media that has really espoused the subject, and it is to them that one must turn to sample the full range (not to mention the deviation and error) of debate and opinion.

Despite all this coverage, however, one soon comes to realize that there is no agreed idea of what actually constitutes a lifestyle drug. Is it, for example, identical to a "lifestyle medicine"? Perhaps the most commonly accepted definition of the latter, as articulated by Gilbert et al. (2000), is a medicine that is used to satisfy a non-health-related goal or is used for treating problems that lie at the margins of health and well-being. Examples that are usually quoted include sildenafil for the treatment of male erectile dysfunction (and, latterly, female sexual dysfunction) and minoxidil or finasteride for the treatment of baldness. By extension, this definition also includes medicines that treat conditions that might be better addressed by changes in lifestyle such as bupropion (to mitigate the nicotine withdrawal syndrome) or antiobesity agents such as orlistat. According to a slightly different interpretation (Gilbert, 1999), lifestyle medicines are used to treat "lifestyle illnesses," these being defined in turn as diseases arising through "lifestyle choices." Smoking is the example *par excellence* of a lifestyle choice leading to lifestyle illness but there are many others (e.g., alcoholism and obesity, among others).

Here, we must distinguish between the all-embracing term "drug" and the rather constrained definition of a "medicine" with its implied therapeutic benefit, because many drugs that are not medicines also fall into the "lifestyle" category. Recognizing this, Young has proposed a more comprehensive classification (Young, 2003) that also incorporates unlicensed "over-the-counter" preparations in addition to recreational drugs. Table 2.1.1 is an attempt at a classification that, although drawing upon the work of earlier authors (Gilbert, 1999; Gilbert et al., 2000; Young, 2003), is not identical to previous schemes. This is not intended to be the last word on the issue and indeed there is no shortage of further definitions. For example, Moffat has described lifestyle drugs as those "taken for pleasure" (Moffat, 2000), whereas other commentators believe: "They are attractive to the popular media; they enhance the quality of life; often address problems of a social or cosmetic nature; and are not conducive to reimbursement" (Kole et al., 2003).

2.1.1 Who Is the "Patient"?

We are accustomed to the notion that the end recipient of a therapeutic transaction is a "patient," but in the case of lifestyle medicines the situation is less straightforward. Are the consumers of these drugs actually "sick" or do they simply have unfulfilled needs or desires? When does a "need" become an "illness" or when does an "aspiration" become a "legitimate" therapeutic goal? And, for that matter, what is an illness? It is a "slippery" term, as Smith pointed out in a recent survey of "non-diseases" in the *British Medical Journal* (Smith, 2002), but then "health" is also a term that is difficult to define. The concept that the latter can be assessed simply in terms of a deviation from the "norm" is flawed because we are all unique and therefore potentially "abnormal" in one way or another. The World Health Organization (WHO) definition of health as "complete physical, psychological and social well-being" is all-embracing and, as has been notably observed (I. Loeffler, quoted in

Smith, 2002), is only likely to be achieved at the moment of simultaneous orgasm.

An illustration might help to explain. In the *British Medical Journal* survey, baldness was number 6 in the top 20 "non-diseases" but one can envisage a case where premature baldness might contribute to psychological problems that could be avoided by the use of a drug that promotes the regrowth of hair. In such a case, this drug might be said to have a therapeutic purpose whereas in others the replacement of hair would be considered a question of personal choice or vanity. Lexchin (2001) quotes another example, that of paroxetine, marketed by SmithKline Beecham in 2000 for the treatment of "social phobia" using the advertising slogan "Imagine being allergic to people." This is a genuine psychiatric condition but, as Lexchin points out, the prescriber's dilemma is that this diagnosis could, with little further extrapolation, be taken to include shyness that is experienced by virtually everyone at some stage in their lives. In other words, we risk turning natural expressions of human behavior into a "disease" that requires—or would benefit from—drug treatment. In such cases, the availability of a "drug" converts a personality trait into an "illness." As Moynihan puts it: "Ever narrowing definitions of 'normal' help turn the complaints of the healthy into the conditions of the sick" (Moynihan, 2003).

This phenomenon, often referred to as *medicalization*, is clearly here to stay. With a growing understanding of the genetic contribution to diseases and other human characteristics, many more of mankind's woes are likely to be perceived by the lay public as being beyond personal control and thus a legitimate target for drug therapy. The gathering pace of pharmaceutical innovation will undoubtedly lead to the discovery of more agents with potential lifestyle applications. Pharmaceutical companies are often demonized for catering to these markets, but society happily endorses the manufacture of all manner of other goods tailored to our lifestyles, so why not pharmaceuticals, too? Indeed, it has been suggested by *Business Week* that the boom in lifestyle drugs "could turn the pharmaceutical industry into an engine of growth for the entire (US) economy" (Webber et al., 1998).

2.1.2 Who Prescribes Them?

Fuelling medicalization is the democratization of medical information brought about by increasing Internet access, disease support groups, patient consumerism, and the direct advertising by the pharmaceutical

Table 2.1.1
What are "lifestyle" drugs and medicines?[a]

Category	Examples[b]	Primary Clinical Use	"Lifestyle" Use
Medicines that have been approved for specific indications and can also be used to satisfy "lifestyle choices" or to treat "lifestyle diseases"	Sildenafil	Erectile dysfunction	Erectile dysfunction
	Oral contraceptives	Preventing conception	Preventing conception
	Orlistat	Obesity	Weight loss
	Sibutramine	Anorectic agent	Weight loss
	Bupropion	Managing nicotine addiction	Managing nicotine addiction
	Methadone	Managing opiate addiction	Opiate substitute
Medicines that have been approved for specific indications but can also be used for other "lifestyle" purposes	Minoxidil	Hypertension	Regrowth of hair
	Finasteride	Prostatic hypertrophy	Regrowth of hair
	Erythropoietin	Chronic anaemia	Enhancing athletic performance
Drugs that have a slight clinical utility but fall mainly into the "lifestyle" category	Alcohol	None as such	Widespread as a component of drinks
	Caffeine	Migraine treatment	Widespread as a component of drinks
	Cannabis	Possibly managing chronic pain	"Recreational" usage
Drugs (generally "illegal") that have no clinical utility but are used to satisfy lifestyle requirements	MDMA ("ecstasy")	None as such	"Recreational" usage
	"Designer steroids"	None as such	Enhancing athletic performance
	Cocaine (some formulations)	None as such	"Recreational" usage

Natural products that are largely unregulated but are associated with many (often anectdotal and unsubstantiated) claims about their action or safety and often cater to lifestyle needs or desires	Fish oils	Slight—perhaps as nutritional supplements	Widespread, for many conditions
	Ascorbic acid	Slight—perhaps as nutritional supplements	Widespread, for many conditions
	Various herbal and other preparations	None	Widespread, for many conditions

[a] An attempt at classification based on the work of Gilbert et al. (2000) and Young (2003).
[b] Abbreviation: MDMA, 3,4-methylenedioxymethamphetamine.

industry to the public, which occurs in some countries. It is characteristic of lifestyle drugs that they encourage self-diagnosis and (where possible) self-prescription, a feature (no doubt gleefully) noted by financial analysts when assessing future market strength (Bogdanovic & Langlands, 2001). Perhaps alerted to a new product by a news item, a mirror is the only diagnostic instrument you will need to reveal that you might benefit from antiobesity treatment (Boseley, 2001), a drug to remove that bald patch or to recolor those unwanted gray streaks (Atkinson, 2002; Etienne, Cony-Makhoul, & Mahon, 2002). An ultimatum from your bank manager might persuade you to take citalopram to curb your urge to shop (Templeton, 2002). A barbed (and obviously unjustified) comment from your partner might send you clicking your way to the online pharmacy houses where sildenafil reportedly retails at £8 to £10 per tablet (Cochrane, 2003). Predictions for the value of the online sales of prescription drugs are assessed at US$15 billion for 2004 and, as investigative journalists have discovered, it is remarkably easy to obtain virtually any drug in this way (Barnett, 2003).

The mere fact that a drug is available often tends to edge out other forms of treatment (Everitt, Avorn, & Baker, 1990) that might actually prove more beneficial. The elderly could be at particular risk (Walley, 2002). As the range of lifestyle drugs increases, we risk forcing physicians to act as social engineers or, at least, arbiters of the social norm. Lexchin (2001) again: "Suppose there was a pill that could make everyone's skin colour exactly the same. If everyone took the medication, discrimination against skin colour would certainly be eliminated."

2.1.3 Are They a New Phenomenon?

If lifestyle drugs resist easy categorization, then we have to look not to the drugs themselves but to our culture to provide the explanation. Our forebears did not distinguish between therapeutic and nontherapeutic objectives in the forensic way that we do. You do not need to be an ethnopharmacologist to know that drugs have been used throughout history for hunting, for cosmetic purposes, and even to further our spiritual aspirations (e.g., the shamanic use of hallucinogenic drugs such as mescaline).

In many instances, the lifestyle drugs of earlier cultures have been "reinvented" as mainstream therapies in today's society. For example, the calabar bean *Physostigma venosum*, which contains eserine, was originally used in West Africa as an "ordeal poison" to determine guilt or innocence; subsequently, its anticholinesterase properties were found to be useful in ophthalmology as a treatment for myasthenia gravis and even as an antidote to some nerve agents. Atropine from *Atropa belladonna* dilates pupils, supposedly making women more attractive, and was thus employed for lifestyle purposes long before its use by Geiger and Hess in the 1830s as an agent to reduce airway secretions. Coca, the "divine plant of the Incas," was also originally used for lifestyle purposes: It "satisfies the hungry, gives new strength to the weary and exhausted and makes the unhappy forget their sorrows," so said Garcilaso de la Vega in 1609. This was long before Alfred Niemann purified "cocaine" in the mid-nineteenth century and Freud, with Koller, began a series of experiments with the drug that culminated in its introduction both as a local anesthetic and, less happily, as a putative treatment for morphine addiction (Weatherall, 1990). Curare and opium are among many other examples and, in the ongoing trials of cannabis for the treatment of neuropathic pain (Young, 2002), one can see the same process of "pharmaceutical emancipation" occurring today.

2.1.4 Who Pays?

Whether you rely on private medical insurance or a social medicine system to underwrite your treatment costs, you can depend on the fact that your provider is worried about the lifestyle drug "revolution" (Gilbert, 1999). When sildenafil was first launched, the *Daily Telegraph* suggested that it could cost the National Health System (NHS) in the United Kingdom an extra £1.25 billion (Hall, 1998). Although in reality the NHS apparently "overspent" by only £5 million to £8 million (Butler, 2000a), there is legitimate concern that a system devised primarily to provide money for the treatment of illness should also be used to pay for drugs for which there is no compelling clinical argument. In the United Kingdom, decisions about the prescription of drugs have been traditionally delegated to physicians and the cost of these reimbursed through the NHS. Some lifestyle drugs, such as the contraceptive pill, are already assimilated into the UK health care budget under these historical conventions, but the culture has now changed. The UK government, sensing an impending problem following the introduction of sildenafil, was quick to intervene (Winyard, 1998) and ultimately restricted its use to a few indications. A similar move followed the introduction of orlistat, although it still costs the NHS a reported £12 million per annum (Boseley, 2001). In the United Kingdom, the National Institute of Clinical Excellence

(NICE), which advises on use of drugs within the NHS, is likely to be faced with increasingly difficult choices in connection with lifestyle medicines and has set up a "citizens council," one of whose tasks is reportedly to advise on the availability of these drugs (Reuters Health Information, 2002).

Health care budgets are, of course, not inexhaustible. Money spent on one drug must usually be saved elsewhere, and one priority must be dropped in favor of another. Rationing is generally considered to be the best way forward, although this has had a mixed success, at least in the United Kingdom (Gilbert, 2000). The whole issue is political dynamite with policies and opinions being divided on both party and factional lines (Butler, 2000b).

Naturally, not everyone agrees that lifestyle drugs should be withheld at all, with some commentators criticizing "the puritan tone" of regulatory decisions and maintaining that "the 'chemicalisation' of happiness, achievable as a controlled event in the body" is not necessarily a bad thing (Kane, 1998). Perhaps they have a valid point, but this argument blurs when it is refocused on the closely related debate on recreational drugs. It is characteristic of our double vision when viewing such issues that the use of recreational drugs other than those that have already been sanctioned by our society (such as alcohol) is frowned upon automatically. Morality is a cultural construct, variable with time and place, but the pharmacology of drugs is not. Hard science tells us that there would be a heavy social price to pay for a complete liberalization of "recreational drugs" but also that there are some instances where legalization could be a reasonable option with probably less damaging consequences than are associated with those recreational drugs that are already "accepted" by society. As Moffat has pointed out in this context (Moffat, 2000), alcohol abuse costs the United Kingdom ~£3.3 billion per annum plus an additional £200 million spent in the NHS on treatment—and these figures are likely to be on the conservative side. It is a stormy debate that will probably rumble on for many years. The recent confusion between MDMA (3,4-methylenedioxymethamphetamine) and methamphetamine in a major neurotoxicity trial (Knight, 2003) will do little to reassure an already bewildered public that even we know what we are doing.

2.1.5 Who Bears the Risk?

Another problem arising from lifestyle drug usage is how to balance the benefit–risk equation. No drug has a single effect, and these secondary actions range from the mildly inconvenient to the frankly dangerous. According to the subtle calculus of human suffering, there is generally some sort of equivalence between the acceptable burden of side-effects and the severity of the illness we are attempting to treat. But when "healthy" people take drugs, who decides what level of side-effects is acceptable, and who bears the cost of treating any ensuing iatrogenic disease? We are again on treacherous ground, but a clue can be found in our attitudes to other "risky" activities. We deplore the number of deaths on our roads, but who has given up motoring? We object to the placing of mobile phone aerials near housing because of a "radiation risk," but nobody would be without their own phone, despite the much greater potential for microwave-induced damage. And what about smoking? I could go on, but the bottom line seems to be that if the intrinsic value of a product to the individual is sufficiently high, and the risk is voluntary, then most people will accept it willingly.

2.1.6 What of the Future?

Perhaps it is time to deconstruct the entire concept and admit that there really is no such thing as a lifestyle drug—only a lifestyle "use." Drugs are just chemical substances that affect our bodies and our minds and that can be put to many different uses, some obviously therapeutic, some more recreational, and some directed toward satisfying other human aspirations. Some drugs might fulfill more than one function depending on culture, context, or formulation.

There is no doubt that lifestyle drugs are here to stay, but how will they shape our society in the future? We can turn to fiction for a suggestion. In *Brave New World*, Huxley (1932) presents us with a vision of a dystopian society where every embryo is preconditioned by exposure to drugs and chemicals and where, as adults, the inhabitants of this world have free access to a range of useful pharmaceutics. For example, a flagging libido can be shored up with "sex-hormone chewing gum," and any ensuing prospect of viviparous conception and birth can be prevented by contraceptives carried by women in their "Malthusian belts." Everyday nutritional deficiencies are easily rectified with "pan-glandular biscuits" or "vitaminized beef-surrogate."

Extraordinary as these ideas must have seemed in 1932 when he wrote the novel, it is a tribute to Huxley's prescience that so many are commonplace to us today. But the pharmacologic jewel in the crown of his imaginary society—and surely the ultimate lifestyle drug—was *Soma*. No doubt developed by "two thousand pharmacologists and biochemists," mentioned in the novel, this was a substance elevated to almost sacramental status: "...there's always *Soma* to give you a holiday from the facts. And there's always *Soma* to calm your anger, to reconcile to your enemies, to make you patient and long suffering," the Controller explains, "Anybody can be virtuous now. You can carry at least half your morality about in a bottle." With the era of lifestyle drugs now seemingly well established, Huxley's vision of a *pax pharmacologica* no longer seems fantastic or even remote. Drug regulators beware!

Notes

1. Editor's note: This reading originally appeared in 2004 in *Trends in Pharmacological Sciences*, volume 25, pages 182–185, and is used with permission. The author is a Principal Fellow of The Wellcome Trust. He would like to thank Lucy Wilkinson for help in searching through the relevant media sites. All websites were accessed during October and November 2003.

References

Atkinson, T. (2002). Lifestyle drug market booming. *Nature Medicine, 8*, 909.

Barnett, A. (2003, 10 August). Deadly cost of the trade in online prescription drugs. *The Observer*, p. 3.

Bogdanovic, S., & Langlands, B. (2001). *Lifestyle drugs patient-initiated prescribing*. Retrieved from http://www.biophoenix.com.

Boseley, S. (2001, 10 March). Anti-obesity drug given licence for NHS use. *The Guardian*. http://www.guardian.co.uk/uk/2001/mar/10/sarahboseley

Butler, P. (2000a, 11 December). NHS overspent on impotence drugs. *The Guardian*. http://www.guardian.co.uk/society/2000/dec/11/health

Butler, P. (2000b, 12 December). Don't let the hypocrites decide. *The Guardian*. http://www.guardian.co.uk/society/2000/dec/12/health.comment

Cochrane, K. (2003, 31 August). Thrusting young men. The new Viagra addicts. *The Times*. http://www.timesonline.co.uk/tol/news/article1154716.ece

Etienne, G., Cony-Makhoul, P., & Mahon, F. X. (2002). Imatinib mesylate and gray hair. *New England Journal of Medicine, 347*(6), 446.

Everitt, D. E., Avorn, J., & Baker, M. W. (1990). Clinical decision-making in the evaluation and treatment of insomnia. *American Journal of Medicine, 89*, 357–362.

Gilbert, D. (1999). *Lifestyle drugs: Who will pay?* London: PJB Publications.

Gilbert, D., Walley, T., & New, B. (2000). Lifestyle medicines. *British Medical Journal, 321*, 1341–1344.

Hall, C. (1998, 8 July). Viagra abuse "will add £1bn to NHS bill." *Daily Telegraph*.

Huxley, A. (1932). *Brave new world*. London: Chatto and Windus.

Kane, P. (1998, 24 September). Lifestyle drugs. *The Herald*.

Knight, J. (2003). Agony for researchers as mix-up forces retraction of ecstasy study. *Nature, 425*, 109.

Kole, P. L., Bhusari, S., Thakurdesai, P. A., & Nagappa, A. N. (2003). *Lifestyle drugs: New avenues*. Retrieved from www.ExpressHealthcareMgmt.com.

Lexchin, J. (2001). Lifestyle drugs: Issues for debate. *Canadian Medical Association Journal, 164*, 1449–1451.

Moffat, A. (2000). Do we need drugs for pleasure? *Pharmaceutical Journal, 265*, 373.

Moynihan, R. (2003). The making of a disease: female sexual dysfunction. *British Medical Journal, 326*, 45–47.

Research and Markets Ltd. (1999). *Lifestyle drugs outlook to 2005*. Retrieved from http://www.researchandmarkets.com.

Reuters Health Information. (2002). UK health agency asks for 'citizen council' for input. Retrieved from http://www.reutershealth.com.

Smith, R. (2002). In search of "non-disease." *British Medical Journal, 324*, 883–885.

Templeton, S. (2002, 22 July). Pill to cure shopping disorder fuels fears over lifestyle drugs. *Sunday Herald*.

Walley, T. (2002). Lifestyle medicines and the elderly. *Drugs Aging, 19*, 163–168.

Weatherall, M. (1990). *In search of a cure*. Oxford: Oxford University Press.

Webber, J., Barrett, A., Mandel, M., & Laderman, J. (1998, 11 May). The new era of lifestyle drugs. *Business Week*, pp. 92–98.

Winyard, G. (1998). Prescribing of sildenafil (Viagra). Retrieved from http://www.doh.gov.uk.

Young, E. (2002, 5 November). Cannabis drugs pass testing "milestone." *New Scientist*. http://www.newscientist.com/article/dn3011-cannabis-drugs-pass-testing-milestone.html

Young, S. N. (2003). Lifestyle drugs, mood, behaviour and cognition. *Journal of Psychiatry & Neuroscience, 28*, 87–89.

Reading 2.2

Neurocognitive Enhancement: What Can We Do and What Should We Do?[1]

Martha J. Farah, Judy Illes, Robert Cook-Deegan, Howard Gardner, Eric Kandel, Patricia King, Eric Parens, Barbara Sahakian, and Paul Root Wolpe

Many are predicting that the twenty-first century will be the century of neuroscience. Humanity's ability to alter its own brain function might well shape history as powerfully as the development of metallurgy in the Iron Age, mechanization in the Industrial Revolution, or genetics in the second half of the twentieth century. This possibility calls for an examination of the benefits and dangers of neuroscience-based technology, or *neurotechnology*, and consideration of whether, when, and how society might intervene to limit its uses.

At the turn of the century, neurotechnology spans a wide range of methods and stages of development. Brain–machine interfaces that allow direct two-way interaction between neural tissue and electronic transducers remain in the "proof of concept" stage but show substantial promise (Donogue, 2002). Neurosurgery is increasingly considered as a treatment for mental illnesses and an array of new procedures are under development, including the implantation of devices and tissue (Malhi & Sachdev, 2002). Noninvasive transcranial magnetic stimulation (TMS) of targeted brain areas is the basis of promising new treatments for depression and other psychopathology (George & Belmaker, 2000).

On the leading edge of neurotechnology is psychopharmacology. Our ability to achieve specific psychological changes by targeted neurochemical interventions, which began through a process of serendipity and trial and error in the mid-twentieth century, is evolving into the science of rational drug design. The psychopharmacopoeia of the early twenty-first century encompasses both familiar, and in some cases highly effective, drugs, and a new generation of more selective drugs that target the specific molecular events that underlie cognition and emotion (Barondes, 2003). For the most part, these drugs are used to treat neurologic and psychiatric illnesses, and there is relatively little controversy surrounding this use. However, psychopharmacology is also increasingly used for "enhancement"; that is, for improving the psychological function of individuals who are not ill.

The enhancement of normal neurocognitive function by pharmacologic means is already a fact of life for many people in our society, from elementary school children to ageing baby boomers. In some school districts in the United States, the proportion of boys taking methylphenidate exceeds the highest estimates of the prevalence of attention deficit–hyperactivity disorder (ADHD) (Diller, 1996), implying that normal childhood boisterousness and distractibility are being targeted for pharmacologic intervention. The use of prescription stimulants (such as methylphenidate and dextroamphetamine) as study aids by high school and college students who do not have ADHD has recently drawn attention and might include as many as 16% of the students on some campuses (Babcock & Byme, 2000). Sales of nutritional supplements that promise improved memory in middle age and beyond have reached a billion dollars annually in the United States alone, despite mixed evidence of effectiveness (Gold, Cahill, & Wenk, 2002). In contrast with the other neurotechnologies mentioned earlier, whose potential use for enhancement is still hypothetical, pharmacologic enhancement has already begun.

2.2.1 What Can We Do?

Many aspects of psychological function are potential targets for pharmacologic enhancement, including memory, executive function, mood, appetite, libido, and sleep (Farah, 2002; Farah & Wolpe, 2004). We will use the first two of these, memory and executive function, as examples to show the state of the art in psychopharmaceutical enhancement, the ethical issues raised by such enhancement, and the policy implications of these ethical issues. A brief review of the state of the art in neurocognitive enhancement is offered here; additional information is freely available to readers of this article at www.nyas.org/ebrief/neuroethics and in recent articles by Rose (2002), Lynch (2002), and Hall (2003).

2.2.1.1 Memory Enhancement

Memory enhancement is of interest primarily to older adults. The ability to encode new memories declines measurably from the third decade of life onwards, and by the fourth decade the decline can become noticeable and bothersome to normal healthy individuals (Craik & Salthouse, 1992). Memory difficulties in middle or old age are not necessarily a harbinger of future dementia but can be part of the normal pattern of cognitive ageing, which does not make it any less inconvenient when we misplace our glasses or forget the name of a recent acquaintance. What can current and imminent neurotechnologies offer us by way of help?

The changes that underlie normal age-related declines in memory probably differ from those that underlie Alzheimer's disease, indicating that the optimal pharmacologic approaches to therapy and enhancement might also differ. Although donepezil, a cholinesterase inhibitor that is used to treat Alzheimer's disease, did enhance performance in one study of healthy middle-aged pilots after flight simulator training (Yesavage et al., 2002), drug companies are looking elsewhere for pharmacologic approaches to memory enhancement in normal individuals. Recent advances in the molecular biology of memory have presented drug designers with many entry points through which to influence the specific processes of memory formation, potentially redressing the changes that underlie both normal and pathologic declines in memory. Most of the candidate drugs fall into one of two categories: those that target the initial induction of long-term potentiation and those that target the later stages of memory consolidation. In the first category are drugs that modulate AMPA (α-amino-3-hydroxy-5-methyl-4-isoxazole propionic acid) receptors to facilitate depolarization, including Cortex Pharmaceuticals' Ampakines (Lynch, 2002). In the second category are drugs that increase CREB (the cAMP response element-binding protein), a molecule that in turn activates genes to produce proteins that strengthen the synapse. One such drug is the molecule MEM1414, which is being tested by Memory Pharmaceuticals (Hall, 2003) (a company co-founded by one of the authors [E.K]).

The pursuit of mastery over our own memories includes erasing undesirable memories as well as retaining desirable ones. Traumatic events can cause lifelong suffering by the intrusive memories of post-traumatic stress disorder (PTSD), and methods are being sought to prevent the consolidation of such memories by pharmacologic intervention immediately after the trauma (Pittman et al., 2002). Drugs whose primary purpose is to block memories are also being developed by the pharmaceutical industry (Hall, 2003). Extending these methods beyond the victims of trauma, to anyone who wishes to avoid remembering an unpleasant event, is another way in which the neural bases of memory could be altered to enhance normal function.

2.2.1.2 Enhancement of Executive Function

Executive function refers to abilities that enable flexible, task-appropriate responses in the face of irrelevant competing inputs or more habitual but inappropriate response patterns. These include the overlapping constructs of attention, working memory, and inhibitory control. Drugs that target the dopamine and noradrenaline neurotransmitter systems are effective at improving deficient executive function, for example in ADHD, and have recently been shown to improve normal executive function as well (Elliott et al., 1997; Mehta et al., 2000).

For example, one of the authors (B.J.S.) found that healthy young volunteers performed the Tower of London problem-solving task more accurately after being given methylphenidate than after being given a placebo when the task was novel (Mehta et al., 2000). Methylphenidate also increased accuracy in a complex spatial working memory task, and this was accompanied by a reduction in the activation of areas of the brain that are related to working memory, as shown by positron emission tomography (PET) (Elliott et al., 1997). For the latter task, the amount of benefit was inversely proportional to the volunteers' working memory capacity as assessed by a different working memory task, digit span, with little or no benefit to those with the highest digit span performances. This is of interest in discussions of enhancement, because it indicates that, for this medication and this cognitive ability at least, those with lower levels of performance are more likely to benefit from enhancement than those with higher levels. Indeed, it is possible that some drugs would compress the normal range of performance in both directions. One of the authors (M.J.F.) found that the dopamine agonist bromocriptine improved performance on various executive function tasks for individuals with lower-than-average working memory capacity but lowered the performance of those with the highest working memory capacities (Kimberg, D'Esposito, & Farah, 1997). Whether enhancement can boost the performance of already high-performing individuals must be determined empirically for each drug and for each type of cognitive ability.

Newer drugs might improve executive function in different ways, influencing different underlying processes and interacting in different ways with individual differences (for example, in working memory capacity) and states (such as restedness). The newest potential neurocognitive enhancer is the drug modafinil, which is approved for the treatment of narcolepsy and is increasingly prescribed off-label for other purposes (Teitelman, 2001). One of the authors (B.J.S.) found that it increases performance among healthy young adults on a set of executive function tasks that differs partly from those that are influenced by methylphenidate, with its effects resulting at least in part from an improved ability to inhibit impulsive responses (Turner et al., 2003).

2.2.2 What Should We Do?

2.2.2.1 Ethical Problems and Policy Solutions

Neurocognitive enhancement raises ethical issues for many different constituencies. These include academic and industry scientists who are developing enhancers and physicians who will be the gatekeepers to them, at least initially. Also included are individuals who must choose to use or not to use neurocognitive enhancers themselves and parents who must choose to give them or not to give them to their children. With the advent of widespread neurocognitive enhancement, employers and educators will also face new challenges in the management and evaluation of people who might be unenhanced or enhanced (for example, decisions to recommend enhancement, to prefer natural over enhanced performance or vice versa, and to request disclosure of enhancement). Regulatory agencies might find their responsibilities expanding into considerations of "lifestyle" benefits and the definition of acceptable risk in exchange for such benefits. Finally, legislators and the public will need to decide whether current regulatory frameworks are adequate for the regulation of neurocognitive enhancement or whether new laws must be written and new agencies commissioned.

To focus our discussion, we will dispense with some ethical issues that are important but not specific to neurocognitive enhancement. The first such issue is research ethics. Research on neurocognitive enhancement, as opposed to therapy, raises special considerations mainly insofar as the potential benefits can be viewed as smaller, and acceptable levels of risk to research subjects would be accordingly lower. This consideration is largely academic for those neurocognitive enhancers that come to market first as therapies for recognized medical conditions, which includes all of the substances that are now available for enhancement, although this might not be true in the future. Another important ethical issue concerns the use of neurocognitive enhancement in the criminal justice system, in which a large proportion of offenders fall in the lower range of cognitive ability in general (Holland, Clare, & Mukhopadhyay, 2002) and executive inhibitory control in particular (Brower & Price, 2001). Although neurocognitive enhancement brings with it the potential for subtle coercion in the office or classroom, "neurocorrection" is more explicitly coercive and raises special issues of privacy and liberty that will not be discussed here. Finally, the ethical problems that are involved in parental decision-making on behalf of minor children are complex and enter into the ethics of neurocognitive enhancement in schoolchildren but will not be discussed here.

The remaining issues can be classified and enumerated in various ways. Four general categories will be used here to organize our discussion of the ethical challenges of neurocognitive enhancement and possible societal responses.

2.2.2.2 Safety

The idea of neurocognitive enhancement evokes unease in many people, and one source of the unease is concern about safety. Safety is a concern with all medications and procedures, but our tolerance for risk is smallest when the treatment is purely elective. Furthermore, in comparison with other comparably elective treatments such as cosmetic surgery, neurocognitive enhancement involves intervening in a far more complex system, and we are therefore at greater risk of unanticipated problems. Would endowing learners with super-memory interfere with their ability to understand what they have learned and relate it to other knowledge? Might today's Ritalin users face an old age of premature cognitive decline? The possibility of hidden costs of neurocognitive enhancement might be especially salient because of our mistrust of unearned rewards and the sense that such opportunities can have Faustian results.

With any drug, whether for therapy or enhancement, we can never be absolutely certain about the potential for subtle, rare, or long-term side effects. Instead, our regulatory agencies determine what constitutes a sufficiently careful search for side effects and what side effects are acceptable in view of a drug's benefits. Although consensus will have to be developed on these issues in connection with neurocognitive enhancement, we see no reason that the same approach cannot be applied here.

2.2.2.3 Coercion

If neurocognitive enhancement becomes widespread, there will inevitably be situations in which people are pressured to enhance their cognitive abilities. Employers will recognize the benefits of a more attentive and less forgetful workforce; teachers will find enhanced pupils more receptive to learning. What if keeping one's job or remaining in one's school depends on practicing neurocognitive enhancement? Such dilemmas are difficult but are not without useful legal precedent. Many of the relevant issues have been addressed in legislation such as Connecticut's statute

"Policies regarding the recommendation of psychotropic drugs by school personnel" (Legislative Commissioners' Office, 2003) and case law such as *Valerie v. Derry Cooperative School District* (United States District Court, 1991).

Of course, coercion need not be explicit. Merely competing against enhanced co-workers or students exerts an incentive to use neurocognitive enhancement, and it is harder to identify any existing legal framework for protecting people against such incentives to compete. But would we even want to? The straightforward legislative approach of outlawing or restricting the use of neurocognitive enhancement in the workplace or in school is itself also coercive. It denies people the freedom to practice a safe means of self-improvement, just to eliminate any negative consequences of the (freely taken) choice not to enhance.

2.2.2.4 Distributive Justice

It is likely that neurocognitive enhancement, like most other things, will not be fairly distributed. Ritalin use by normal healthy people is highest among college students, an overwhelmingly middle-class and privileged segment of the population. There will undoubtedly be cost barriers to legal neurocognitive enhancement and possibly social barriers as well for certain groups. Such barriers could compound the disadvantages that are already faced by people of low socioeconomic status in education and employment. Of course, our society is already full of such inequities, and few would restrict advances in health or quality of life because of the potential for inequitable distribution. Unequal access is generally not grounds for prohibiting neurocognitive enhancement, any more than it is grounds for prohibiting other types of enhancement, such as private tutoring or cosmetic surgery, that are enjoyed mainly by the wealthy. Indeed, in principle there is no reason that neurocognitive enhancement could not help to equalize opportunity in our society. In comparison with other forms of enhancement that contribute to gaps in socioeconomic achievement, from good nutrition to high-quality schools, neurocognitive enhancement could prove easier to distribute equitably.

2.2.2.5 Personhood and Intangible Values

Enhancing psychological function by brain intervention is in some ways like improving a car's performance by making adjustments to the engine. In both cases the goal is to improve function, and to the extent that we succeed without compromising safety, freedom of choice, or fairness, we can view the result as good. But in other ways the two are very different, because modifying brains, unlike engines, affects persons. The fourth category of ethical issue encompasses the many ways in which neurocognitive enhancement intersects with our understanding of what it means to be a person, to be healthy and whole, to do meaningful work, and to value human life in all its imperfection. The recent report of the President's Council on Bioethics (Kass, 2003) emphasizes these issues in its discussion of enhancement.

Attempts to derive policies from these considerations must contend with the contradictory ways in which different values are both challenged and affirmed by neurocognitive enhancement. For example, we generally view self-improvement as a laudable goal. At the same time, improving our natural endowments for traits such as attention span runs the risk of commodifying them. We generally encourage innovations that save time and effort, because they enable us to be more productive and to direct our efforts toward potentially more worthy goals. However, when we improve our productivity by taking a pill, we might also be undermining the value and dignity of hard work, medicalizing human effort, and pathologizing a normal attention span. The self-transformation that we effect by neurocognitive intervention can be seen as self-actualizing or as eroding our personal identity. Neither the benefits nor the dangers of neurocognitive enhancement are trivial.

In weighing the dangers of neurocognitive enhancement against its benefits, it is important to note the many ways in which similar trade-offs are already present in our society. For example, the commodification of human talent is not unique to Ritalin-enhanced executive ability. It is probably more baldly on display in books and classes that are designed to prepare preschoolers for precocious reading, music, or foreign language skills, but many loving parents seek out such enrichment for their children. Americans admire the effort that was expended in Abraham Lincoln's legendary four-mile walk to school every day, but no one would do that (or want their child to do that) if a bus ride were available. Medicalization has accompanied many improvements in human life, including improved nutrition and family planning. And if we are not the same person on Ritalin as off, neither are we the same person after a glass of wine as before, or on vacation as before an exam. As these examples show, many of our "lifestyle" decisions end up on the right side of one value and the wrong side of another, but this does not necessarily mean that these decisions are wrong.

2.2.2.6 Disentangling Moral Principle and Empirical Fact

Since pre-Socratic times, philosophers have sought ways of systematizing our ethical intuitions to identify a set of guiding principles that could be applied in any situation to dictate the right course of action. All of us have ethical intuitions about most situations; one goal of ethics is to replace case-by-case intuitions with principled decisions. A practical social advantage of ethical principles is that they can provide guidance when intuitions are unclear or inconsistent from person to person. The success of an ethical discussion depends on the discussants' ability to articulate the relevant principles as well as the relevant facts about a situation to which the principles apply.

In the ethics of neurocognitive enhancement, we are still feeling our way toward the relevant principles, and we still have much to learn about the relevant facts. Is it a matter of principle that "medicalization" is bad or that hard work confers "dignity"? Or are these moral heuristics, rules of thumb that might be contradicted in some cases? And is it a matter of fact that Ritalin reduces our opportunities to learn self-discipline, or could it in fact have no effect or even help us in some way? Until we have disentangled the *a priori* from the empirical claims and evaluated the empirical claims more thoroughly, we are at risk of making wrong choices.

2.2.2.7 When Not to Decide Is to Decide

Neurocognitive enhancement is already a fact of life for many people. Market demand, as measured by sales of nutritional supplements that promise cognitive enhancement, and ongoing progress in psychopharmacology portend a growing number of people practicing neurocognitive enhancement in the coming years. In terms of policy, we will soon reach the point where not to decide is to decide. Continuing our current *laissez-faire* approach, with individuals relying on their physicians or illegal suppliers for neurocognitive enhancement, risks running afoul of public opinion, drug laws, and physicians' codes of ethics. The question is therefore not whether we need policies to govern neurocognitive enhancement but rather what kind of policies we need. The choices range from minimal measures, such as raising public awareness of the potential practical and moral difficulties of neurocognitive enhancement, to the wholesale enacting of new laws and the creation of new regulatory agencies. In between these extremes lie a host of other options, for example the inclusion of neurocognitive enhancement policies in codes of ethics of the professional organizations of physicians, scientists, human resource managers, and educators, and short-term moratoria on neurocognitive enhancement.

Francis Fukuyama (2002) has argued for new legislation to control the use of neurocognitive enhancement, among other biotechnologies. He characterizes the work of groups such as the President's Council on Bioethics in the United States and the European Group on Ethics in Science and New Technology as the "intellectual spade work of thinking through the moral and social implications of biomedical research" and suggests that "it is time to move from thinking to acting, from recommending to legislating. We need institutions with real enforcement powers."

We admit to being less certain about the right course of action. With respect to the first three categories of issue, concerning safety, freedom, and fairness, current laws and customs already go a long way toward protecting society. With respect to the fourth category of issue, we believe that there is much more "spade work" (in Fukuyama's words) to be done in sorting out the moral and social implications of neurocognitive enhancement before we move from recommendations to legislation. We should draw an object lesson from the history of federal stem cell legislation in the United States, which was enacted hastily in the wake of reported attempts at human reproductive cloning with limited public understanding of the issues. That legislation is now viewed by many as a setback for responsible biomedical research, and two states have now enacted their own laws to permit a wider range of research activity.

The need for more discussion of the issues is a predictable conclusion for an article like this one but nevertheless a valid conclusion. One urgent topic for discussion is the role of physicians in neurocognitive enhancement (Chatterjee, 2004). Although Western medicine has traditionally focused on therapy rather than enhancement, exceptions are well established. Cosmetic surgery is the most obvious example, but dermatology, sports medicine, and fertility treatments also include enhancement among their goals. Enabling a young woman to bank her eggs to allow later childbearing, for example, is not therapeutic but enhancing. Will neurocognitive enhancement join these practices? If so, will it be provided by specialists or family practitioners? What responsibility will physicians take for the social and psychological impact of the enhancements they prescribe, and by what means (for example, informal or formal psychological screening as used by cosmetic surgeons or fertility specialists)?

Beyond these immediate practical issues, we must clarify the intangible ethical issues that apply to neurocognitive enhancement. This requires interdisciplinary discussion, with neuroscientists available to identify the factual assumptions that are implicit in the arguments for and against different positions and ethicists available to articulate the fundamental moral principles that apply. As a society, we are far from understanding the facts and identifying the relevant principles. With many of our college students already using stimulants to enhance executive function and the pharmaceutical industry soon to be offering an array of new memory-enhancing drugs, the time to begin this discussion is now.

Notes

1. Editor's note: This reading originally appeared in 2004 in *Nature Reviews Neuroscience,* volume 5, pages 421–425, and is used with permission. This paper is based, in part, on a meeting held at the New York Academy of Sciences in June 2003, supported by a grant to J.I. from the National Science Foundation with co-sponsorship of a Mushett Family Foundation grant to the Academy. The writing of this paper was supported by NSF and NIH grants to M.J.F. and an NIH grant and a Greenwald Foundation grant to J.I.

References

Babcock, Q., & Byrne, T. (2000). Student perceptions of methylphenidate abuse at a public liberal arts college. *Journal of American College Health, 49,* 143–145.

Barondes, S. (2003). *Better than Prozac: Creating the next generation of psychiatric drugs.* New York: Oxford University Press.

Brower, M. C., & Price, B. H. (2001). Neuropsychiatry of frontal lobe dysfunction in violent and criminal behaviour: A critical review. *Journal of Neurolology, Neurosurgery and Psychiatry, 71,* 720–726.

Chatterjee, A. (2004). Cosmetic neurology: The controversy over enhancing movement, mentation and mood. *Neurology, 63,* 968–974.

Craik, F. I. M., & Salthouse, T. A. (1992). *The handbook of aging and cognition.* Hillsdale, NJ: Lawrence Erlbaum.

Diller, L. H. (1996). The run on Ritalin. Attention deficit disorder and stimulant treatment in the 1990s. *Hastings Center Report, 26,* 12–18.

Donogue, J. (2002). Connecting cortex to machines: Recent advances in brain interfaces. *Nature Neuroscience, 5*(Suppl), 1085–1088.

Elliott, R., Sahakian, B. J., Matthews, K., Bannerjea, A., Rimmer, J., & Robbins, T. W. (1997). Effects of methylphenidate on spatial working memory and planning in healthy young adults. *Psychopharmacology, 131,* 196–206.

Farah, M. (2002). Emerging ethical issues in neuroscience. *Nature Neuroscience, 5,* 1123–1129.

Farah, M. J., & Wolpe, P. R. (2004). Monitoring and manipulating brain function: New neuroscience technologies and their ethical implications. *Hastings Center Report, 34,* 35–45.

Fukuyama, F. (2002). *Our posthuman future.* New York: Farrar, Strauss and Giroux.

George, M. S., & Belmaker, R. H. (2000). *Transcranial magnetic stimulation in neuropsychiatry.* Washington, DC: American Psychiatric Press.

Gold, P. E., Cahill, L., & Wenk, G. L. (2002). Ginkgo biloba: A cognitive enhancer? *Psychological Science in the Public Interest, 3*, 2–11.

Hall, S. S. (2003, September). The quest for a smart pill. *Scientific American, 289*(3), 54–65.

Holland, T., Clare, I. C., & Mukhopadhyay, T. (2002). Prevalence of criminal offending by men and women with intellectual disability and the characteristics of offenders. *Journal of Intellectual Disabilities Research, 46*(Suppl), 6–20.

Kass, L. (2003). *Beyond therapy—Biotechnology and the pursuit of happiness.* New York: HarperCollins.

Kimberg, D. Y., D'Esposito, M., & Farah, M. J. (1997). Effects of bromocriptine on human subjects depend on working memory capacity. *Neuroreport, 8*, 3581–3585.

Legislative Commissioners' Office. (1 January 2003). *General Statutes of Connecticut.* Title 10, Ch. 169, Sect. 10–212b.

Lynch, G. (2002). Memory enhancement: the search for mechanism-based drugs. *Nature Neuroscience, 5*, 1035–1038.

Malhi, G. S., & Sachdev, P. (2002). Novel physical treatments for the management of neuropsychiatric disorders. *Journal of Psychosomatic Research, 53*, 709–719.

Mehta, M. A., Owen, A. M., Sahakian, B. J., Mavaddat, N., Pickard, J. D., & Robbins, T. W. (2000). Methylphenidate enhances working memory by modulating discrete frontal and parietal lobe regions in the human brain. *Journal of Neuroscience, 20*, RC65.

Pittman, R. K., Sanders, K. M., Zusman, R. M., Healy, A. R., Cheema, F., & Lasko, N. B. (2002). Pilot study of secondary prevention of posttraumatic stress disorder with propranolol. *Biological Psychiatry, 15*, 189–192.

Rose, S. P. R. (2002). "Smart drugs": Do they work? Are they ethical? Will they be legal? *Nature Reviews Neuroscience, 3*, 975–979.

Teitelman, E. (2001). Off-label uses of modafinil. *American Journal of Psychiatry, 158*, 1341.

Turner, D. C., Robbins, T. W., Clark, L., Aron, A. R., Dowson, J., & Sahakian, B. J. (2003). Cognitive enhancing effects of modafinil in healthy volunteers. *Psychopharmacology (Berl.), 165*, 260–269.

United States District Court, New Hampshire (1 August 1991). Case No. C-88-412-L.

Yesavage, J. A., Mumenthaler, M. S., Taylor, J. L., Friedman, L., O'Hara, R., & Sheikh, J. (2002). Donepezil and flight simulator performance: Effects on retention of complex skills. *Neurology, 59*, 123–125.

Reading 2.3

The Run on Ritalin: Attention Deficit Disorder and Stimulant Treatment in the 1990s[1]

Lawrence H. Diller

Stimulants were first reported as a pharmacologic treatment for children's behavioral problems in 1937 (Bradley, 1937). Methylphenidate, a derivative of piperidine, was synthesized in the 1940s and marketed as Ritalin in the 1960s (Greenhill, 1992). It is structurally related to the older drug still used for the treatment of hyperactivity, *d*-amphetamine. Their pharmacologic actions are essentially the same.

Stimulant treatment for children became more common in the 1960s when its short-term benefits for what was then called hyperactivity were documented in controlled trials. In 1970, it was estimated that 150,000 children were taking stimulant medication in the United States (Gadow, 1981).

A furor over stimulants began in 1970. The reaction stemmed from an article in the popular press charging that 10% of the children in the Omaha school district in Nebraska were being medicated with Ritalin (Maynard, 1970). While ultimately shown to contain inaccuracies, the article spurred other reports of "mind control" over children and led to congressional hearings about stimulants that same year (Gallagher, 1970). Numerous articles in newspapers and magazines and one book attacked Ritalin and the "myth" of the hyperactive child (Rogers, 1971) Subsequently it was found that some of the criticism appeared to be led by supporters of the Scientology movement, who have consistently challenged mainstream psychiatry's use of psychoactive medications (Sappell & Welkos, 1990). Yet the negative publicity struck a nerve with the general public, which by the mid-1970s made it quite difficult to convince parents and teachers in many communities to attempt a trial of Ritalin.

The U.S. Drug Enforcement Agency (DEA) began monitoring the amounts of methylphenidate and amphetamine produced in this country in 1971. Both became Schedule II controlled drugs partly in response to an epidemic of methylphenidate abuse occurring in Sweden and the illegal use of stimulants in this country. Estimates on the number of children using stimulants have varied widely. In 1980 it was estimated that from 270,000 to 541,000 elementary school children were receiving stimulants (Gadow, 1981). In 1987 a national estimate of 750,000 children was made (Safer & Krager, 1988). Both estimates were guesses extrapolated from local surveys.

More precise than national estimates of children taking stimulants are the records of production quotas maintained by the DEA that show a steady output of approximately 1700 kg of legal methylphenidate through the 1980s followed by a sharp increase in production in 1991 (Office of Public Affairs, 1995). From 1990 through May 1995, the annual U.S. production of methylphenidate has increased by 500% to 10,410 kg (Federal Register, 1995), "an increase rarely seen for any other Schedule II Controlled Substance," according to the DEA (Office of Public Affairs, 1993). A national survey of physicians' diagnoses and practices based upon data collected in 1993 found that of the 1.8 million persons receiving medication for attention deficit–hyperactivity disorder, 1.3 million were taking methylphenidate (Swanson, Lemer, & Williams, 1995). A comparison of 1993 Ritalin production with the latest figures available for 1995 suggests that 2.6 million people currently are taking Ritalin, the vast majority of who are children ages 5 through 12.

Who is taking all of this Ritalin, and why? To get at the answers to these questions, we need to look at changes in professional and lay attitudes regarding psychoactive drugs, the brain, and children's behavior. Six hypotheses are suggested to explain the sudden increase in the demand for this drug.

2.3.1 Changes in Diagnostic Criteria

As more children's behavior is viewed as abnormal, more treatment is offered. The American Psychiatric Association distinguishes deviancy from normalcy in its *Diagnostic and Statistical Manual of Mental Disorders* (*DSM*). With the introduction of the *DSM–III* in 1980, mainstream psychiatry officially changed its view from a diagnosis of hyperactivity, highlighting physical movement, to one where problems with attention, attention deficit disorder (ADD), were of primary concern. This change reflected research that suggested the primary problem for children was one of focus and distractibility. Hyperactivity, as a reflection of motoric impulsivity, was still important but not critical to the diagnosis. Thus, one could meet the criteria for ADD without being overly motorically active at all. The name of the condition was changed again in *DSM–III–R* to attention deficit–hyperactivity disorder, and in the *DSM–IV* (1994) separate subtypes of inattention and hyperactivity/ impulsivity were restored.

There have been additional interpretative changes to the diagnosis. One need not demonstrate symptoms in every situation. Rather, one need only display symptoms in at least two environments. Similarly one may concentrate satisfactorily at a number of tasks, perhaps even overfocus, yet still meet criteria for diagnosis if concentration and focus are problems for important tasks (for example, "selective inattention or attentional inconsistency").

The changes in diagnostic criteria and interpretation have greatly broadened the group of children *and adults* who might qualify for the diagnosis. The line between children with "normal" variations of temperament, lively or spontaneous children who are sensitive to stimuli, and those who have a "disorder" has become increasingly blurred. The *sine qua non* for the diagnosis of hyperactivity in the mid-1970s was a demonstration of motoric overactivity and/or distractibility in nearly all settings including the doctor's office. Some children who may have benefited from identification and treatment were undoubtedly missed under these criteria, but this is less likely today. Now children who sit quietly and perform well in social situations or in one-on-one psychometric testing can still be candidates for the diagnosis and treatment of attention deficit–hyperactivity disorder (ADHD) if their parents or teachers report poor performance in completing tasks at school or at home.

Prior to *DSM–III*, etiologic factors were important in the diagnosis of psychiatric disorders. Since 1980 diagnosis has been descriptive, based primarily on observed behavior and self-report. While the multiaxial codings of the *DSM–III* presumably account for medical factors and social stressors on the patient, less emphasis is placed on psychosocial influences, such as family, school, or work environments. In addition, the ascendancy of biological psychiatry, with its emphasis on the genetic and neurochemical factors directing behavior, implicitly diminishes the significance of development, learning disabilities, emotional status, family interaction, classroom size, and other environmental factors that may be relevant. Meeting ADHD criteria, which strictly speaking involves demonstrating a group of behaviors, has come to mean "having" ADHD, a neurologic condition, such as Pervasive Developmental Disorder or Tourette syndrome. Research purported to support a biological basis for ADHD, a brain scan of the cerebral cortex, or a survey of family epidemiology cannot conclusively distinguish between biological or environmental etiologies (Ross & Pam, 1995).

Environmental factors can be seen either as contributing to the etiology or maintaining the symptomatic behavior. Indeed, a strongly stated case for neurologic factors has been useful to counterbalance beliefs that such behaviors were attributable to lazy children and disorganized adults. However, if the symptoms of ADHD are to be viewed within the biopsychosocial model, calling ADHD a neurologic disorder can mislead some into discounting psychosocial factors as unimportant.

2.3.2 The "Lean and Mean" 1990s

As professional viewpoints have changed, so too have societal pressures and public attitudes toward attention and behavior problems in children and adults. Over the past two decades, the pressure on children to perform has increased while support needed to help maximize performance has declined. Twenty-five years ago, 3- and 4-year-old children were not expected to know the alphabet and numbers. Community programs like Head Start and television shows such as *Sesame Street*, while benefiting millions, have also led to expectations that children can learn at an earlier age. Yet over the concurrent past 20 years, poverty rates for children, as a measure of their general well-being, have increased from 15% to 20% nationwide, and children comprise 40% of all those who live in poverty (Roberts, 1994).

More families are requiring two incomes to maintain their standard of living, and the increasing number of women in the workforce has led to large-scale preschool enrollment of children, requiring that younger children adhere to a more organized and less flexible social structure. Many children adapt easily to preschool and thrive in that group environment. Yet some children are not developmentally or socially ready for preacademic learning and a more demanding social structure. These children, had they stayed at home, would not be exposed to community scrutiny or come to the attention of teachers and physicians at an early age. At age 3 or 4, their behavior may qualify them for an ADHD diagnosis.

At the elementary school level, funding pressures on school systems have led to increased classroom sizes and higher student-to-teacher ratios. Also, more stringent criteria exist to qualify for special education services, which are often inadequately funded.

Similar conditions exist for high school and college students, especially in public education. The pressure to do well academically is immense.

Inexorable pressures have developed to maintain a high grade point average in order to gain entry into a "good" college or graduate school. "Cs" have become unacceptable to many middle class families. This trend can be seen in the popularity of bolstering SAT scores through extracurricular preparatory classes. Increasing attention through the use of medication may be seen as just another method to improve performance and results.

A declining standard of living undergirds many of these conditions. With corporations "downsizing," there is greater competition for fewer choice positions. In the current climate of job insecurity, the perception is "Perform or else!" The human gene pool cannot change for cultural or economic reasons in 25 or 30 years. Thus relatively greater numbers of children and adults may be found wanting in their abilities to concentrate given the current pressures of their academic and work environments.

2.3.3 Pressures on Physicians and Educators

Physicians are also under pressure. Even before the managed care era, the time and economic constraints on the primary care physician were great. When presented with a potentially complex child behavioral problem, the physician may be attracted to the option of prescribing a medication rather than addressing the thornier and more time-consuming issues of emotions, family relationships, or school environment. Even with genuine concern for a multimodal evaluation and treatment plan, often little else is done on the primary care level (Wolraich et al., 1990).

Specialists, such as behavioral–developmental pediatricians and child psychiatrists, should be capable of spending more time and lending greater expertise to the resolution of the intricacies of the child, family, and school situation. These specialists are concerned, however, that the cost-containment measures of managed care will increasingly permit referrals only when medication is being considered for the child.

Increasing pupil–teacher ratios and diminishing special education services also have an effect. These conditions make it "easier" to medicate a child than to work with a dysfunctional family, decrease the size of the classroom, or augment funding for special education services. Because stimulants "work" more quickly, they are more attractive not only to families and physicians but also to managed care companies and financially strapped educational systems. It is unlikely that either would insist on medication in lieu of counseling or special education services, but neither would protest if medication allowed the child to function better without either service.

2.3.4 The Disability Issue

Society increasingly has interpreted performance problems as disease, which then become defined as a disability. People with defined disabilities cannot legally face discrimination and are entitled to the benefits of special services. The increasing numbers of children and adults who meet the broader ADHD criteria are beginning to have an impact in the classroom and workplace. Parents find the only way to get extra help for their children is to have them labeled with a disorder. The Individuals with Disabilities Education Act of 1990 and recent interpretations of Section 504 of the 1973 Rehabilitation Act have become broad and potent legal tools for families of children with ADHD seeking special services from their school districts (Reid & Katsiyannis, 1995). Unfortunately, funding of the legislation has been fragmented and never fully appropriated by Congress. State educational systems have been left to provide funding for these entitlements. Court actions have forced one school district to scramble and allocate funds disproportionately to services for the disabled. In New York City, where nearly a quarter of the school budget is spent on 13% of the students categorized as disabled, the situation has reached crisis proportions. With nearly six times as much money spent on the special student, spending on regular students is among the lowest in the nation (Dillon, 1994). Whether this skimping on regular classrooms paradoxically will push more children struggling at the edge of "normalcy" to meet criteria for a "disorder" is speculative but provocative and disturbing.

Categorizing ADHD as a disability has created other dilemmas. Typically someone with a disability is provided special circumstances or allowances for optimal performance, for example, more time provided in a college entrance examination. However, at least at one large university, ADHD diagnosis has become so common that a doctor's diagnosis is no longer sufficient as the basis for special consideration (H. Beck, personal communication). Instead, the school psychologists assess the student's performance in real-life situations of test taking and studying to see how and why performance may be affected. In the workplace, more and more employers are being asked to make changes for their workers who are affected with ADHD. It is only a matter of time until an employer balks and a suit is filed. The trend is being followed closely by the business community.

Attention deficit–hyperactivity disorder has also contributed to a crisis in the disability insurance industry. Claims have soared for a host of ill-defined conditions. Insurers have fled the business of providing disability insurance, making it prohibitively expensive or impossible to obtain (Quint, 1994). As more people meet diagnostic criteria, many will attempt to gain services. It remains to be seen whether broadly defined disabilities ultimately trivialize suffering or make it more difficult for the more severely impaired person to obtain urgently needed recognition and services.

2.3.5 The Culture of Prozac

Prozac (fluoxetine), the first of the serotonin reuptake inhibitors, went on the market in 1988. With its low side-effect profile compared with the earlier generation of antidepressants, it widened the range of individuals who might tolerate a psychotropic drug for depression and led to widespread popular debate on the subject. Peter Kramer's best-selling book, *Listening to Prozac*, reflected and further encouraged popular interest in the use of psychiatric medication to enhance mood and performance. The overall prevalence of antidepressant use in certain communities has quadrupled in a 10-year period (Hume et al., 1995). It has become much more acceptable to take a psychotropic medication. This new atmosphere has also increased acceptance of stimulant use for behavioral problems in children and attentional problems in adults.

2.3.6 The Role of Mass Media

The effects of mass media on the practice of medicine and concerns of patients are well documented. As television news, talk shows, and print journalism have highlighted the use of psychotropic medication to cope with one's problems, a corresponding public interest in Ritalin and ADHD has developed. Personal and affecting testimonies of dramatic improvements after using Ritalin have been reported on national television broadcasts and many syndicated talk shows. A book on ADHD in adults and children has edged toward the best-seller list for over a year (Hallowell & Ratey, 1994b). Prominent local and national news weeklies have made ADHD their cover stories (Wallis, 1994).

In both the professional and lay media, ADHD is routinely referred to as a neurologic disorder. While most experts agree that genetic-biochemical factors influence behavior to some degree, the general public tends to transform this view into a biological determinism in which only heredity and brain chemistry determine behavior rather than in interactions with the environment. This interpretation may be comforting to some perplexed and worried parents who feel responsible for their children's difficulties and help overburdened teachers gain assistance in teaching children with this "disability." Psychotherapeutic strategies can help "externalize" the disease as separate from the child. Yet when behavior is regarded as stemming from biological pathology, interventions like stimulant medication become more easily justified and emphasized, while others become less valued. Indeed, ADHD has become the somewhat dubious leading self-diagnosis as the "biological cause... for job failure, divorce, poor motivation, lack of success, and chronic mild depression" (Shaffer, 1994).

2.3.7 Cosmetic Ritalin

"Cosmetic psychopharmacology," as Kramer puts it, is the elective use of medically prescribed drugs to enhance mood or improve behavior. Currently, it is considered medically and ethically justifiable to prescribe stimulants only when behavior meets criteria for a medical or psychiatric disorder. It is not known how much Ritalin currently is being prescribed for those on the indistinct line between "disease" and the general struggle for success. It is known, however, that Ritalin improves the focus and performance of those who do not meet ADHD criteria (normal, nonreferred children [Rapoport et al., 1978]) and that the drug is prescribed for such use.

There remains no definitive "test" for ADHD. The ambiguities of the ADHD diagnosis were highlighted in a study on stimulant medication and primary care (Wolraich et al., 1990). Over one quarter of children diagnosed with ADHD by their physician failed to meet criteria for the diagnosis when the cases were compared with structured psychiatric interviews with the parents. The number of children who failed to meet criteria increased to half when compared with structured interviews with the children's teachers. While the overall number of children medicated was not seen as high by the investigators, they noted the nonspecificity of the behavioral symptoms in the children that responded to stimulants. Thus, the ADHD diagnosis was seen as a diagnostic cover, albeit inaccurate, for the use of stimulants in a range of behavioral and performance problems in children.

2.3.8 Questioning Ritalin

Stimulants can be used in an effective and sensible way, especially when other modalities of treatment for attentional problems are addressed concurrently. Undoubtedly many parents of children with ADHD and adults with ADHD feel Ritalin has been of immense benefit to their children or themselves. However, important questions remain unanswered, and a pending request to decrease DEA controls on methylphenidate production and physician prescription practices makes them all the more urgent (Drug Enforcement Agency, 1995).

Ritalin's reemergence as a popular "fix" overlooks adverse side effects, a dearth of long-term studies, and a host of other ethical questions concerning unwitting coercion, fairness, informed consent, and potentially inadequate treatment of patients. Larger societal questions also should be asked: Should society use a biological fix to address problems that have roots in social and environmental factors? If it consistently does, how might society be affected?

If elective treatments are to warrant consideration, their side effects must be minimal. The short- and long-term physical side effects of Ritalin are generally considered minor on the basis of 15- to 20-year follow-up studies involving children who took stimulants for several years up until early adolescence (Jacobvitz et al., 1990). The effects of continuous Ritalin use through adolescence and into adulthood have not, as yet, been studied. The drug's immediate side effects, brief appetite suppression and possible insomnia, are generally well tolerated by children. Some reports suggest Ritalin unmasks the tics of Tourette syndrome, but this remains controversial (Gadow et al., 1995). Long-term growth suppression has been attributed to Ritalin, but this effect can be minimized through the scheduling of drug "holidays" (Klein et al., 1988).

Although one reason for the much greater use of Ritalin compared with amphetamine for ADHD has been the erroneous belief that it has less abuse potential, there exists a possibility of abusing Ritalin. The Swedish experience of the late 1960s and very recent examples of Ritalin abuse by teenagers in this country belie this myth of safety (Jaffe, 1991). However, there is little evidence of physical addiction to or abuse of Ritalin when used *appropriately* for ADHD. Despite the possibilities of abuse, Ritalin appears relatively safe from a strictly physical standpoint.

Evaluation of the emotional and psychological consequences of Ritalin use is more complex. There is still a strong cultural belief that it is better to cope by using one's inherent resources and interacting with people than by resorting to medication. This "pharmacological Calvinism" may lead to feelings of inadequacy in the child who takes the drug (Sleator, Ullman, & von Neumann, 1982) despite the physician's and family's view that Ritalin is necessary or benign. Teenagers, particularly sensitive about their identity, are especially vulnerable to issues of competence and biologic integrity. These beliefs can be overcome but remain a potential downside.

While physical addiction does not occur when Ritalin is used as prescribed for ADHD, psychological dependence is possible for the child, the family, the adolescent, or adult. When queried, children attributed most of their success in a "vigilance" assignment to their own efforts rather than medication (Milich et al., 1989). It is family members and teachers who more often notice the child performing suboptimally and ask, "Did you take your pill today?" The question expresses an underlying message to the child about the drug's important contribution to performance and behavior, and, ultimately, this message may undermine the child's confidence. This sense of dependency is highlighted when the medication is used "as necessary," in event-driven dosing, for example, when studying for an exam or attending a weekend family gathering. It is even possible that event-driven dosing may promote or exacerbate the often disorganized ADHD lifestyle by allowing the procrastinating individual to "catch up" at the last minute. The teenager and adult may also be tempted to stretch the normal wake–sleep cycle in order to achieve even greater performance, which could ultimately lead to an abuse pattern. The long-term consequences of self-administered stimulants by teenagers and adults for ADHD have not been studied to determine the likelihood of such a pattern developing. Thus, while achievements made under the influence of stimulants can enhance a sense of competence, self-esteem, and independence, the specter of psychological dependence, altered self-image, and potential abuse remains, especially in a society that paradoxically continues to be somewhat critical of psychotropic drugs while demanding greater performance.

Ritalin should be questioned further because no long-term studies prove its efficacy. Numerous reports show the stimulants to be of value in short-term memory and performance. In long-term studies, benefits to children formally classified under the hyperactivity diagnosis have not been demonstrated (Swanson et al., 1993). For children with ADHD without hyperactivity or for teenagers and adults, there are *no* long-term studies of Ritalin's efficacy.

Long-term controlled studies are difficult to run and fund, and one can question the ethics of withholding a potentially effective treatment until there is more definitive proof of benefit. However, a single study in which children received Ritalin along with child–family counseling and special education services is the *only* research demonstrating long-term improvements (Satterfield & Schell, 1987).[2] Most child behavior experts advocate a multimodal approach to treatment despite the lack of definitive evidence of improvement. In actual practice, though, the follow through for behavioral recommendations is poor (Wolraich et al., 1990). The multimodal model of treatment is also suggested for adults (Nadeau, 1995). Yet, here too, the emphasis in professional and lay articles is on the pharmacologic interventions.

The increasing availability and use of Ritalin to enhance performance also raise questions of subtle coercion and fairness. As more children and adults use Ritalin to work more efficiently at school or in the office, will those who are also struggling to perform feel pressured to consider medication? Will there be an impetus to keep up with others, to compete for the good grade, bonus, or job promotion by whatever means necessary, medication or otherwise? Moreover, is it fair to use the same performance criteria for those who use Ritalin as for those who do not? In athletic competition, stimulants remain banned precisely because of fairness issues. Yet, recently, the case has been made that athletes with ADHD be allowed to compete while taking Ritalin because of their "handicap" (Dyment, 1990). Somehow viewing behavior as neurologically based makes it more acceptable to use medication.

Because many Ritalin users are children, issues concerning informed consent also arise. Although the treatment of undesired nonpathologic conditions in adult medicine is not uncommon (for example, plastic surgery, topical minoxidil for baldness, estrogens for menopause, treatment for infertility, and contraceptives), elective therapies for children have been more controversial because it is the parents, not the children, who decide upon treatment. For example, growth hormone for constitutional short stature has been hotly debated (Lantos, Siegler, & Cuttler, 1989). Who decides for whom in these cases? And how high should the standard be?

One last question concerns the tendency for genetic contributions and neurochemical influences on behavior to be understood deterministically by society, such as the media and the courts. Such an interpretation can have the effect of eclipsing other treatment options. Even "good" psychopharmacology decreases the need to scrutinize the child's social environment and may permit a poor situation to continue or grow worse. Should dysfunctional family patterns and overcrowded classrooms be tolerated just because Ritalin improves the child's behavior? An effort is under way to determine which combination of treatments is most effective. The National Institutes of Mental Health has funded a multisite ADHD study involving several thousand children, with the goal of comparing treatment efficacies with a variety of approaches and combinations (Richters et al., 1995). Yet in the absence of confirmed, effective long-term treatment for ADHD and the general recommendations for a multimodal approach, will medication-only treatment produce persistent problems later in a child's life?

Furthermore, this bioreductionistic interpretation of the neurobiological components of ADHD behavior attributes less power to free will and individual choice. Thus, the popular viewpoint of maladaptive behavior as disease conflicts with another historically strong cultural perspective: accountability and responsibility. This clash of views is likely to be resolved ultimately in the civil and criminal court systems and by the economic imperatives of the workplace. It is worth noting how recent court decisions on recovered memory of child sexual abuse are influencing psychiatric technique and practice and the frequency of diagnosis of multiple personality disorder (Saks, 1994). Similar court guidelines are likely to emerge for those on the borderline of an ADHD diagnosis.

2.3.9 Responses to the Epidemic

The main response to date over the epidemic of ADHD and the use of stimulants in America has been further efforts at informing professionals and the public about the "new" ADHD (without hyperactivity) (Hallowell & Ratey, 1994). For many physicians, psychologists, and educators, the identification of potential ADHD and consequent stimulant treatment are meeting an important need of the community. Further education about the benefits of diagnosis and stimulants is the current goal. Academic medicine remains primarily focused on substantiating a biological substrate for ADHD. A notable exception is a recent study on the effects of family stressors in the development of ADHD (Biederman et al., 1995).

However, another view of ADHD diagnosis and the rise in stimulant use is far more sobering. As suggested earlier, the ADHD/stimulant phenomenon may reflect how the demands on children and families have increased as the social network supporting them has declined. The rise in the use of stimulants is alarming and signals an urgent need for American society to reevaluate its priorities.

On a clinical level, physicians treating children and adults may be locked into a "social trap" (Whalen & Henker, 1980). Though it may make sense to medicate individuals so they can function more effectively and competently within a certain environment, do doctors unwittingly permit and support a long-term collective negative outcome for the society? Are they unintentionally promoting an antihumanistic, competitive environment that demands performance at any cost? Should they more aggressively promote a general redistribution of society's resources to children and families? Some say there is no choice but to offer medication; it is not up to physicians to address society's ills. Peter Kramer in *Listening to Prozac* seems rather sanguine about a society that copes with newer, safer, improved psychopharmacologic agents. Whether individually beneficial or societally dangerous, it behooves the physician to at least raise these questions about ADHD and stimulants with parents, teachers, and colleagues.

Notes

1. Editor's note: This reading originally appeared in 1996 in *The Hastings Center Report*, volume 26, pages 12–18, and is used with permission. The author gratefully thanks John Jacobs, Lane Tanner, Denise Bostrom, Glenn Elliott, and Tom Boyce for their review and comments on the manuscript, Fred Gardner for editorial assistance, and Helen Reyes, librarian, John Muir Medical Center, for research support.

2. Editor's note: Since the original publication of Diller's article, the results of the landmark "Multimodal Treatment Treatment Study of Children with ADHD" were published, confirming the superiority of medication to behavioral therapy alone in the treatment of ADHD with a 14-month follow-up period. Details can be found in: MTA Cooperative Group (1999). A 14-month randomized clinical trial of treatment strategies for attention-deficit/hyperactivity disorder. *Archives of General Psychiatry,* 56, 1073–1086. On even longer-term follow-up, 6 to 8 years later, differences among the treatments administered in childhood disappeared. For details see: MTA Cooperative Group (2009). The MTA at 8 Years: Prospective Follow-up of Children Treated for Combined-Type ADHD in

a Multisite Study *Journal of the American Academy of Child & Adolescent Psychiatry, 48*, 484–500.

References

American Psychiatric Association. (1994). *Diagnostic and statistical manual of mental disorders* (4th ed.) (pp. 577–585). Washington, DC: Author.

Beck, H., Disabled Students' Program, University of California, Berkeley, personal communication.

Biederman, J., Milberger, S., Faraone, S. V., Kiely, K., Guite, J., & Mick, E. (1995). Family-environment risk factors for attention deficit hyperactivity disorder. *Archives of General Psychiatry, 52*, 464–470.

Bradley, C. (1937). The behavior of children receiving benzedrine. *American Journal of Psychiatry, 94*, 577–585.

Dillon, S. (1994, April 4). Special education absorbs school resources. *New York Times*, A1.

Drug Enforcement Agency, Drug and Chemical Evaluation Section. (1995). Methylphenidate (a background paper). Office of Diversion Control, U.S. Department of Justice, Drug Enforcement Administration. Washington, D.C.: U.S. Government Printing Office.

Dyment, P. G. (1990). Hyperactivity, stimulants, and sports. *The Physician and Sportsmedicine, 18*, 22.

Gadow, K. D. (1981). Prevalence of drug treatment for hyperactivity and other childhood behavior disorders. In K. D. Gadow & J. Loney (Eds.), *Psychosocial aspects of drug treatment for hyperactivity*. Boulder, CO: Westview Press.

Gadow, K. D., Nolan E, Sprafkin J, & Sverd J. (1995). School observations of children with attention deficit hyperactivity disorder and comorbid tic disorder: Effects of methylphenidate treatment. *Journal of Developmental and Behavioral Pediatrics, 16*, 167–176.

Gadow, K. D. (1981). Prevalence of drug treatment for hyperactivity. In K. D. Gadow and J. Loney (Eds.), *Psychosocial aspects of drug treatment for hyperactivity*. Boulder, CO: Westview Press.

Gallagher, C. E. (1970). Federal involvement in the use of behavior modification drugs on grammar school children of the right to privacy inquiry. Committee on Governmental Operations. House of Representatives 91st Congress, 2nd session, no. 52-268. Washington DC: US Government Printing Office.

Greenhill, L. L. (1992). Pharmacologic treatment of attention deficit disorder. *Psychiatric Clinics of North America, 15*, 1–27.

Hallowell, E. M., & Ratey, J. J. (1994a). *Answers to distractions*. New York: Pantheon Books.

Hallowell, E. M., & Ratey, J. J. (1994b). *Driven to distraction*. New York: Pantheon Books.

Hume, A., Barbour, M. N., Lapane, K. L., & Carleton, R. A. (1995). Is antidepressant use changing? Prevalence and clinical correlates in two New England communities. *Pharmacotherapy, 15*, 78–84.

Jacobvitz, D., Sroufe, L. A., Stewart, M., & Leffert, N. (1990). Treatment of attentional and hyperactivity problems in children with sympathomimetic drugs: A comprehensive review. *Journal of the American Academy of Child and Adolescent Psychiatry, 29*, 677–688.

Jaffe, S. L. (1991). Intranasal abuse of prescribed methylphenidate by an alcohol and drug abusing adolescent with ADHD. *Journal of the American Academy of Child and Adolescent Psychiatry, 30*, 773–775.

Klein, R., Landa, B., Mattes, J. A., & Klein, D. F. (1988). Methylphenidate and growth in hyperactive children. *Archives of General Psychiatry, 45*, 1127–1130.

Lantos, J., Siegler, M., & Cuttler, M. (1989). Ethical issues in growth hormone therapy. *Journal of the American Medical Association, 261*, 1020–1024.

Maynard, R. (1970, June 29). Omaha pupils given "behavior drugs." *Washington Post.*

Milich, R., Licht, B. G., Murphy, D. A., & Pelham, W. E. (1989). Attention deficit hyperactivity disordered boys' evaluation of and attributions for task performance on medication versus placebo. *Journal of Abnormal Psychology, 98*, 280–284.

Nadeau, K. G. (Ed.). (1995). *A comprehensive guide to attention deficit disorder in adults: Research, diagnostic, and treatment.* New York: Brunner/Mazel.

Office of Public Affairs, Drug Enforcement Administration, Department of Justice. (1975–1995). *Methylphenidate yearly production quota.* Washington, DC: Author.

Office of Public Affairs, Drug Enforcement Administration, Department of Justice. (1993). Background information on methylphenidate (Ritalin). Washington, DC: U.S. Government Printing Office.

Quint, M. (1994, November 28). New ailments: Bane of insurers. *New York Times*, D1.

Rapoport, J. L., Buchsbaum, M. S., Zahn, T. P., Weingartner, H., Ludlow, C., & Mikkelsen, E. J. (1978). Dextroamphetamine: Cognitive and behavioral effects in normal prepubertal boys. *Science, 199*, 560–562.

Reid, R., & Katsiyannis, A. (1995). Attention-deficit/hyperactivity disorder and section 504. *Remedial and Special Education, 16*, 44–52.

Richters, J. E., Arnold, L. E., Jensen, P. S., Abikoff, H., Conners, C. K., & Greenhill, L. L. (1995). NIMH collaborative multisite multimodal treatment study of children with ADHD: Background and rational. *Journal of the American Child and Adolescent Psychiatry, 34*, 987–1000.

Roberts, S. (1994). *Who are we? A portrait of America today based on the latest United States census* (p. 199). New York: Times Books.

Rogers, J. M. (1971). Drug abuse—just what the doctor ordered. *Psychology Today, 9*, 16–24.

Ross, C. A., & Pam, A. (1995). *Pseudoscience in biological psychiatry: Blaming the body.* New York: Wiley.

Safer, D. J., & Krager, J. M. (1988). A survey of medical treatment for hyperactive/inattentive students. *The Journal of the American Medical Association, 260,* 2256–2258.

Saks, E. R. (1994). Does multiple personality disorder exist? The beliefs, the data, and the law. *International Journal of Law and Psychiatry, 17,* 43–78.

Sappell, J., & Welkos, R. (1990, June 6). Suits, protests fuel a campaign against psychiatry. *Los Angeles Times,* A48:1.

Satterfield, B. T., & Schell, A. M. (1987). Therapeutic interventions to prevent delinquency in hyperactive boys. *Journal of the American Academy of Child and Adolescent Psychiatry, 26,* 56–64.

Shaffer, D. (1994). Attention deficit hyperactivity disorder in adults. *American Journal of Psychiatry, 151,* 633–638.

Sleator, E. K., Ullman, R. K., & von Neumann, A. (1982). How do hyperactive children feel about taking stimulants and will they tell the doctor? *Clinical Pediatrics, 21,* 474–479.

Swanson, J. M., McBurnett, K., Wigal, T., Pfiffner, L. J., Lerner, M. A., & Williams, L. (1993). Effect of stimulant medication on children with attention deficit disorder: A "review of reviews." *Exceptional Children, 60,* 154–162.

Swanson, J. M., Lemer, M., & Williams, L. (1995). More frequent diagnosis of attention deficit-hyperactivity disorder [Letter]. *The New England Journal of Medicine, 333,* 94.

Wallis, C. (1994, July 18). Life in overdrive *Time,* 42–50.

Whalen, C. K., & Henker, B. (1980). The social ecology of psychostimulant treatment: A model for conceptual and empirical analysis. In C. K. Whalen & B. Henker (Eds.), *Hyperactive children: The social ecology of identification and treatment.* New York: Academic Press.

Wolraich, M. L., Lindgren, S., Stromquist, A., Milich, R., Davis, C., & Watson, D. (1990). Stimulant medication use by primary care physicians in the treatment of attention deficit hyperactivity disorder. *Pediatrics, 86,* 95–101.

Reading 2.4

Beyond Therapy: Essential Sources of Concern[1]

President's Council on Bioethics

Our familiar worries about issues of safety, equality, and freedom, albeit very important, do not exhaust the sources of reasonable concern. When richly considered, they invite us to think about the deeper purposes for the sake of which we want to live safely, justly, and freely. And they enable us to recognize that even the safe, equally available, noncoerced and nonfaddish uses of biomedical technologies to pursue happiness or self-improvement raise ethical and social questions, questions more directly connected with the essence of the activity itself: the use of technological means to intervene into the human body and mind, not to ameliorate their diseases but to change and improve their normal workings. Why, if at all, are we bothered by the voluntary *self*-administration of agents that would change our bodies or alter our minds? What is disquieting about our attempts to improve upon human nature, or even our own particular instance of it?

The subject being relatively novel, it is difficult to put this worry into words. We are in an area where initial revulsions are hard to translate into sound moral arguments. Many people are probably repelled by the idea of drugs that erase memories or that change personalities, or of interventions that enable 70-year-olds to bear children or play professional sports, or, to engage in some wilder imaginings, of mechanical implants that would enable men to nurse infants or computer–brain hookups that would enable us to download the *Oxford English Dictionary*. But can our disquiet at such prospects withstand rational, anthropologic, or ethical scrutiny? Taken one person at a time, with a properly prepared set of conditions and qualifications, it will be hard to say what is wrong with any biotechnical intervention that could improve our performances, give us (more) ageless bodies, or make it possible for us to have happier souls. Indeed, in many cases, we ought to be thankful for or pleased with the improvements our biotechnical ingenuity is making possible.

If there are essential reasons to be concerned about these activities and where they may lead us, we sense that it may have something to do with challenges to what is naturally human, what is humanly dignified, or to attitudes that show proper respect for what is naturally and dignifiedly human. As it happens, there are at least four such considerations: appreciation of and respect for "the naturally given," threatened by hubris; the dignity of human activity, threatened by "unnatural" means; the preservation of identity, threatened by efforts at self-transformation; and full human flourishing, threatened by spurious or shallow substitutes.

2.4.1 Hubris or Humility: Respect for "the Given"

A common, man-on-the-street reaction to the prospects of biotechnological engineering beyond therapy is the complaint of "man playing God." If properly unpacked, this worry is in fact shared by people holding various theological beliefs and by people holding none at all. Sometimes the charge means the sheer prideful presumption of trying to alter what God has ordained or nature has produced, or what should, for whatever reason, not be fiddled with. Sometimes the charge means not so much usurping God-like powers but doing so in the absence of God-like knowledge: the mere playing at being God, the hubris of acting with insufficient wisdom.

Over the past few decades, environmentalists, forcefully making the case for respecting Mother Nature, have urged upon us a "precautionary principle" regarding all our interventions into the natural world. Go slowly, they say, you can ruin everything. The point is certainly well taken in the current context. The human body and mind, highly complex and delicately balanced as a result of eons of gradual and exacting evolution, are almost certainly at risk from any ill-considered attempt at "improvement." There is not only the matter of unintended consequences, a concern even with interventions aimed at therapy. There is also the matter of uncertain goals and absent natural standards, once one proceeds "beyond therapy." When a physician intervenes therapeutically to correct some deficiency or deviation from a patient's natural wholeness, he acts as a servant to the goal of health and as an assistant to nature's own powers of self-healing, themselves wondrous products of evolutionary selection. But when a bioengineer intervenes for nontherapeutic ends, he stands not as nature's servant but as her aspiring master, guided by nothing but his own will and serving ends of his own devising. It is far from clear that our delicately integrated natural bodily powers will take kindly to such impositions, however desirable the sought-for change may seem to the intervener. And there is the further question of the unqualified goodness of the goals being sought, a matter to which we shall return.[2] One revealing way to formulate the problem of hubris is what one of our Council Members has called the temptation to "hyper-agency," a Promethean aspiration to remake nature, including human nature, to serve our purposes and to satisfy our desires. This attitude is to be faulted not only because it can lead to bad, unintended consequences; more fundamentally, it also represents a false understanding of, and an improper disposition toward, the naturally given world. The root of the difficulty seems to be both cognitive and moral: the failure properly to appreciate and respect the "giftedness" of the world. Acknowledging the giftedness of life means recognizing that our talents and powers are not wholly our own doing, nor even fully ours, despite the efforts we expend to develop and to exercise them. It also means recognizing that not everything in the world is open to any use we may desire or devise. Such an appreciation of the giftedness of life would constrain the Promethean project and conduce to a much-needed humility. Although it is in part a religious sensibility, its resonance reaches beyond religion.

Human beings have long manifested both wondering appreciation for nature's beauty and grandeur and reverent awe before nature's sublime and mysterious power. From the elegance of an orchid to the splendor of the Grand Canyon, from the magnificence of embryologic development to the miracle of sight or consciousness, the works of nature can still inspire in most human beings an attitude of respect, even in this age of technology. Nonetheless, the absence of a respectful attitude is today a problem in some—though by no means all—quarters of the biotechnical world. It is worrisome when people act toward, or even talk about, our bodies and minds—or human nature itself—as if they were mere raw material to be molded according to human will. It is worrisome when people speak as if they were wise enough to redesign human beings, improve the human brain, or reshape the human life cycle. In the face of such hubristic temptations, appreciating that the given world—including our natural powers to alter it—is not of our own making could induce a welcome attitude of modesty, restraint, and humility. Such a posture is surely recommended for anyone inclined to modify human beings or human nature for purposes beyond therapy.

Yet the respectful attitude toward the "given," while both necessary and desirable as a restraint, is not by itself sufficient as a guide. The "giftedness of nature" also includes smallpox and malaria, cancer and Alzheimer's disease, decline and decay. Moreover, nature is not equally generous with her gifts, even to man, the most gifted of her creatures. Modesty born of gratitude for the world's "givenness" may enable us to recognize that not everything in the world is open to any use we may desire or devise, but it will not *by itself* teach us *which* things can be tinkered with and which should be left inviolate. Respect for the "giftedness" of things cannot tell us which gifts are to be accepted as is, which are to be improved through use or training, which are to be housebroken through self-command or medication, and which opposed like the plague.

To guide the proper use of biotechnical power, we need something in addition to a generalized appreciation for nature's gifts. We would need also a particular regard and respect for the special gift that is our own given nature. For only if there is a *human* "givenness," or a given humanness, that is also good and worth respecting, either as we find it or as it could be perfected *without ceasing to be itself*, will the "given" serve as a *positive* guide for choosing what to alter and what to leave alone. Only if there is something precious in our given human nature—beyond the fact of its giftedness—can what is given guide us in resisting efforts that would degrade it. When it comes to human biotechnical engineering beyond therapy, only if there is something inherently good or dignified about, say, natural procreation, the human life cycle (with its rhythm of rise and fall), and human erotic longing and striving; only if there is something inherently good or dignified about the ways in which we engage the world as spectators and appreciators, as teachers and learners, leaders and followers, agents and makers, lovers and friends, parents and children, citizens and worshippers, and as seekers of our own special excellence and flourishing in whatever arena to which we are called—only then can we begin to see why those aspects of our nature need to be defended against our deliberate redesign.

We must move, therefore, from the danger of hubris in the powerful designer to the danger of degradation in the designed, considering how any proposed improvements might impinge upon the nature of the one being improved. With the question of human nature and human dignity in mind, we move to questions of means and ends.

2.4.2 "Unnatural" Means: The Dignity of Human Activity

Until only yesterday, teaching and learning or practice and training exhausted the alternatives for acquiring human excellence, perfecting our natural gifts through our own efforts. But perhaps no longer: Biotechnology may be able to do nature one better, even to the point of requiring less teaching, training, or practice to permit an improved nature to shine forth.... The insertion of the growth-factor gene into the muscles of rats and mice bulks them up and keeps them strong and sound without the need for nearly as much exertion. Drugs to improve alertness (today) or memory and amiability (tomorrow) could greatly relieve the need for exertion to acquire these powers, leaving time and effort for better things. What, if anything, is disquieting about such means of gaining improvement?

The problem cannot be that they are "artificial," in the sense of having man-made origins. Beginning with the needle and the fig leaf, man has from the start been the animal that uses art to improve his lot by altering or adding to what nature alone provides.[3] Ordinary medicine makes extensive use of similar artificial means, from drugs to surgery to mechanical implants, in order to treat disease. If the use of artificial means is absolutely welcome in the activity of healing, it cannot be their unnaturalness alone that disquiets us when they are used to make people "better than well."

Still, in those areas of human life in which excellence has until now been achieved only by discipline and effort, the attainment of similar results by means of drugs, genetic engineering, or implanted devices looks to many people (including some Members of this Council) to be "cheating" or "cheap." Many people believe that each person should work hard for his achievements. Even if we prefer the grace of the natural athlete or the quickness of the natural mathematician—people whose performances deceptively appear to be effortless—we admire also those who overcome obstacles and struggle to try to achieve the excellence of the former. This matter of character—the merit of disciplined and dedicated striving—is surely pertinent. For character is not only the source of our deeds but also their product.... Healthy people whose disruptive behavior is "remedied" by pacifying drugs rather than by their own efforts are not learning self-control[4]; if anything, they may be learning to think it unnecessary. People who take pills to block out from memory the painful or hateful aspects of a new experience will not learn how to deal with suffering or sorrow. A drug that induces fearlessness does not produce courage.

Yet things are not so simple. Some biotechnical interventions may assist in the pursuit of excellence without in the least cheapening its attainment. And many of life's excellences have nothing to do with competition or overcoming adversity. Drugs to decrease drowsiness, increase alertness, sharpen memory, or reduce distraction may actually help people interested in their natural pursuits of learning or painting or performing their civic duty. Drugs to steady the hand of a neurosurgeon or to prevent sweaty palms in a concert pianist cannot be regarded as "cheating," for they are in no sense the source of the excellent activity or achievement. And, for people dealt a meager hand in the dispensing of nature's gifts, it should not be called cheating or cheap if biotechnology could assist them in becoming better equipped—whether in body or in mind.

Nevertheless, as we suggested at some length in Chapter Three of the Report from which this excerpt is taken, there remains a sense that the "naturalness" of means matters. It lies not in the fact that the assisting drugs and devices are artifacts, but in the danger of violating or deforming the nature of human agency and the dignity of the naturally human way of activity. In most of our ordinary efforts at self-improvement, whether by practice, training, or study, we sense the relation between our doings and the resulting improvement, between the means used and the end sought. There is an experiential and intelligible connection between means and ends; we can see how confronting fearful things might eventually enable us to cope with our fears. We can see how curbing our appetites produces self-command. Human education ordinarily proceeds by speech or symbolic deeds, whose meanings are at least in principle directly accessible to those upon whom they work.

In contrast, biotechnical interventions act directly on the human body and mind to bring about their effects on a passive subject, who plays little or no role at all. He can at best *feel* their effects *without understanding their meaning in human terms*. Thus, a drug that brightened our mood would alter us without our understanding how and why it did so—whereas a mood brightened as a fitting response to the arrival of a loved one or to an achievement in one's work is perfectly, because humanly, intelligible. And not only would this be true about our states of mind. All of our encounters with the world, both natural and interpersonal, would be mediated, filtered, and altered. Human experience under biological intervention becomes increasingly mediated by unintelligible forces and vehicles, separated from the human significance of the activities so altered. The relations between the knowing subject and his activities, and between his activities and their fulfillments and pleasures, are disrupted.

The importance of human effort in human achievement is here properly acknowledged: the point is less the exertions of good character against hardship, but the manifestation of an alert and self-experiencing agent making his deeds flow intentionally from his willing, knowing, and embodied soul. If human flourishing means not just the accumulation of external achievements and a full curriculum vitae but a lifelong being-at-work exercising one's *human* powers *well* and without great impediment, our genuine happiness requires that there be little gap, if any, between the dancer and the dance.[5]

2.4.3 Identity and Individuality

With biotechnical interventions that skip the realm of intelligible meaning, we cannot really own the transformations nor can we experience them as genuinely ours. And we will be at a loss to attest whether the resulting conditions and activities of our bodies and our minds are, in the fullest sense, our own as human. But our interest in identity is also more personal. For we do not live in a generic human way; we desire, act, flourish, and decline *as ourselves*, as individuals. To be human is to be someone, not anyone—with a given nature (male or female), given natural abilities (superior wit or musical talent), and, most important, a real history of attachments, memories, and experiences, acquired largely by living with others.

In myriad ways, new biotechnical powers promise (or threaten) to transform what it means to be an individual: giving increased control over our identity to others, as in the case of genetic screening or sex selection of offspring by parents; inducing psychic states divorced from real life and lived experience; blunting or numbing the memories we wish to escape; and achieving the results we could never achieve unaided, by acting as ourselves alone.

To be sure, in many cases, biomedical technology can restore or preserve a real identity that is slipping away: keeping our memory intact by holding off the scourge of Alzheimer's disease; restoring our capacity to love and work by holding at bay the demons of self-destroying depression. In other cases, the effect of biotechnology on identity is much more ambiguous. By taking psychotropic drugs to reduce anxiety or overcome melancholy, we may become the person we always wished to be—more cheerful, ambitious, relaxed, content. But we also become a different person in the eyes of others, and in many cases we become dependent on the continued use of psychotropic drugs to remain the new person we now are.

As the power to transform our native powers increases, both in magnitude and refinement, so does the possibility for "self-alienation"—for losing, confounding, or abandoning our identity. I may get better, stronger, and happier—but I know not how. I am no longer the agent of self-transformation, but a passive patient of transforming powers. Indeed, to the extent that an achievement is the result of some extraneous intervention, it is detachable from the agent whose achievement it purports to be. "Personal achievements" impersonally achieved are not truly the achievements of persons. That I can use a calculator to do my arithmetic does not make *me* a knower of arithmetic; if computer chips in my brain were to "download" a textbook of physics, would that make *me* a knower of physics? Admittedly, the relation between biological boosters and personal identity is much less clear: if I make myself more alert through Ritalin, or if drugs can make up for lack of sleep, I may be able to learn more using my unimpeded native powers while it is still unquestionably *I* who am doing the learning. And yet, to find out that an athlete took steroids before the race or that a test-taker (without medical disability) took Ritalin before the test is to lessen our regard for the achievement of the doer. It is to see not just an acting self, but a dependent self, one who is less himself for becoming so dependent.

In the deepest sense, to have an identity is to have limits: my body, not someone else's—even when the pains of aging might tempt me to become young again; my memories, not someone else's—even when the traumas of the past might tempt me to have someone else's memories; my achievements and potential, not someone else's—even when the desire for excellence might tempt me to "trade myself in" for a "better model." We seek to be happy—to achieve, perform, take pleasure in our experiences, and catch the admiring eye of a beloved. But we do not, at least self-consciously, seek such happiness at the cost of losing our real identity.

2.4.4 Partial Ends, Full Flourishing

Beyond the perils of achieving our desired goals in a "less-than-human way" or in ways "not fully our own," we must consider the meaning of the ends themselves: better children, superior performance, ageless bodies, and happy souls. Would their attainment in fact improve or perfect our lives as human beings? Are they—always or ever—reasonable and attainable goals?

Everything depends, as we have pointed out in each case, on how these goals are understood, on their specific and concrete content. Yet, that said, the first two human ends—better children and superior performance—do seem reasonable and attainable, sometimes if not always, to some degree if not totally. When asked what they wish for their children, most parents say: "We want them to be happy," or "We want them to live good lives"—in other words, to be better and to do better. The desire is a fitting one for any loving parent. The danger lies in misconceiving what "better children" really means, and thus coming to pursue this worthy goal in a misguided way, or with a false idea of what makes for a good or happy child.

Likewise, the goal of superior performance—the desire to be better or do better in all that we do—is good and noble, a fitting human aspiration. We admire excellence whenever we encounter it, and we properly seek to excel in those areas of life, large and small, where we ourselves are engaged and at-work. But the danger here is that we will become better in some area of life by diminishing ourselves in others, or that we will achieve superior results only by compromising our humanity, or by corrupting those activities that are not supposed to be "performances" measured in terms of external standards of "better and worse."

In many cases, biotechnologies can surely help us cultivate what is best in ourselves and in our children, providing new tools for realizing good ends, wisely pursued. But it is also possible that the new technological means may deform the ends themselves. In pursuit of better children, biotechnical powers risk making us "tyrants"; in pursuit of superior performance, they risk making us "artifacts." In both cases, the problem is not the ends themselves but our misguided idea of their attainment or our false way of seeking to attain them. And in both cases, there is the ubiquitous problem that "good" or "superior" will be reconceived to fit the sorts of goals that the technological interventions can help us attain. We may come to believe that genetic predisposition or brain chemistry holds the key to helping our children develop and improve, or that stimulant drugs or bulkier muscles hold the key to excellent human activity. If we are equipped with hammers, we will see only those things that can be improved by pounding.

The goals of ageless bodies and happy souls—and especially the ways biotechnology might shape our pursuit of these ends—are perhaps more complicated.[6] The case for ageless bodies seems at first glance to look pretty good. The prevention of decay, decline, and disability, the avoidance of blindness, deafness, and debility, the elimination of feebleness, frailty, and fatigue, all seem to be conducive to living fully as a human being at the top of one's powers—of having, as they say, a "good quality of life" from beginning to end. We have come to expect organ transplantation for our worn-out parts. We will surely welcome stem-cell–based therapies for regenerative medicine, reversing by replacement the damaged tissues of Parkinson's disease, spinal cord injury, and many other degenerative disorders. It is hard to see any objection to obtaining a genetic enhancement of our muscles in our youth that would not only prevent the muscular feebleness of old age but would empower us to do any physical task with greater strength and facility throughout our lives. And, should aging research deliver on its promise of adding not only extra life to years but also extra years to life, who would refuse it?

But . . . there may in fact be many human goods that are inseparable from our aging bodies, from our living in time, and especially from the natural human life cycle by which each generation gives way to the one that follows it. Because this argument is so counterintuitive, we need to begin not with the individual choice for an ageless body, but with what the individual's life might look like in a world in which everyone made the same choice. We need to make the choice universal, and see the meaning of that choice in the mirror of its becoming the norm.

What if everybody lived life to the hilt, even as they approached an ever-receding age of death in a body that looked and functioned—let's not be too greedy—like that of a 30-year-old? Would it be good if each and all of us lived like light bulbs, burning as brightly from beginning to end, then popping off without warning, leaving those around us suddenly in the dark? Or is it perhaps better that there be a shape to life, everything in its due season, the shape also written, as it were, into the wrinkles of our bodies that live it—provided, of course, that we do not suffer years of painful or degraded old age and that we do not lose our wits? What would the relations between the generations be like if there never came a point at which a son surpassed his father in strength or vigor? What incentive would there be for the old to make way for the young, if the old slowed down little and had no reason to think of retiring—if Michael could play basketball until he were not 40 but 80? Might not even a moderate prolongation of life span with vigor lead to a prolongation in the young of functional immaturity—of the sort that has arguably already accompanied the great increase in average life expectancy experienced in the past century?[7] Going against both common intuition and native human desire, some commentators have argued that living with full awareness and acceptance of our finitude may be the condition of many of the best things in human life: engagement, seriousness, a taste for beauty, the possibility of virtue, the ties born of procreation, the quest for meaning.[8] This might be true not just for immortality—an unlikely achievement, likely to produce only false expectations—but even for more modest prolongations of the maximum life span, especially in good health, that would permit us to live as if there were always tomorrow. The pursuit of perfect bodies and further life-extension might deflect us from realizing more fully the aspirations to which our lives naturally point, from living well rather than merely staying alive. A concern with one's own improving agelessness might finally be incompatible with accepting the need for procreation and human renewal. And far from bringing contentment, it might make us increasingly anxious over our health or dominated by the fear of death. Assume, merely for the sake of the argument, that even a few of these social consequences would follow from a world of much greater longevity and vigor: What would we then say about the simple goodness of seeking an ageless body?

What about the pursuit of happy souls, and especially of the sort that we might better attain with pharmacologic assistance? Painful and shameful memories are disturbing; guilty consciences trouble sleep; low self-esteem, melancholy, and world-weariness besmirch the waking hours. Why not memory-blockers for the former, mood-brighteners for the latter, and a good euphoriant—without risks of hangovers or cirrhosis—when celebratory occasions fail to be jolly? For let us be clear: If it is imbalances of neurotransmitters that are largely responsible for our state of soul, would it not be sheer priggishness to refuse the help of pharmacology for our happiness, when we accept it guiltlessly to correct for an absence of insulin or thyroid hormone?

And yet, . . . there seems to be something misguided about the pursuit of utter and unbroken psychic tranquility or the attempt to eliminate all shame, guilt, and painful memories. Traumatic memories, shame, and guilt are, it is true, psychic pains. In extreme doses, they can be crippling. Yet, short of the extreme, they can also be helpful and fitting. They are appropriate responses to horror, disgraceful conduct, injustice, and sin, and, as such, help teach us to avoid them or fight against them in the future. Witnessing a murder should be remembered as horrible; doing a beastly deed should trouble one's soul. Righteous indignation at injustice depends on being able to feel injustice's sting. And to deprive oneself of one's memory—including and especially its truthfulness of feeling—is to deprive oneself of one's own life and identity.

These feeling states of soul, though perhaps accompaniments of human flourishing, are not its essence. Ersatz pleasure or feelings of self-esteem are not the real McCoy. They are at most shadows divorced from the underlying human activities that are the essence of flourishing. Most people want both to feel good and to feel good about themselves, but only as a result of being good and doing good.

At the same time, there appears to be a connection between the possibility of feeling deep unhappiness and the prospects for achieving genuine happiness. If one cannot grieve, one has not truly loved. To be capable of aspiration, one must know and feel lack. As Wallace Stevens put it: Not to have is the beginning of desire. In short, if human fulfillment depends on our being creatures of need and finitude and therewith of longings and attachment, there may be a double-barreled error in the pursuit of ageless bodies and factitiously happy souls: far from bringing us what we really need, pursuing these partial goods could deprive us of the urge and energy to seek a richer and more genuine flourishing.

Looking into the future at goals pursuable with the aid of new biotechnologies enables us to turn a reflective glance at our own version of the human condition and the prospects now available to us (in principle) for a flourishing human life. For us today, assuming that we are blessed with good health and a sound mind, a flourishing human life is not a life lived with an ageless body or an untroubled soul, but rather a life lived in rhythmed time, mindful of time's limits, appreciative of each season and filled first of all with those intimate human relations that are ours only because we are born, age, replace ourselves, decline, and die—and know it. It is a life of aspiration, made possible by and born of experienced lack, of the disproportion between the transcendent longings of the soul and the limited capacities of our bodies and minds. It is a life that stretches toward some fulfillment to which our natural human soul has been oriented, and, unless we extirpate the source, will always be oriented. It is a life not of better genes and enhancing chemicals but of love and friendship, song and dance, speech and deed, working and learning, revering and worshipping.

If this is true, then the pursuit of an ageless body may prove finally to be a distraction and a deformation. And the pursuit of an untroubled and self-satisfied soul may prove to be deadly to desire, if finitude recognized spurs aspiration and fine aspiration acted upon *is itself* the core of happiness. Not the agelessness of the body, nor the contentment of the soul, nor even the list of external achievements and accomplishments of life, but the engaged and energetic being-at-work of what nature uniquely gave to us is what we need to treasure and defend. All other "perfections" may turn out to be at best but passing illusions, at worst a Faustian bargain that could cost us our full and flourishing humanity.

Summing up these "essential sources of concern," we might succinctly formulate them as follows:

In wanting to become more than we are, and in sometimes acting as if we were already superhuman or divine, we risk despising what we are and neglecting what we have.
In wanting to improve our bodies and our minds using new tools to enhance their performance, we risk making our bodies and minds little different from our tools, in the process also compromising the distinctly human character of our agency and activity.
In seeking by these means to be better than we are or to like ourselves better than we do, we risk "turning into someone else," confounding the identity we have acquired through natural gift cultivated by genuinely lived experiences, alone and with others.
In seeking brighter outlooks, reliable contentment, and dependable feelings of self-esteem in ways that bypass their usual natural sources, we risk flattening our souls, lowering our aspirations, and weakening our loves and attachments.
By lowering our sights and accepting the sorts of satisfactions that biotechnology may readily produce for us, we risk turning a blind eye to the objects of our natural loves and longings, the pursuit of which might be the truer road to a more genuine happiness.

To avoid such outcomes, our native human desires need to be educated against both excess and error. We need, as individuals and as a society, to find these boundaries and to learn how to preserve and defend them. To do so in an age of biotechnology, we need to ponder and answer questions like the following:

When does parental desire for better children constrict their freedom or undermine their long-term chances for self-command and genuine excellence?
When does the quest for self-improvement make the "self" smaller or meaner?

When does a preoccupation with youthful bodies or longer life jeopardize the prospects for living *well*?
When does the quest for contentment or self-esteem lead us away from the activities and attachments that prove to be essential to these goals when they are properly understood?

Answers to these questions are not easily given in the abstract or in advance. Boundaries are hard to define in the absence of better knowledge of the actual hazards. Such knowledge will be obtainable only in time and only as a result of lived experience. But centrally important in shaping the possible future outcomes will be the cultural attitudes and social practices that shape desires, govern expectations, and influence the choices people make, now and in the future. This means reflecting more specifically on how biotechnology beyond therapy might affect and be affected by American society.

Notes

1. Editor's note: This reading is excerpted from the report *Beyond therapy: Biotechnology and the pursuit of happiness* (President's Council on Bioethics, 2003) available at www.bioethics.gov/reports/beyondtherapy.

2. The question of the knowledge and goodness of goals is often the neglected topic when people use the language of "mastery," or "mastery and control of nature," to describe what we do when we use knowledge of how nature works to alter its character and workings. Master of the means of intervention without knowing the goodness of the goals of intervening is not, in fact, mastery at all. In the absence of such knowledge of ends, the goals of the "master" will be set rather by whatever it is that happens to guide or move his will—some impulse or whim or feeling or desire—in short, by some residuum of nature still working within the so called master or controller. To paraphrase C. S. Lewis, what looks like man's mastery of nature turns out, in the absence of guiding knowledge, to be nature's master of man. (See his *The abolition of man,* New York: Macmillan, 1965, paperback edition, pp. 72–80.) There can, in truth, be no such thing as the *full* escape from the grip of our own nature. To pretend otherwise is indeed a form of hubristic and dangerous self-delusion. For reasons given in the text, therapeutic medicine, though it may use the same technologies, should not be regarded as "mastery of nature," but as service to nature, as we come to know, through medical science, how it might best be served.

3. By his very nature, man is the animal constantly looking for ways to better his life through artful means and devices; man is the animal with what Rousseau called "perfectability."

4. We have also noted that other people suffering from certain neuropsychiatric disorders become capable of learning self control only with the aid of medication addressed to their disorders.

5. This is not merely to suggest that there is a disturbance of human agency or freedom, or a disruption of activities that will confound the assignment of personal responsibility or undermine the proper bestowal of praise and blame. To repeat: most of life's activities are noncompetitive; most of the best of them—loving and working and savoring and learning—are self-fulfilling beyond the need for praise and blame or any other external reward. In these activities, there is at best no goal beyond the activity itself. It is the possibility of natural, unimpeded, for itself human activity that we are eager to preserve against dilution and distortion.

6. The discussion that follows depends heavily on a paper by Leon R. Kass, "Beyond Therapy: Biotechnology and the Pursuit of Human Improvement," prepared for the President's Council on Bioethics, Washington, DC, January 16, 2003. Copy available at the Council's website at www.bioethics.gov.

7. The gift of added years of expected future life is surely a great blessing for the young. But is the correlative perception of a seemingly limitless future an equal blessing? How preciously do people regard each day of life when its limits are out of sight?

8. See, for example, Jonas, H. (January/February 1992). The Blessings and Burdens of Mortality. *Hastings Center Report*; Kass, L. (2002). *L'Chaim* and its limits: Why not immortality. In *Life, liberty, and the defense of dignity: The challenge for bioethics*. San Francisco: Encounter Books.

Reading 2.5

Toward Responsible Use of Cognitive-Enhancing Drugs by the Healthy: Policy Suggestions[1]

Henry Greely, Barbara Sahakian, John Harris, Ronald C. Kessler, Michael Gazzaniga, Philip Campbell, and Martha J. Farah

The new methods of cognitive enhancement are "disruptive technologies" that could have a profound effect on human life in the twenty-first century. A *laissez-faire* approach to these methods will leave us at the mercy of powerful market forces that are bound to be unleashed by the promise of increased productivity and competitive advantage. Concerns about safety, freedom, and fairness may well seem less important than the attractions of enhancement, for sellers and users alike.

In consideration of these dangers, Fukuyama (2002) has proposed the formation of new laws and regulatory structures to protect against the harms of unrestrained biotechnological enhancement. In contrast, we suggest a policy that is neither *laissez-faire* nor primarily legislative. We propose to use a variety of scientific, professional, educational, and social resources, in addition to legislation, to shape a rational, evidence-based policy informed by a wide array of relevant experts and stakeholders. Specifically, we propose four types of policy mechanism.

2.5.1 Research

The first mechanism is an accelerated program of research to build a knowledge base concerning the usage, benefits, and associated risks of cognitive enhancements. Good policy is based on good information, and there is currently much we do not know about the short- and long-term benefits and risks of the cognitive-enhancement drugs currently being used, and about who is using them and why. For example, what are the patterns of use outside of the United States and outside of college communities? What are the risks of dependence when used for cognitive enhancement? What special risks arise with the enhancement of children's cognition? How big are the effects of currently available enhancers? Do they change "cognitive style," as well as increasing how quickly and accurately we think? And given that most research so far has focused on simple laboratory tasks, how do they affect cognition in the real world? Do they increase the total knowledge and understanding that students take with them from a course? How do they affect various aspects of occupational performance?

We call for a program of research into the use and impacts of cognitive-enhancing drugs by healthy individuals.

2.5.2 Leadership from Relevant Professions

The second mechanism is the participation of relevant professional organizations in formulating guidelines for their members in relation to cognitive enhancement. Many different professions have a role in dispensing, using, or working with people who use cognitive enhancers. By creating policy at the level of professional societies, it will be informed by the expertise of these professionals and their commitment to the goals of their profession.

One group to which this recommendation applies is physicians, particularly in primary care, pediatrics, and psychiatry, who are most likely to be asked for cognitive enhancers. These physicians are sometimes asked to prescribe for enhancement by patients who exaggerate or fabricate symptoms of attention deficit–hyperactivity disorder (ADHD), but they also receive frank requests, as when a patient says "I know I don't meet diagnostic criteria for ADHD, but I sometimes have trouble concentrating and staying organized, and it would help me to have some Ritalin on hand for days when I really need to be on top of things at work." Physicians who view medicine as devoted to healing will view such prescribing as inappropriate, whereas those who view medicine more broadly as helping patients live better or achieve their goals would be open to considering such a request (Chatterjee, 2004). There is certainly a precedent for this broader view in certain branches of medicine, including plastic surgery, dermatology, sports medicine, and fertility medicine.

Because physicians are the gatekeepers to medications discussed here, society looks to them for guidance on the use of these medications and devices, and guidelines from other professional groups will need to take into account the gatekeepers' policies. For this reason, the responsibilities that physicians bear for the consequences of their decisions are particularly sensitive, being effectively decisions for all of us. It would therefore be helpful if physicians as a profession gave serious consideration to the ethics of appropriate prescribing of cognitive enhancers and consulted widely as to how to strike the balance of limits for patient benefit and protection in a liberal democracy. Examples of such limits in other areas of enhancement medicine include the psychological screening of candidates for cosmetic surgery or tubal ligation, and upper bounds on maternal age or number of embryos transferred in fertility treatments. These examples of limits may not be specified by law, but rather by professional standards.

Other professional groups to which this recommendation applies include educators and human-resource professionals. In different ways, each of these professions has responsibility for fostering and evaluating cognitive performance and for advising individuals who are seeking to improve their performance, and some responsibility also for protecting the interests of those in their charge. In contrast with physicians, these professionals have direct conflicts of interest that must be addressed in whatever guidelines they recommend: Liberal use of cognitive enhancers would be expected to encourage classroom order and raise standardized measures of student achievement, both of which are in the interests of schools; it would also be expected to promote workplace productivity, which is in the interests of employers.

Educators, academic admissions officers, and credentials evaluators are normally responsible for ensuring the validity and integrity of their examinations and should be tasked with formulating policies concerning enhancement by test-takers. Laws pertaining to testing accommodations for people with disabilities provide a starting point for discussion of some of the key issues, such as how and when enhancements undermine the validity of a test result and the conditions under which enhancement should be disclosed by a test-taker.

The labor and professional organizations of individuals who are candidates for on-the-job cognitive enhancement make up our final category of organization that should formulate enhancement policy. From assembly line workers to surgeons, many different kinds of employee may benefit from enhancement and want access to it, yet they may also need protection from the pressure to enhance.

We call for physicians, educators, regulators, and others to collaborate in developing policies that address the use of cognitive-enhancing drugs by healthy individuals.

2.5.3 Public Education

The third mechanism is education to increase public understanding of cognitive enhancement. This would be provided by physicians, teachers, college health centers, and employers, similar to the way that information about nutrition, recreational drugs, and other public-health information is now disseminated. Ideally it would also involve discussions of different ways of enhancing cognition, including through adequate sleep, exercise, and education, and an examination of the social values and pressures that make cognitive enhancement so attractive and even, seemingly, necessary.

We call for information to be broadly disseminated concerning the risks, benefits, and alternatives to pharmaceutical cognitive enhancement.

2.5.4 Legislation

The fourth mechanism is legislative. Fundamentally new laws or regulatory agencies are not needed. Instead, existing law should be brought into line with emerging social norms and information about safety. Drug law is one of the most controversial areas of law, and it would be naive to expect rapid or revolutionary change in the laws governing the use of controlled substances. Nevertheless, these laws should be adjusted to avoid making felons out of those who seek to use safe cognitive enhancements. And regulatory agencies should allow pharmaceutical companies to market cognitive-enhancing drugs to healthy adults provided they have supplied the necessary regulatory data for safety and efficacy.

We call for careful and limited legislative action to channel cognitive-enhancement technologies into useful paths.

Like all new technologies, cognitive enhancement can be used well or poorly. We should welcome new methods of improving our brain function. In a world in which human work-spans and life spans are increasing, cognitive enhancement tools—including the pharmacological—will be increasingly useful for improved quality of life and extended work productivity, as well as to stave off normal and pathologic age-related cognitive declines (Beddington et al., 2008). Safe and effective cognitive enhancers will benefit both the individual and society.

But it would also be foolish to ignore problems that such use of drugs could create or exacerbate. With this, as with other technologies, we need to think and work hard to maximize its benefits and minimize its harms.

Notes

1. Editor's note: This reading was excerpted from a substantially longer article entitled "Toward responsible use of cognitive-enhancing drugs by the healthy" that appeared in 2008 in *Nature*, volume 456, pages 702–705, and is used with permission. The section reprinted here presents our policy suggestions. The omitted sections summarize the social and ethical issues surrounding cognitive enhancement more generally. The original article is the result of a seminar held by the authors at Rockefeller University. Funds for the seminar were provided by Rockefeller University and *Nature*.

References

Chatterjee, A. (2004). Cosmetic neuology: The controversy over enhancing movement, mentation and mood. *Neurology, 63*, 968–974.

Beddington, J., Cooper, C. L., Field, J., Goswami, U., Huppert, F. A., Jenkins, R., Jones, H. S., Kirkwood, T. B., Sahakian, B. J., & Thomas, S. M. (2008). The mental wealth of nations. *Nature, 455*, 1057–1060.

Fukuyama, F. (2002). *Our posthuman future: Consequences of the biotechnology revolution.* New York: Farrar, Straus and Giroux.

3

Brain, Self, and Authenticity

Academics may argue over whether it makes sense to talk of a fixed or unitary "self," but in everyday life we constantly invoke just such a concept. From Shakespeare's exhortation "to thine own self be true" to the fitting room chatter "this dress is just not me," we take for granted that there is an enduring and characteristic set of values and dispositions that defines each of us. Of course, this set is not entirely fixed. We are familiar with changes that occur in development from childhood to maturity, and we also say things like "he's a new man" or "I'm not myself today." These figures of speech mean that aspects of a person's characteristic ways of viewing or reacting to the world can change. But substantial transformations of an adult's personality or self-concept are unusual and are difficult for us to assimilate into our traditional understanding of human nature. That such transformations could be effected by a pill makes them all the more jarring. As Peter Kramer so aptly put it in his landmark book, *Listening to Prozac* (1997, p. 332): "The self, and how it fares in a world where personality is understood as 'biological' and subject to biological influence, is a central issue for our time.... Who are we, if we can be so altered by medication? And why should the medicated self on occasion feel more 'true' than the unmedicated?"

The readings in this section reflect on the ways in which psychopharmacology can alter core aspects of our personal identity, through manipulation of our memories and our feelings about them, and through manipulation of our personalities. They highlight both the hopeful and worrisome potentials of such technologies, for the individual using them and for society as a whole. We begin with the topic of memory, that neural autobiography that writers and nonwriters alike create and carry with them through life.

Identity, Memory, and the Brain

By virtue of what is someone the same person despite the passage of time? Philosophers generally dismiss answers that depend on physical identity or continuity. If we imagine the admittedly unrealistic scenario of having body parts all gradually replaced or being reconstituted at a distance by a Star Trek–style transporter, our intuitions tell us that we would remain the same person and hence that personal identity is not a matter of physical stuff. Starting with John Locke, philosophers have tended to focus on psychological continuity (see Parfit, 1984, for an influential contemporary example of this approach). According to this view, what yesterday's me and today's me have in common is that today's me remembers the experiences of yesterday's me.

Brain interventions to manipulate memories would enable us to edit our mental autobiography at will. This has a number of implications for personal identity, from the perspective of both authenticity and moral responsibility, as the President's Council on Bioethics and Kolber argue at length. To what extent can memories be erased with current or foreseeable neurotechnology? The answer is that some degree of memory modification appears to be possible, but we are nowhere near the capabilities that worry the President's Council on Bioethics.

One approach to memory erasure is based on "reconsolidation," which is the controversial idea that every time a memory is recalled, it is rendered temporarily vulnerable and must again undergo the process of consolidation into a more stable and enduring form, just as it did when it was first acquired (Nader, 2003). If this were true, then memories could be erased by evoking them and applying a treatment that interferes with consolidation. Something like this procedure was envisioned by Michel Gondry in his film *The Eternal Sunshine of the Spotless Mind*. Although there is laboratory evidence for a period of vulnerability for memories after each act of recall, they are not literally returned to their preconsolidation state: Whatever degree of erasure is accomplished after recall appears to be temporary, so that the erased memory gradually returns over time (Eisenberg & Dudai, 2004).

Another way of manipulating memory requires intervening at or near the time the memory is being formed and has so far been limited to influencing the strength of emotional memories. Highly emotional experiences tend to be remembered especially vividly (Phelps, 2006). This feature of our memory system has obvious survival value. Once you've seen your neighbor from the next cave eaten by a saber-toothed tiger, an especially vivid memory of this will help you to avoid saber-toothed tigers in the future. The neural mechanisms responsible for the special staying power of emotional memories are fairly well worked out.

Detection of a threat triggers a chain of neurochemical events leading to the secretion of epinephrine and norepinephrine (also called adrenaline and noradrenaline). These molecules affect many bodily systems, including the heart, gut, and muscles as well as many parts of the brain, causing a reaction known as the "fight or flight" response. Within the brain, the amygdala has receptors for these chemicals and, when activated by them, it shifts its functioning and that of the hippocampus toward consolidating more enduring memories with strong associated feelings. By administering drugs that block the receptors for adrenaline and noradrenaline, we do not prevent the memory from being formed, but we weaken it and the accompanying emotion (relative to what they would have been). The name of the class of drugs that has been most explored for memory dampening is "beta blockers," because it is the beta-adrenergic receptors that play a role in memory. Propranolol is a beta-blocker and has been used experimentally to prevent post-traumatic stress disorder (PTSD). If administered shortly after a traumatic experience, people are less likely to suffer from the intrusive and disturbing memories that characterize PTSD (Pitman et al., 2002).

The Neurochemistry of Personality

Another area of neuroscience with direct relevance to personal identity is the neuroscience of personality. Contemporary personality research is largely concerned with the identification and understanding of dimensions of personality; that is, the handful of major ways in which people differ from one another. The most widely used framework is called the "Big Five" (Costa & McCrae, 1985) because it identifies five relatively independent traits on which people vary: Neuroticism (a tendency to experience changing mood and especially negative emotion), Conscientiousness (a concern with order and the regulation of impulses), Extraversion (engagement with the world, especially positive experiences and other people), Openness to Experience (distinguishing the flexible and imaginative from the more conventional and down-to-earth people), and Agreeableness (valuing harmony and cooperation). There is of course much more to a person than a position in a five-dimensional space. Yet for purposes of discussing personal identity, the Big Five theory and others of its ilk provide a useful framework. For example, the melancholic temperament discussed in the reading by Kramer is generally viewed as a combination of high Neuroticism and low Extraversion.

We know from twin studies and other methods of behavioral genetics research that personality, as measured by levels of these traits, is partly genetically determined (Parens, 2004). Recent research has sought to identify some of the contributing genes, and not surprisingly, the focus is on genes that govern neurotransmitter function (Munafò et al., 2003). Molecular genetics suggests that individual personality is at least partly a function of individual neurochemistry. This suggests that drugs that alter neurotransmitter functions may also alter personality. Not surprisingly, when patients who are clinically depressed or anxious are successfully treated with medication, certain personality traits change. Less obvious is the fact that such medications can subtly alter personality in normal healthy individuals.

In the earliest study of antidepressant effects on the personality of normal subjects, Knutson and colleagues (1998) administered either a selective serotonin reuptake inhibitor (SSRI, discussed in the previous chapter) or a placebo to different groups of healthy young adults for 4 weeks. The subjects on the SSRI rated themselves as feeling less negative emotion, particularly hostility, and behaved more cooperatively in a collaborative problem-solving situation. Similar effects on social behavior were observed by Tse and Bond (2002), although in this shorter-duration study no effects on mood were found. In a study of mood reactivity, Furlan et al. (2004) administered SSRI or placebo to healthy older subjects and had them keep diaries recording daily events as well as their moods. Although they did not find an overall change in positive or negative mood, they found a specific change in reactivity to negative events. As the authors put it, the "hassles" of life evoked less negative emotion in the subjects on SSRI than in those on placebo, which is a fairly good operational definition of reduced Neuroticism. In sum, our natural personalities are at least in part determined by characteristics of neurotransmitter activity under genetic control, and psychopharmacology can alter our personality by altering neurotransmitter activity.

Neuroethics of Personal Identity

The Readings

The readings in this section revisit some of the same neuroethical issues discussed in the previous section, such as the elusive line between therapy and enhancement and the medicalization of life's ups and downs. In addition, however, they focus on a specific concern that applies to the modification of our memories and our personalities, the concern that an enhanced self is a less authentic self.

The link between memory and authenticity is very much at the heart of the *President's Council on Bioethics'* critique of memory-dampening drugs. They enumerate several ways in which editing memory could undermine our personal authenticity and our individual and societal conscience and make us into happy but shallow beings. In their words, memory dampening could lead us to pursue "happiness by willingly abandoning or compromising our own truthful identities: instead of integrating, as best we can, the troubling events of our lives into a more coherent whole, we might just prefer to edit them out..." A number of commentators have opposed the Council's position on relatively pragmatic grounds, for example by citing the need to balance the social risks of medicalization against the health risks of PTSD (Henry, Fishman, & Youngner, 2007). In contrast, *Kolber* engages the Council's arguments one by one, including the relatively philosophical worries about personal identity, authenticity, and other ethical and societal concerns. He carefully unpacks the arguments, identifying unstated assumptions and offering analogies, precedents, and counterexamples to assuage many of the Council's concerns.

Elliott puts the idea of pharmacologic self-transformation into historical context and reflects on the ways that Prozac and similar drugs provide a new kind of identity for those using them today. He notes that decades before Prozac came onto the market, antianxiety medications were being touted as a new means to self-improvement, allowing people to achieve their potential in life unimpeded by fears and other negative emotions. He highlights the contradictory values that our society holds concerning pharmacologic enhancement of the self, with our respect for authenticity at odds with our admiration for adaptability and self-improvement. Elliott also calls our attention to the role that pharmaceutical company marketing has had on our self-concepts. By popularizing new and expanded diagnostic categories that encroach on seemingly normal life problems, and offering medical explanations and solutions, these companies have provided us with new personal identities based on those categories.

Kramer tackles some deeply ingrained cultural assumptions about temperament, art, and authenticity, which Elliott has written about elsewhere (e.g., Elliott, 1998). Kramer asks whether the elimination of melancholy would make us more superficial as individuals or would rob us of great art and literature as a society. Can we be fully human and fully ourselves without being fully vulnerable to dread and pain? With choice quotations and allusions to literature and art, Kramer demonstrates the reality of what could be called melancholic chauvinism and goes on to argue persuasively that it rests on assumptions rather than proven facts.

Selected Cross-cutting Issues

Although the readings in this section cover two distinct sets of neurotechnologies, memory dampening and personality enhancement, they converge in addressing broader common issues of self and authenticity. In what sense are these broader issues really the same? Let us examine the concepts of "authentic self" discussed by the President's Council on Bioethics, Kolber, Elliot, and Kramer and find out how they are related.

In the case of memory dampening, the word *authenticity* means something very close to its most literal sense, that is, "true" or "veridical." We are our authentic self if we retain the memories of what we have truly seen, heard, and done. If we edit them out at will, we have disconnected our sense of self from reality, undermining the veridicality or "authenticity" of that self.

In the case of personality enhancement, authenticity involves seeing the world as it is, without rose-colored glasses, and therefore also involves the concept of veridicality. Elliott points out that the SSRIs, and the tranquilizers that preceded them, have been treated as solutions to problems that may originate in the pressures and constraints of modern life. If our life does not feel right, it may be easier to conclude that our brain chemistry needs a little help than to confront the problematic realities of our existence. In this vein he quotes Marya Mannes on the use of medication to soothe anxieties created by the media and consumer culture. Elliot has elsewhere written of the importance of being able to experience alienation when appropriate; that is, when the values of the surrounding community are worthy of rejection (Elliott, 1998). In both cases, it is possible to pharmacologically "enhance" our personalities to exist more contentedly in a screwed-up world, but we are sacrificing the authenticity of our experience of the world if we do so. As Elliott points out in reading 3.3, the interpretation of life's anxieties and pains as medical problems with pharmaceutical solutions can give rise to a new kind of identity, inauthentic and indeed scripted by the marketing departments of drug companies, whereby we are all patients in need of treatment.

One might expect the notion of veridicality to be less relevant to understanding authenticity in the artistic temperament, because art is not about literal truth. Yet as Kramer reminds us, art is a response to the world, and reflects our perceptions, hopes, and fears. For reasons unknown, a disproportionate amount of great literature has been written by melancholics, and hence a fundamentally cheery view of the world seems inauthentic to those with a literary worldview. But if the talented writers of the world all took Prozac, perhaps this cheer would come to seem authentic and gloom would seem silly and melodramatic.

The value of pain is another theme that runs through the readings of this section. Painful memories of our own suffering, or that of others, may prevent us from causing pain in the future. The dampening of such memories might lead us to care less about the welfare of others. And painful memories may play a role in establishing a personal identity, because in coming to terms with them we forge a stronger personality and construct a life narrative that helps give meaning to our lives. These opinions are expressed most strongly by the President's Council on Bioethics, but even Kolber grants that there is truth to them and argues only that they are frequently outweighed by other considerations. The relation of personal authenticity to pain is raised by Elliott, who points out the contradiction between the ideals of both "being yourself" and striving to change for the better. Although he discusses the changes wrought by Prozac in fairly general terms, including increased energy, confidence, and flexibility, these more general effects are at least partly the result of reduced anxiety and psychic pain. Hence Elliott in effect asks if we are living inauthentically when we use medication to dull the pains we would normally suffer from life's insecurities and losses. Finally, as both Kolber and Kramer note, the value of pain is implicit in the lore of artistic creation; the chronic pain of the melancholic has been credited with making his art better and more profound. Of course, Kramer points out that there is little evidence to support or deny this.

Questions for Discussion

1. Pharmacologic transformation of the self will obviously affect individual selves, but the authors in this section also reflect on the effects of memory and personality manipulation on society. What are some of the ways in which the neurotechnologies discussed here could affect the legal system, culture, and politics? What other spheres of society would be affected?

2. Imagine that you are about to undergo a long, drawn out and horrific experience; for example, a year-long prison term during which you will be tortured frequently. Also imagine two hypothetical drugs that you could take during this period of your life: Although neither drug would change the experience itself, one would engender strong optimism and hope throughout the ordeal (a kind of super-Prozac) and one would render the memories of the torture sessions very abstract and faint as soon as each is over (a kind of super-propranolol). How would each affect the person you would be during the year and afterward, and which person would you rather be? In what ways are their effects similar and in what ways different? Which would you choose and why?
3. The neurotechnologies discussed in this section have been anticipated by novelists and film-makers, who have expressed some of the same concerns over their likely impact on our self-concepts, our relationships with others, and our very humanity. Many of the stories of Phillip K. Dick and the movies they inspired (e.g. *Total Recall*, *Paycheck*) explore the idea of memory manipulation, and Michel Gondry's (2004) film *Eternal Sunshine of the Spotless Mind* is a moving, if wacky, look at what we stand to lose by altering our memories for the sake of living more comfortably. Stories premised on drugs that reduce negative affect are generally morality tales, focusing on bad outcomes. Examples include Aldous Huxley's *Brave New World* with its mood-enhancing "Soma," Walker Percy's *Thanatos Syndrome*, and Robin Cook's *Acceptable Risk*. What other books and films explore these issues? Focus on one, named here or of your own choosing, and identify which of the concerns raised in the readings are played out in this work of fiction. Do other ethical or social issues arise in the fictional work that were not mentioned in the readings?

References

Costa, P. T., Jr., & McCrae, R. R. (1985). *The NEO personality inventory manual.* Odessa, FL: Psychological Assessment Resources.

Eisenberg, M., & Dudai, Y. (2004). Reconsolidation of fresh, remote, and extinguished fear memory in Medaka: Old fears don't die. *The European Journal of Neuroscience, 20*(12), 3397–3403.

Elliott, C. (1998). The tyranny of happiness: Ethics and cosmetic psychopharmacology. In E. Parens (Ed.), *Enhancing human traits*. Washington DC: Georgetown University Press.

Furlan, P. M., Kallan, M. J., Have, T. T., Lucki, I., & Katz, I. (2004). SSRIs do not cause affective blunting in healthy elderly volunteers. *The American Journal of Geriatric Psychiatry, 12*(3), 323–330.

Gondry, M. (Director). (2004). *The eternal sunshine of the spotless mind* [Motion picture]. Los Angeles: Focus Features.

Henry, M., Fishman, J. R., & Youngner, S. J. (2007). Propranolol and the prevention of post-traumatic stress disorder: Is it wrong to erase the "string" of bad memories? *American Journal of Bioethics, 7,* 12–20.

Knutson, B., Wolkowitz, O. M., Cole, S. W., Chap, T., & Moore, E. A. (1998). Selective alteration of personality and social behavior by seretonergic intervention. *American Journal of Psychiatry, 155,* 373–379.

Kramer, P. D. (1993). *Listening to prozac.* New York: Viking.

Munafò, M. R., Clark, T. G., Moore, L. R., Payne, E., Walton, R., & Flint, J. (2003). Genetic polymorphisms and personality in healthy adults: A systematic review and meta-analysis. *Molecular Psychiatry, 8,* 471–484.

Nader, K. (2003). Memory traces unbound. *Trends in Neurosciences, 26,* 65–72.

Parens, E. (2004). Genetic differences and human identities: On why talking about behavioral genetics is important and difficult. *The Hastings Center Report,* 34, S1–S36.

Parfit, D. (1984). *Reasons and persons.* Oxford: Oxford University Press.

Phelps, E. A. (2006). Emotion and cognition: Insights from studies of the human amygdale. *Annual Review of Psychology, 57,* 27–53.

Pitman, R. K., Sanders, K. M., Zusman, R. M., Healy, A. R., Cheema, F., Lasko, N. B., et al. (2002). A pilot study of secondary prevention of post-traumatic stress disorder with propranolol. *Biological Psychiatry, 51,* 189–192.

Tse, W. S., & Bond, A. J. (2002). Difference in serotonergic and noradrenergic regulation of human social behaviors. *Psychopharmacology, 159,* 216–221.

Reading 3.1

Memory Blunting: Ethical Analysis[1]

President's Council on Bioethics

If we had the power, by promptly taking a memory-altering drug, to dull the emotional impact of what could become very painful memories, when might we be tempted to use it? And for what reasons should we yield to or resist the temptation?

At first glance, such a drug would seem ideally suited for the prevention of post-traumatic stress disorder (PTSD), the complex of debilitating symptoms that sometimes afflict those who have experienced severe trauma. These symptoms—which include persistent re-experiencing of the traumatic event and avoidance of every person, place, or thing that might stimulate the horrid memory's return[2]—can so burden mental life as to make normal everyday living extremely difficult, if not impossible.[3] For those suffering these disturbing symptoms, a drug that could separate a painful memory from its powerful emotional component would appear very welcome indeed.

Yet the prospect of preventing (even) PTSD with beta-blockers or other memory-blunting agents seems to be, for several reasons, problematic. First of all, the drugs in question appear to be effective only when administered during or shortly after a traumatic event—and thus well before any symptoms of PTSD would be manifested. How then could we make, and make on the spot, the *prospective* judgment that a particular event is sufficiently terrible to warrant preemptive memory-blunting? Second, how shall we judge *which* participants in the event merit such treatment? After all, not everyone who suffers through painful experiences is destined to have pathologic memory effects. Should the drugs in question be given to everyone or only to those with an observed susceptibility to PTSD, and, if the latter, how will we know who these are? Finally, in some cases merely witnessing a disturbing event (for example, a murder, rape, or terrorist attack) is sufficient to cause PTSD-like symptoms long afterwards. Should we then, as soon as disaster strikes, consider giving memory-altering drugs to all the witnesses, in addition to those directly involved?

These questions point to other troubling implications. Use of memory-blunters at the time of traumatic events could interfere with the normal psychic work and adaptive value of emotionally charged memory. A primary function of the brain's special way of encoding memories for emotional experiences would seem to be to make us remember important events longer and more vividly than trivial events. Thus, by blunting the emotional impact of events, beta-blockers or their successors would concomitantly weaken our recollection of the traumatic events we have just experienced. Yet often it is important, in the aftermath of such events, that at least someone remembers them clearly. For legal reasons, to say nothing of deeper social and personal ones, the wisdom of routinely interfering with the memories of trauma survivors and witnesses is highly questionable.

If the apparent powers of memory-blunting drugs are confirmed, some might be inclined to prescribe them liberally to all who are involved in a sufficiently terrible event. After all, even those not destined to come down with full-blown PTSD are likely to suffer painful recurrent memories of an airplane crash, an incident of terrorism, or a violent combat operation. In the aftermath of such shocking incidents, why not give everyone the chance to remember these events without the added burden of painful emotions? This line of reasoning might, in fact, tempt us to give beta-blockers liberally to soldiers on the eve of combat, to emergency workers en route to a disaster site, or even to individuals requesting prophylaxis against the shame or guilt they might incur from future misdeeds—in general, to anyone facing an experience that is likely to leave lasting intrusive memories.

Yet on further reflection it seems clear that not every intrusive memory is a suitable candidate for prospective pharmacologic blunting. As Daniel Schacter has observed, "attempts to avoid traumatic memories often backfire":

> Intrusive memories need to be acknowledged, confronted, and worked through, in order to set them to rest for the long term. Unwelcome memories of trauma are symptoms of a disrupted psyche that requires attention before it can resume healthy functioning. Beta-blockers might make it easier for trauma survivors to face and incorporate traumatic recollections, and in that sense could facilitate long-term adaptation. Yet it is also possible that beta-blockers would work against the normal process of recovery: traumatic memories would not spring to mind with the kind of psychological force that demands attention and perhaps intervention. Prescription of beta-blockers could bring about an effective

trade-off between short-term reductions in the sting of traumatic memories and long-term increases in persistence of related symptoms of a trauma that has not been adequately confronted.[4]

The point can be generalized: in the immediate aftermath of a painful experience, we simply cannot know either the full meaning of the experience in question or the ultimate character and future prospects of the individual who experiences it. We cannot know how this experience will change this person at this time and over time. Will he be cursed forever by unbearable memories that, in retrospect, clearly should have been blunted medically? Or will he succeed, over time, in "redeeming" those painful memories by actively integrating them into the narrative of his life? By "rewriting" memories pharmacologically we might succeed in easing real suffering at the risk of falsifying our perception of the world and undermining our true identity.

Finally, the decision whether or not to use memory-blunting drugs must be made in the absence of clearly diagnosable disease. The drug must be taken right after a traumatic experience has occurred, and thus before the different ways that different individuals handle the same experience has become clear. In some cases, these interventions will turn out to have been preventive medicine, intervening to ward off the onset of PTSD before it arrives—though it is worth noting that we would lack even post hoc knowledge of whether any particular now-unaffected individual, in the absence of using the drug, would have become symptomatic.[5] In other cases, the interventions would not be medicine at all: altering the memory of individuals who could have lived well, even with severely painful memories, without pharmacologically dulling the pain. Worse, in still other cases, the use of such drugs would inoculate individuals in advance against the psychic pain that *should* accompany their commission of cruel, brutal, or shameful deeds. But in all cases, from the defensible to the dubious, the use of such powers changes the character of human memory, by intervening directly in the way individuals "encode," and thus the way they understand, the happenings of their own lives and the realities of the world around them. Sorting out how and why this matters, and especially what it means for our idea of human happiness, is the focus of the more particular—albeit brief—ethical reflections that follow.

3.1.1 Remembering Fitly and Truly

Altering the formation of emotionally powerful memories risks severing what we remember from how we remember it and distorting the link between our perception of significant human events and the significance of the events themselves. It risks, in a word, falsifying our perception and understanding of the world. It risks making shameful acts seem less shameful, or terrible acts less terrible, than they really are.

Imagine the experience of a person who witnesses a shocking murder. Fearing that he will be haunted by images of this event, he immediately takes propranolol (or its more potent successor) to render his memory of the murder less painful and intrusive. Thanks to the drug, his memory of the murder gets encoded as a garden-variety, emotionally neutral experience. But in manipulating his memory in this way, he risks coming to think about the murder as more tolerable than it really is, as an event that should not sting those who witness it. For our opinions about the meaning of our experiences are shaped partly by the feelings evoked when we remember them. If, psychologically, the murder is transformed into an event our witness can recall without pain—or without *any* particular emotion—perhaps its moral significance will also fade from consciousness. If so, he would in a sense have ceased to be a genuine witness of the murder. When asked about it, he might say, "Yes, I was there. But it wasn't so terrible."

This points us to a deeper set of questions about bad memories: Would dulling our memory of terrible things make us too comfortable with the world, unmoved by suffering, wrongdoing, or cruelty? Does not the experience of hard truths—of the unchosen, the inexplicable, the tragic—remind us that we can never be fully at home in the world, especially if we are to take seriously the reality of human evil? Further, by blunting our experience and awareness of shameful, fearful, and hateful things, might we not also risk deadening our response to what is admirable, inspiring, and lovable? Can we become numb to life's sharpest sorrows without also becoming numb to its greatest joys?

These questions point to what might be the highest cost of making our memory of intolerable things more tolerable: Armed with new powers to ease the suffering of bad memories, we might come to see all psychic pain as unnecessary and in the process come to pursue a happiness that is less than human: an unmindful happiness, unchanged by time and events, unmoved by life's vicissitudes. More precisely, we might come to pursue such happiness by willingly abandoning or compromising our own truthful identities: instead of integrating, as best we can, the troubling events of our lives into a more coherent whole, we might just prefer to edit them out or make them less difficult to live with than they really are.

There seems to be little doubt that some bitter memories are so painful and intrusive as to ruin the possibility for normal experience of much of life and the world. In such cases the impulse to relieve a crushing burden and restore lost innocence is fully understandable: If there are some things that it is better never to have experienced at all—things we would avoid if we possibly could—why not erase them from the memory of those unfortunate enough to have suffered them? If there are some things it is better never to have known or seen, why not use our power over memory to restore a witness's shattered peace of mind? There is great force in this argument, perhaps especially in cases where children lose prematurely that innocence that is rightfully theirs.

And yet, there may be a great cost to acting compassionately for those who suffer bad memories, if we do so by compromising the truthfulness of how they remember. We risk having them live falsely in order simply to cope, to survive by whatever means possible. Among the larger falsehoods to which such practices could lead us, few are more problematic than the extreme beliefs regarding the possibility—and impossibility—of human control. Erring on the one side, we might come to imagine ourselves as having more control over our memories and identities than we really do, believing that we can be authors and editors of our memories while still remaining truly—and true to—ourselves. Erring on the other side, we might come to imagine that we are impotently in the grip of the past as we look to the future, believing that we can never learn to live with this particular memory or give it new meaning. And so we ease today's pain, but only by foreclosing, in a certain way, the possibility of being the kind of person who can live well with the whole truth—both chosen and unchosen—and the kind of person who can live well as himself.

3.1.2 The Obligation to Remember

Having truthful memories is not simply a personal matter. Strange to say, our own memory is not merely our own; it is part of the fabric of the society in which we live. Consider the case of a person who has suffered or witnessed atrocities that occasion unbearable memories; for example, those with firsthand experience of the Holocaust. The life of that individual might well be served by dulling such bitter memories,[6] but such a humanitarian intervention, if widely practiced, would seem deeply troubling: Would the community as a whole—would the human race—be served by such a mass numbing of this terrible but indispensable memory? Do those who suffer evil have a duty to remember and bear witness, lest we all forget the very horrors that haunt them? (The examples of this dilemma need not be quite so stark: the memory of being embarrassed is a source of empathy for others who suffer embarrassment; the memory of losing a loved one is a source of empathy for those who experience a similar loss.) Surely, we cannot and should not force those who live through great trauma to endure its painful memory *for the benefit of the rest of us*. But as a community, there are certain events that we have an obligation to remember—an obligation that falls disproportionately, one might even say unfairly, on those who experience such events most directly.[7] What kind of people would we be if we did not "want" to remember the Holocaust, if we sought to make the anguish it caused simply go away? And yet, what kind of people are we, especially those who face such horrors firsthand, that we can endure such awful memories?

The answer, in part, is that those who suffer terrible things cannot or should not have to endure their own bad memories alone. If, as a people, we have an obligation to remember certain terrible events truthfully, surely we ought to help those who suffered through those events to come to terms with their worst memories. Of course, one might see the new biotechnical powers, developed precisely to ease the psychic pain of bad memories, as the mark of such solidarity: perhaps it is our new way of meeting the obligation to aid those who remember the hardest things, those who bear witness to us and for us. But such solidarity may, in the end, prove false: for it exempts us from the duty to suffer-with (literally, to feel *com*-passion for) those who remember; it does not demand that we preserve the truth of their memories; it attempts instead to make the problem go away, and with it the truth of the experience in question.

3.1.3 Memory and Moral Responsibility

The question of how responsible we are or should be held for our memories, especially our memory failures, is a complicated one: Are remembering and forgetting voluntary or involuntary acts? To what extent should a man who forgets his child in a car, by mistake, be held "morally accountable" for his forgetting? Is remembering "something we do" or "something that happens to us"?

Hard as these questions are, this much seems clear: Without memory, both our own and that of others, the notion of moral responsibility would largely unravel. In particular, the power to numb or eliminate the psychic sting of certain memories risks eroding the responsibility we take for our own actions—as we would never have to face the harsh judgment of our own conscience (Lady Macbeth) or the memory of others. The risk applies both to self-serving uses of such a power (for example, drugs taken after a criminal act and before the next one) and to more ambiguous "social" uses (for example, drugs taken after killing in war and before killing again). Without truthful memory, we could not hold others or ourselves to account for what we do and who we are. Without truthful memory, there could be no justice or even the possibility of justice; without memory, there could be no forgiveness or the possibility of forgiveness—all would simply be *forgotten.*

The desire for powers that numb our most painful memories is largely a personal desire: to have such drugs for myself, in the service of my own peace of mind and happiness. Yet we cannot be blind to the potentially coercive and immoral uses—by other individuals and by the state—of biotechnical interventions that alter how we remember and what we forget and that indirectly affect our well-being. Just as drugs that dull the emotional sting of certain memories might be desired by the victim to ease his trauma, so they might be useful to the assailant to dull his victim's sense of being wronged. Perhaps no one has a greater interest in blocking the painful memory of evil than the evildoer. We also cannot ignore the potentially coercive nature of normalizing the use of such drugs in certain occupations; that is, by making chemically aided desensitization part of the "job description" (augmenting or replacing existing nonchemical means of desensitization). Nor can we forget the central place of manipulating memory in totalitarian societies, both real and imagined, and the way such manipulation made living truthfully—and living happily—impossible.

3.1.4 The Soul of Memory, the Remembering Soul

Perhaps more than any other subject in this report, memory is puzzling. It is both central to who we are as individuals and as a society, yet very hard to pin down—so variable in its many meanings and many manifestations. Jane Austen may have captured this complexity best:

> If any one faculty of our nature may be called more wonderful than the rest, I do think it is memory. There seems something more speakingly incomprehensible in the powers, the failures, the inequalities of memory, than in any other of our intelligences. The memory is sometimes so retentive, so serviceable, so obedient—at others, so bewildered and so weak—and at others again, so tyrannical, so beyond control!—We are to be sure a miracle every way—but our powers of recollecting and of forgetting, do seem peculiarly past finding out.[8]

On the one hand, when considering the meaning of human memory, we need to face the fact that there are limits to our control over who we are and what we become. We are not free to decide everything that happens to us; some experiences, both great joys and terrible misfortunes, simply befall us. These experiences become part of who we are, part of our own life as truthfully lived. And yet, we do have some measure of freedom in *how* we live with such memories—the meaning we assign them, the place we give them in the larger narrative of our lives. But this meaning is not simply arbitrary; it must connect the truth or significance of the events themselves, as they really were and really are, with our own continuing pursuit of a full and happy life. In doing so, we might often be tempted to sacrifice the accuracy of our memories for the sake of easing our pain or expanding our control over our own psychic lives. But doing so means, ultimately, severing ourselves from reality and leaving our own identity behind; it risks making us false, small, or capable of great illusions, and thus capable of great decadence or great evil, or perhaps simply willing to accept a phony contentment. We might be tempted to alter our memories to preserve an open future—to live the life we wanted to live before a particular experience happened to us. But in another sense, such interventions assume that our own future is not open—that we cannot and could never redeem the unwanted memory over time, that we cannot and could never integrate the remembered experience with our own truthful pursuit of happiness.

In the end, we must wonder what life would be like—and what kind of a people we would become—with only happy memories, with everything difficult, uncertain, and hard edited out of our lives as we remembered and understood them. We would suffer no loss, but perhaps only because we loved feebly and cared little for what we had. We would never shudder at life's injustices, but perhaps only because we had little interest in justice. We would little relish our own achievements, as we would achieve them without any memory of hardship along the way and with no recollection of achieving in spite of the odds. To have only happy memories would be a blessing—and a curse. Nothing would trouble us, but we would probably be shallow people, never falling to the depths of despair because we have little interest in the heights of human happiness or in the complicated lives of those around us. In the end, to have only happy memories is not to be happy in a truly human way. It is simply to be free of misery—an understandable desire given the many troubles of life, but a low aspiration for those who seek a truly human happiness.

Notes

1. Editor's note: This reading is excerpted from the report *Beyond Therapy: Biotechnology and the Pursuit of Happiness* (President's Council on Bioethics, 2003) available at www.bioethics.gov/reports/beyondtherapy.

2. There is no definitive diagnostic criterion for PTSD, but the core symptoms are thought to include persistent re-experiencing of the traumatic event, avoidance of associated stimuli, and hyperarousal. See *Diagnostic and Statistical Manual of Mental Disorders, Fourth Edition, text revision*, Washington, DC: American Psychiatric Association, 2000, pp. 463–486.

3. These symptoms are observed especially among combat veterans; indeed, PTSD is the modern name for what used to be called *shell shock* or *combat neurosis*. Among veterans, PTSD is frequently associated with recurrent nightmares, substance abuse, and delusional outbursts of violence. There is controversy about the prevalence of PTSD, with some studies finding that up to 8% of adult Americans have suffered the disorder, as well as a third of all veterans of the Vietnam War. See Kessler, R. C., et al., "Post-traumatic stress disorder in the national comorbidity survey," *Archives of General Psychiatry* 52(12): 1048–1060, 1995; Kulka, R. A., et al., *Trauma and the Vietnam War Generation: Report of Findings from the National Vietnam Veterans Readjustment Study*, New York: Brunner/Mazel, 1990.

4. Schacter, D., *The Seven Sins of Memory: How the Mind Forgets and Remembers*, New York: Houghton Mifflin, 2001, p. 183.

5. There is already ongoing controversy about excessive diagnosis of PTSD. Many psychotherapists believe that a patient's psychic troubles are generally based on some earlier (now repressed) traumatic experience that must be unearthed and dealt with if relief is to be found. True PTSD is, however, generally transient, and the search for treatment is directed against the symptoms of its initial (worst) phase—the sleeplessness, the nightmares, the excessive jitteriness.

6. Of course, many Holocaust survivors managed, without pharmacologic assistance, to live fulfilling lives while never forgetting what they lived through. At the same time, many survivors would almost certainly have benefited from pharmacologic treatment.

7. For a discussion of memory-altering drugs and the meaning of "bearing witness," see the essay by Cohen, E., "Our psychotropic memory," *SEED*, no. 8, Fall 2003, p. 42.

8. Austen, Jane, *Mansfield Park* (1814), Ch. 22. London: J. M. Dent & Co.

Reading 3.2

Ethical Implications of Memory Dampening[1]

Adam J. Kolber

Although true memory erasure is still the domain of science fiction,[2] less dramatic means of dampening the strength of a memory may have already been developed. Some experiments suggest that propranolol, an FDA-approved drug, can dull the emotional pain associated with the memory of an event when taken within 6 hours *after* the event occurs.[3] Furthermore, by reducing the emotional intensity of a memory, propranolol may be capable of dampening its factual richness as well.[4] Together, the research holds out the possibility that, under some circumstances, propranolol may dampen both emotional and factual components of memory.

Researchers are now conducting larger studies with propranolol to confirm these preliminary results[5] and to test whether propranolol might alleviate traumatic memories from the more distant past.[6] Meanwhile, even though propranolol was originally granted FDA approval to treat hypertension, clinicians may already use it to treat traumatic memories because doctors are permitted to prescribe it for off-label purposes.[7] Whether or not further research supports the use of propranolol to treat traumatic memories or focuses on some more potent successor, the quest for drugs to "therapeutically forget" is under way, and the search is starting to show promise.[8]

The President's Council on Bioethics (the "Council") explored many of these issues in a series of hearings in 2002 and 2003.[9] By and large, the Council was skeptical of the merits of memory dampening, raising concerns that memory dampening may (1) prevent us from truly coming to terms with trauma, (2) tamper with our identities, leading us to a false sense of happiness, (3) demean the genuineness of human life and experience, (4) encourage us to forget memories that we are obligated to keep, and (5) inure us to the pain of others. Although the Council did not make policy recommendations concerning memory-dampening drugs, one might ask whether the kinds of concerns raised by the Council could justify prohibiting or broadly restricting their use.

3.2.1 Specific Responses to the Prudential Concerns

3.2.1.1 The Tough Love Concern

The Council claims that memory dampening, by offering us a solution in a bottle, allows us to avoid the difficult but important process of coming to terms with emotional pain. There are two ways to understand the concern. The first is that there is something false or undeserved about the manner in which memory dampening eases distress. Gilbert Meilaender makes this point in his essay on memory dampening where he claims that, rather than erasing traumatic experiences, "it might still be better to struggle—with the help of others—to fit them into a coherent story that is the narrative of our life." "Our task," according to Meilaender, "is not so much to erase embarrassing, troubling, or painful moments, but, as best we can and with whatever help we are given, to attempt to redeem those moments by drawing them into a life whose whole transforms and transfigures them."[10]

People have divergent views, however, about what it means to transform and transfigure our experiences into "a coherent story." It seems quite plausible that one could craft a coherent life narrative that is punctuated by periods of dampened memories. Moreover, it is open to debate how important it is that one's life story be coherent or otherwise neatly packaged. Some recent research suggests that those with narcissistic, self-enhancing personalities tend to be particularly resilient after traumatic experiences.[11] Yet, whereas such personality traits may make it easier to cope with traumatic events, they do not necessarily do well for us in other aspects of our lives. Thus, it is at least a complicated matter whether we should seek to develop those aspects of our personalities that help us rebound after trauma.

Furthermore, even if one shares Meilaender's preference to redeem and transform our experiences without memory dampeners, two additional responses are suggested. First, many experiences are simply tragic and terrifying, offering virtually no opportunity for redemption or transformation. For example, after a 1978 plane crash in San Diego, desk clerks and baggage handlers were assigned to retrieve dead bodies and clean up the crash site.[12] Emotionally unprepared for this task, many of them were so distraught that they were unable to return to work. In such cases, it seems unlikely that the traumatized employees should, in Meilaender's words, "redeem those moments by drawing them into a life whose whole transforms and transfigures them."[13] Most would agree that such employees should not have participated in the cleanup in the first place, and, hence, they should not be required or expected to bear the emotional burden of having done so.[14]

Second, even if it is better to weave traumatic events into positive, life-affirming narratives, many people are never able to do so. Memory-dampening drugs may enable such people to make life transformations that they would be *incapable* of making in the absence of the drugs. For others, pharmaceuticals may drastically shorten the time it takes to recover from a traumatic experience. Suppose a person spends 10 years coming to terms with a traumatic event that he could have come to terms with in 2 years with pharmaceutical assistance. Whereas he might be viewed as heroic by Meilaender, others might view him as extremely obstinate. Therefore, even in those instances when positive human transformation should accompany traumatic experience, there may well be a role for memory dampening to facilitate the process.

The more modest version of the "tough love" concern merely states that "[p]eople who take pills to block from memory the painful or hateful aspects of a new experience will not learn how to deal with suffering or sorrow."[15] This concern, however, merely fights the hypothetical existence of effective memory-dampening drugs. If a memory-dampening drug increases the overall psychological distress of patients, by being addictive or by otherwise leading them to make poor life choices, it will be unappealing to doctors and patients, not as a matter of ethics, but as a matter of science. Such drugs would not be deemed effective psychiatric tools. To even launch the interesting policy questions related to memory dampening, we must assume the existence of a drug that is not highly addictive and that satisfies basic requirements of medical efficacy and safety.

Assuming that we identify such a drug, legitimate but manageable concerns may arise about overuse. If the drug is used principally for victims of motor vehicle accidents and violent crimes, the drug is not likely to be used often by the same people. Furthermore, many of those with good coping skills have never had a motor vehicle accident nor been the victim of a violent crime; thus, working through these experiences cannot be critical to the development of these skills. If, however, a person frequently dampens his memory for comparatively insignificant events, then the Council's fear seems more plausible. Yet, virtually every medication runs a risk of overuse, and barring evidence that a medication is addictive, we usually manage that risk with our ordinary restrictions on prescription medications.

3.2.1.2 The Personal Identity Concern

Memory and identity are closely linked.[16] We feel a special connection to our past selves largely because we remember having our past experiences. For example, when I get out of bed in the morning, I consider myself the same person who went to sleep there the night before, in part, because I remember doing so. Those with extreme memory disorders, like advanced Alzheimer's disease, may lack such memories and may lose a stable sense of self.[17] Although memory is not the sole constituent of personal identity, it creates much of the psychological continuity that makes us aware of our continuing existence over time.[18]

John Locke deemed memory and identity to be so closely connected that he claimed that we should not punish a person for a crime he no longer remembers committing.[19] According to Locke, the person who cannot recall the crime is a different person than the perpetrator because the two lack an essential connection through memory, and the former should not be punished for the crime of the latter.

A glimmer of the Lockean view may be found in various places in the law of insanity where we are disinclined to hold people responsible for actions taken by their psychologically discontinuous alter egos. For example, in a case of dissociative identity disorder (formerly known as multiple personality disorder), the court held that the defendant—more specifically, the dominant personality of the defendant—could not be held responsible for the crimes of an alternate personality when the dominant personality was unaware of those crimes at the time they were committed, even if the alternate personality was legally sane.[20] In addition, the Supreme Court has held it unconstitutional to execute an insane death row inmate, even if the inmate was sane at the time of the murder.[21] Our unwillingness to execute the insane may recognize, in some measure, the psychological discontinuity between an insane inmate and his sane counterpart who committed the crime.[22]

Recognizing the important connection between memory and identity, the Council suggests that memory dampening may weaken our sense of identity by dissociating memories of our lives from those lives as they were actually lived. Selectively altering our memories, according to the Council, can distort our identity, "subtly reshap[ing] who we are, at least to ourselves."[23] "[W]ith altered memories," the Council writes, "we might feel better about ourselves, but it is not clear that the better-feeling 'we' remains the same as before."[24]

Yet, even in the absence of memory dampeners, we cannot help but selectively remember. Memories have a natural rate of decay and are far more a synthesis and reconstruction of our past than a verbatim transcript. Just to process the tremendous amount of information that is presented to our senses, we must constantly abstract away from the "real" world. As the Council acknowledges, "individuals 'naturally' edit their memory of traumatic or significant events—both giving new meaning to the past in light of new experiences and in some cases distorting the past to make it more bearable."[25] In fact, such selective reconstruction of our lives seems to be at the very heart of the creation of a coherent life story that Gilbert Meilaender advocates. Nevertheless, we do not worry whether our better-feeling naturally reconstructed selves remain the same as before.

It is, thus, not at all clear why we ought to revere the selective rewriting of our lives that we do without pharmaceuticals, yet be so skeptical of pharmaceutically assisted rewriting. In fact, memory dampening may strengthen our sense of identity. By preventing traumatic memories from consuming us, memory dampeners may allow us to pursue our own life projects, rather than those dictated by bad luck or past mistakes. As David Wasserman has noted, "pharmacologically assisted authorship may strengthen rather than reduce narrative identity," by allowing one to "edit his autobiography, instead of having it altered only by the vagaries of neurobiology."[26] Thus, to the extent that people voluntarily make changes to their mental processes, such changes may be perceived as bolstering self-identity. In fact, many people who begin taking antidepressants report feeling like themselves for the first time.[27] This suggests that some deliberate shifts in identity may not seem alienating at all.

3.2.1.3 The Genuine Experiences Concern

The Council also worries that a memory-dampened life, chemically altered as it is, is somehow a less genuine life.[28] According to the Council, "we might often be tempted to sacrifice the accuracy of our memories for the sake of easing our pain or expanding our control over our own psychic lives. But doing so means, ultimately, severing ourselves from reality and leaving our own identity behind."[29] This, according to the Council, "risks making us false, small, or capable of great illusions."[30] It also risks making us "capable of great decadence or great evil."[31]

Unfortunately, the Council never explains what makes a life genuine and truthful (nor how leading a life that is otherwise makes us capable of great evil). Is a memory-dampened life thought less genuine simply because some of the memories associated with it decay at a faster rate than they otherwise would have? Given that memories never precisely replicate our past experiences, do undampened memories provide a standard of genuineness? How important is it to lead a "genuine" life, whatever that means?[32]

In the case of those who are emotionally traumatized, memories of the trauma can be overwhelming and trigger exaggerated responses to harmless stimuli associated with a traumatic memory. Such overreactions are themselves divorced from reality. Memory dampeners, by preventing people from being overtaken by trauma, may actually make them more genuine, more true to what they take their lives to be, than they would be if they were gripped by upsetting memories.

Furthermore, we are not always troubled by discrepancies between our perceptions and the world as it "genuinely" is. It has been widely observed that in many areas of life, people systematically overestimate their abilities and prospects relative to others:

> People (nondepressed people, at least) rate themselves as better—friendlier, more likely to have gifted children, more in control of their own lives, more likely to quickly recover from illness, less likely to get ill in the first place, better leaders, and better drivers—than they really are.... There is evidence associating the above sorts of positive illusions with increased happiness, "ability to care for others," "motivation, persistence," and "the capacity for creative, productive work."[33]

Suppose there were a pill that eliminated these systematic self-enhancing biases. On the one hand, one could argue, those who took such pills would lead less genuine lives, as they would no longer understand the world in the way that they would in the absence of the pill. Their lives would be less genuine in the sense that they would lack a characteristically human understanding of the world. On the other hand, those who took the pill might lead more genuine lives, freed from the ruby-colored lenses that nature has given us.

No doubt, as a general life strategy, we do well to firmly commit ourselves to reality and to discovering the truth about ourselves and the world around us. Yet, such a strategy might, at times, be worse for us all things considered; or, at least, the Council has not shown otherwise. To make the case that memory-dampening drugs will harmfully affect our lives, the Council must be much more specific about what makes a life genuine, how these drugs make lives less genuine, and why that should matter so much to us that we ought to suffer in distress to preserve our unadulterated memories.

3.2.2 General Response to the Prudential Concerns

In the preceding section, I argued that many of the Council's concerns about memory dampening are founded on controversial premises. Not all of us will agree with the Council about how we ought to cope with emotional pain, what changes to our memory will damage our sense of self, and what makes one set of experiences more genuine and, therefore, better than another. Whereas the concerns expressed by the Council and some of its members may prove insightful to like-minded patients or medical professionals, they are insufficiently developed to provide a basis for broad restrictions on memory dampening.

Each of the concerns presented reflects a bias for our natural, pharmaceutical-free mechanisms of responding to trauma. The Council implicitly or explicitly defended (1) our natural ability to surmount difficult life obstacles, (2) our natural memories as the desirable basis for our sense of identity, and (3) our natural memories as more genuine and more desirable than those that are pharmaceutically altered.

There are two reasons commonly given for this preference for the status quo. The first is that we doubt that human intervention can improve upon our natural endowments when it comes to responding to difficult memories. We generally do an astonishingly good job of remembering what we need to remember and forgetting what we can do without. If millions of years of evolution have tended to select for brains that optimally balance retained and deleted memories, then we may find it very difficult indeed to improve upon our natural endowment.

However, whereas evolution has made the human brain remarkably adept at balancing our needs to retain and to forget memories, it surely did not lead each of us to an optimal balance. The conditions and needs of modern society differ substantially from those during most of our evolution. Furthermore, some people have better memories than others, and some are more susceptible to PTSD than others. It is very unlikely that we each have a brain optimized for our individual needs, especially because our needs can change during the course of a lifetime.

A second reason to defend our natural balance of retention and forgetting is that, with such a balance, we lead distinctively human lives and perhaps doing so is itself valuable. In a concluding section of its report, the Council expresses a similar sentiment, acknowledging that its concerns with memory dampening and certain other new technologies "may have something to do with challenges to what is naturally human, what is humanly dignified, or to attitudes that show proper respect for what is naturally and dignifiedly human."[34]

A running theme in the Council's report is that memory dampening dehumanizes us by giving us too much control over our life experiences. According to the Council, "We are not free to decide everything that happens to us; some experiences, both great joys and terrible misfortunes, simply befall us. These experiences become part of who we are," part of our lives "as truthfully lived."[35]

Yet the Council acknowledges exactly what makes this view so unpersuasive: "The 'giftedness of nature' also includes smallpox and malaria, cancer and Alzheimer [sic] disease, decline and decay."[36] Surely we are not expected to accept everything in the world that is "given." The Council, however, offers no principled basis for deciding when to intervene, insisting that a "respectful attitude toward the 'given'" is "both necessary and desirable as a restraint,"[37] even though "[r]espect for the 'giftedness' of things cannot tell us which gifts are to be accepted as is, which are to be improved through use or training, which are to be housebroken through self-command or medication, and which opposed like the plague."[38] At some point, one must wonder whether this distinction actually serves to distinguish. Indeed, what is "given" may itself be dynamic, for our "given" nature might be to transcend our boundaries and constantly improve ourselves. At one point, the Council makes exactly that suggestion.[39] It is, therefore, very difficult to understand why human enhancement should be restrained by our "given" nature.

The weaknesses of a status quo preference can be illustrated by imagining a world called Dearth, where the inhabitants are very much like us except that, on average, they are less likely than we are to suffer from traumatic memories. Perhaps Dearthlings are less emotionally aroused by traumatic experiences than humans typically are. One day, the government of Dearth establishes a commission that holds hearings on an emerging technology, called traumatic memory *enhancement.* Using memory-enhancing drugs, Dearthlings can make their traumatic memories more vivid, more persistent, and otherwise more like those of typical humans. Should Dearthlings enhance their responses to trauma to make them more like the responses of typical humans?

With limited facts, it is difficult to say. Without the drug, Dearthlings suffer less; on the other hand, they might, in some sense, experience a richer, more meaningful life with the drug. Most would agree, however, that a Dearthling should not be forced to take a drug that will create a significant risk that he will develop upsetting memories from a recent traumatic experience. Similarly, a human being with a significant risk of developing upsetting memories from a recent traumatic experience should be permitted to use memory-dampening drugs to prevent those memories from forming. The only difference between a Dearthling at risk from traumatic-memory enhancement and a human at risk from refraining from memory dampening is whether the risk comes from taking a pill or from not taking it. If the Dearthling is permitted to avoid a bad state of affairs by not taking a pill, the human should be able to avoid that same bad state of affairs by taking one. Otherwise, the preference for the status quo begins to seem like an unprincipled taboo on pill taking.

Some Council members might respond by saying that there is a very important difference between these two individuals—namely, one is a human and one is a Dearthling—and the human ought to deal with traumatic memories in characteristically human rather than Dearthling ways. In response, however, I must present the scary news that there are Dearthlings among us, for some humans are quite resilient in the face of traumatic experiences whereas others are prone to PTSD.[40] In fact, one sibling may be quite sensitive to trauma whereas another is the human equivalent of a Dearthling. Given the amount of variation among humans, appeals to human nature tell us little about whether we must respond to trauma like a Dearthling or like a statistically typical human.

At this point, the Council might reiterate that our human nature may require each of us to accept his own personal "given" response to trauma whatever it might be. Yet, the Council encourages us to change our "given" response to traumatic memories so long as we do so the old-fashioned way. It is difficult, however, to see why the method of change matters if it leads to the same end point. Perhaps the Council doubts that a pharmaceutical intervention will get us to the same end point as a non-pharmaceutical intervention. That, however, would merely serve as a critique of some particular imperfect form of memory dampening rather than a critique of memory dampening in general.

To recap, two potential reasons were considered for preferring our status quo methods of dealing with trauma to those using memory dampening. The first was that our status quo methods are simply the best methods possible. I argued that this is highly implausible as an empirical matter. The second was that our status quo methods are best because they are, in some sense, given to us as part of our human nature. I argued that there is little reason to prefer some state of affairs simply because it is the status quo, and it is virtually impossible to determine when human nature dictates that we leave some state of affairs alone and when it dictates that we do whatever we can to change it.

Another reason why the Council's concerns about memory dampening do not translate well into legal restrictions on memory dampening is that the concerns discussed so far are not quintessentially ethical in nature. For example, the Council advises each of us to lead a genuine life because such a life is valuable to the person living it. To the extent that there is an ethical obligation to lead such a life, it is an obligation one has to one's self. Yet the notion of having an obligation to one's self is controversial. If A has an obligation to B, then, ordinarily, B can choose to release A from that obligation. Now suppose that A has an obligation to himself. Can A release himself from an obligation to himself? If so, it is not clear that A is obligated in any meaningful way.[41]

Restrictions based on what I call the Council's prudential concerns are paternalistic in nature. Paternalistic limitations on our freedom may "serve ... the reflective values of the actor," or "impose ... values that the actor rejects."[42] The "soft" paternalism that is consistent with our own values is usually thought less invasive and more respectful of individual autonomy than the "hard" paternalism that imposes values foreign to the actor. To the extent that I have shown that the Council's concerns in the past section are founded on controversial premises and do not reflect quintessentially ethical obligations, I have thereby suggested that interventions based on those concerns are of the more suspect variety.

The Council's prudential concerns provide little ground for doubting the ability of individual patients and their doctors to collectively decide when to use memory-dampening drugs, much as they would collectively decide to use any other physical or psychiatric medical treatment. The possibility remains, however, that the concerns described here could be reconfigured in terms of the effects that they would have on others. In that case, perhaps one could formulate nonpaternalistic reasons for restrictions. Indeed, in the next two sections, I describe concerns of the Council that I take to be somewhat stronger because they do identify more widespread societal effects of memory dampening.

3.2.2.1 Obligations to Remember

In the Supreme Court's most influential "right to die" case, *Cruzan v. Director, Missouri Department of Health*,[43] Nancy Cruzan's family failed in its effort to obtain a court order to disconnect Nancy from the artificial feeding and hydration equipment that kept her alive in a persistent vegetative state. Writing in dissent, Justice John Paul Stevens emphasized that "[e]ach of us has an interest in the kind of memories that will survive [us] after death." Stevens dissented, in part, because Nancy Cruzan may have had "an interest in being remembered for how she lived rather than how she died," and he feared that "the damage done to those memories by the prolongation of her death is irreversible."[44]

Stevens suggests that people have strong interests in being remembered in certain ways for who they are and what they do. If Stevens is correct, then we may have obligations to satisfy these interests by appropriately remembering people and events. Because memory dampeners may facilitate violations of these obligations, we arguably have grounds to heavily restrict their use. In this section, I will suggest otherwise. First, I will describe the concerns of Council members that memory dampening may violate obligations to remember. Then, I will argue that even if we sometimes have ethical obligations to others to remember, these obligations cannot by themselves justify broad restrictions on memory dampening.

Council member Gilbert Meilaender suggests, albeit meekly, that we may have ethical obligations to remember those "treated unjustly . . . to remember the evil done them," which "might be necessary not just for the sake of the victims themselves but for our common humanity."[45] Whereas Meilaender merely "suspect[s] we can imagine circumstances in which we might think that there is indeed an obligation not to forget,"[46] I think that such obligations, at least where understood as *prima facie* obligations, are quite common, stemming perhaps from interests in respect, honor, or justice.[47]

In a world without memory dampening, it may seem that one cannot possibly be responsible for failing to remember, as we have limited control over our memories, and voluntary control is often thought to be a prerequisite to responsibility. On further examination, however, we clearly hold people responsible for failing to remember. For example, we blame those who forget an important birthday or anniversary, and we penalize those who forget to file a timely tax return. Some of the most tragic instances of failed memory occur when parents unintentionally cause the death of their young children by leaving them stranded in the backseats of automobiles on hot days, sometimes leading to criminal punishment.[48]

The nature of our obligations to remember are radically underexplored, however, partly because, prior to the realistic possibility of memory dampening, there was relatively little one could do to consciously alter one's memories, and there was correspondingly little one could do to consciously fulfill or escape obligations to remember. One explanation for the observation that we do, in fact, hold people responsible for forgetting is that, in the examples given above—failing to commemorate a special occasion, to file tax returns, and to care for one's children—we are actually faulting people, not for their involuntary forgetfulness, but for some intentional failure at an earlier point in time.[49] For example, perhaps the neglectful taxpayer intentionally decided not to record his filing deadline on his calendar or made other deliberate choices not to develop those attributes that would have prevented his memory failure. In a world with memory-altering drugs (either enhancing or dampening), we would have more opportunities to consciously alter our inclinations to remember or forget, leading perhaps to more responsibility for whatever memories we keep or discard.

Even if we can have obligations to remember, however, it is easy to overestimate the strength of these obligations. Perhaps the Council does so when it states that it may have been inappropriate for those with firsthand experiences of the Holocaust to dampen their traumatic memories:

> Consider the case of a person who has suffered or witnessed atrocities that occasion unbearable memories: for example, those with firsthand experience of the Holocaust. The life of that individual might well be served by dulling such bitter memories, but such a humanitarian intervention, if widely practiced, would seem deeply troubling: Would the community as a whole—would the human race—be served by such a mass numbing of this terrible but indispensable memory? Do those who suffer evil have a duty to remember and bear witness, lest we all forget the very horrors that haunt them?[50]

There is something harsh about expecting trauma sufferers to bear the additional burden of carrying forward their traumatic memories for the benefit of others. The Council, recognizing this, goes on to soften its perspective somewhat, stating that "we cannot and should not force those who live through great trauma to endure its painful memory *for the benefit of the rest of us*."[51] Yet, even for those who suffer from the most tragic of memories, the Council is ambivalent about the ethics of pharmaceutical dampening:

> [A]s a community, there are certain events that we have an obligation to remember—an obligation that falls disproportionately, one might even say unfairly, on those who experience such events most directly. What kind of people would we be if we did not "want" to remember the Holocaust, if we sought to make the anguish it caused simply go away? And yet, what kind of people are we, especially those who face such horrors firsthand, that we can endure such awful memories?

According to the Council, we are sometimes obligated to remember some person or set of events because doing so pays respect to that person or set of events. For example, we may have obligations to remember great sacrifices that others make on our behalf, not because these memories will guide our actions, but rather because retaining the memory demonstrates a kind of respect or concern for these others.

The case for legally restricting memory dampening is particularly weak when it comes to such "homage" memories. What makes the retention of a traumatic homage memory significant is that the person who bears the traumatic memory has chosen to identify with it in some way. In fact, memory-dampening drugs, by giving us the opportunity to consciously choose to keep a memory intact, may actually facilitate our identification with it. On the other hand, if an individual retains a homage memory simply because he has no choice—because the tragic memory was indelibly imprinted into his brain by stress hormones or because memory dampening has been prohibited—the holding of the homage memory loses much of its significance. Such memories are not truly homages at all.

Nevertheless, we can easily imagine situations where our obligations to remember are much stronger. For example, suppose a bystander is the only person to see the face of a serial rapist fleeing the home of his latest victim. Though the bystander may find the memory of the perpetrator's appearance quite upsetting, virtually everyone would agree that the bystander ought to retain the memory if doing so will ultimately help prosecute the perpetrator and protect potential future victims. Such a conclusion would be much less likely, however, if we consider instead the point of view, not of a mere bystander-witness, but of the traumatized victim who, let us now suppose, is the only one to see the perpetrator's face. In that case, we might still expect the victim to experience even this more intense trauma for, say, an hour until a police sketch artist can preserve the memory. It is much less clear, however, if the victim should be obligated to wait more than 6 hours to begin memory dampening in a world (like ours today, perhaps) where memory dampening would no longer be effective. At a minimum, however, it is clear that some people have obligations to remember because there are strong societal interests in preserving certain memories.

Translating ethical obligations to remember into legal restrictions on memory dampening is no simple matter. Memory dampening is a kind of medical treatment, and we do not ordinarily limit a person's access to medical resources simply to further police investigations. On the other hand, memory dampening can destroy evidence, and we have plenty of laws prohibiting that. It, therefore, seems plausible that some balancing of interests should occur when a person wishes to dampen memories that hold substantial instrumental value to society.

Yet, even if we sometimes have ethical obligations to retain memories that ought sometimes be backed by legal sanctions, there is little reason to think that broad restrictions on memory dampening are needed. So, for example, an expansion of obstruction of justice statutes could further limit the use of memory-dampening drugs when patients have memories that are needed to protect societal interests in justice and safety. Alternatively, physicians could be required to make certain inquiries before prescribing memory-dampening drugs and could perhaps be obliged to notify authorities if a patient seeks to dampen or erase memories, where doing so may endanger someone else's life. Limited restrictions like these derive from concerns about memory dampening that, unlike those previously discussed, are based on ethical obligations we have to others and do not rely on much-disputed conceptions of human nature or controversial preferences for what is deemed natural.

3.2.2.2 Coarsening to Horror

The Council also expressed concern that memory dampening will coarsen our reactions to horror and tragedy. If we see the world from a chemically softened, affect-dulled perspective, we may grow inured to trauma and its associated distress, "making shameful acts seem less shameful, or terrible acts less terrible, than they really are."[52] As an example, the Council describes a hypothetical witness to a violent crime who dampens his memory and eventually perceives the crime as less severe than he would have without pharmaceutical assistance:

> Imagine the experience of a person who witnesses a shocking murder. Fearing that he will be haunted by images of this event, he immediately takes propranolol (or its more potent successor) to render his memory of the murder less painful and intrusive. Thanks to the drug, his memory of the murder gets encoded as a garden-variety, emotionally neutral experience. But in manipulating his memory in this way, he risks coming to think about the murder as more tolerable than it really is, as an event that should not sting those who witness it. For our opinions about the meaning of our experiences are shaped partly by the feelings evoked when we remember them. If, psychologically, the murder is transformed into an event our witness can recall without pain—or without any particular emotion—perhaps its moral significance will also fade from consciousness.[53]

One concern suggested by this example is that memory dampening will make it more difficult to accurately convey evidence and other kinds of information to each other. According to the Council, the person described above "would in a sense have ceased to be a genuine witness of the murder," and when later asked about the event, "he might say, 'Yes, I was there. But it wasn't so terrible.'"[54] Though the Council asks whether this person was a "genuine witness of the murder," the implicit reference to the natural is more appropriate here than it was with respect to the Council's prudential concerns. If this person were to appear before a jury, his description of the events surrounding the murder will be interpreted by listeners against a backdrop of *natural* linguistic conventions that help connect a speaker's affect to the events he describes. Similarly, in the military context, some worry that memory-dampened soldiers will come back from battle with unnatural affect-reduced descriptions of their experiences, making combat seem less horrific than it would otherwise. Against a standard backdrop of communicative conventions, we would understandably be puzzled by a flat, lifeless description of human tragedy.

Indeed, if memory dampening has a tendency to alter our perceptions and our understanding of events in the world, then, as the Council's example suggests, it may affect more than just the ways we communicate. A deeper concern is that memory dampening will coarsen our feelings and make us less willing to respond to tragic situations. Along these lines, one can imagine a would-be-famous civil rights leader in the 1960s who, in order to combat the memory of childhood injustices, would have gone on to revolutionize our social institutions but, due to his use of memory dampeners, instead pursues a more mundane life plan and is never so much as mentioned in the history books.

Not only might our coarsened emotions disincline us to take positive action, but also it has been suggested that memory dampeners could reduce our inhibitions to engage in socially destructive action. Thus, violent criminals could use memory dampeners to ease feelings of guilt, making them more likely to recidivate.[55] In addition, it has been claimed, memory-dampened soldiers, freed from burdens of conscience, may be more effective at killing.[56] These examples suggest that fear and remorse or expectations of fear and remorse inhibit certain antisocial behaviors and that memory dampening may interfere with this desirable control mechanism. It may warrant studying whether any proposed memory-dampening agent actually has such effects.

Even if there is some empirical basis for these concerns, however, it is important not to overstate their importance. For even if memory dampening does make some trauma *seem* less horrible, this happens in part because memory dampening can *actually make* trauma less horrible. That is, much of what is bad about traumatic experience is that it traumatizes those who survive it. So, for example, to the extent that we can ease the traumatic memories of those involved in military conflict (without leading to a significant increase in total military conflict), then memory dampening makes combat somewhat better than it would otherwise be. Furthermore, when soldiers are injured in battle, we heal their physical wounds using advanced technology, even if doing so makes war seem less horrible; so it is unclear why their emotional wounds should be treated any differently.

Though the coarsening concern is far from overwhelming, it at least shows how the widespread use of memory dampeners can potentially affect the lives of those who do not use them. Nevertheless, this concern cannot alone justify broad restrictions on memory dampening, at least not if such restrictions are consistent with our typical policies of drug regulation. For example, people consume alcohol to relieve themselves of the pain of traumatic events. Whether or not this leads to some general inurement to tragedy in society (which seems doubtful), most would not address the problem with a comprehensive prohibition of alcohol. Similarly, even if antidepressants are used for relief from the pain of traumatic experiences, we would not generally prohibit them for fear that society will be less compassionate. Likewise, the world may benefit from the inspired artwork of a Vincent van Gogh, yet few would deprive a tortured soul of antidepressants in order to foster artistic creation.

We likely permit the use of such drugs, despite whatever minimal effects they may have on our reactions to tragedy, because their costs are outweighed by other benefits. So even if data someday support the Council's concern that memory-dampening drugs can have negative effects on soldiers' battlefield reactions or on societal reactions more generally, we can surely tailor limits on their use in particular contexts. And if the testimony of memory-dampened witnesses has a different emotional tone than that of ordinary witnesses, experts can explain the differences to jurors.

Whereas memory dampening has its drawbacks, such may be the price we pay in order to heal intense emotional suffering. In some contexts, there may be steps that ought to be taken to preserve valuable factual or emotional information contained in a memory, even when we must delay or otherwise impose limits on access to memory dampening. None of these concerns, however, even if they find empirical support, are strong enough to justify broad-brushed restrictions on memory dampening.

Notes

1. Editor's note: This reading was excerpted from a substantially longer article that appeared in 2006 in the *Vanderbilt Law Review*, volume 59, pages 1561–1626, entitled "Therapeutic Forgetting: The Legal and Ethical Implications of Memory Dampening," and is used with permission. The reprinted sections address the arguments against memory blunting raised by the President's Council on Bioethics in their report *Beyond Therapy: Biotechnology and the Pursuit of Happiness* and also reprinted in this volume. Other parts of Kolber's published article review the technology of memory dampening and relevant legal issues such as difficulties in obtaining informed consent for memory dampening and the ways in which the availability of memory dampening could affect legal practices surrounding claims of emotional distress. The excerpt itself has also been abridged with the permission of the author. The author's acknowledgments follow here: "For helpful comments, I thank Jeremy Blumenthal, Rebecca Dresser, Donald Dripps, James DuBois, Adam Elga, David Fagundes, Jesse Goldner, Kent Greenawalt, Tracy Gunter, Steven Hartwell, Orin Kerr, Ivy Lapides, David Law, Orly Lobel, Elizabeth Loftus, David McGowan, Camille Nelson, Richard Redding, Alan Scheflin, Larry Solum, Graham Strong, and Mary Jo Wiggins, as well as colloquia and conference participants at Hofstra Law School (faculty workshop), New York University (Joseph Le Doux Lab & Elizabeth Phelps Lab), Saint Louis University Law School (Health Law Scholars Workshop), University of California, San Diego (Biomedical Ethics Seminar Series), University College London (Law, Mind & Brain Colloquium), University of Maryland (Health Law Teachers Conference), University of San Diego School of Law (faculty workshop), and Vanderbilt University Law School (S.E.A.L. Conference). I also thank my research assistants Jane Ong, Samuel Park, Sarah Pinkerton, and Michelle Webb. This project was generously supported by a summer research grant from the University of San Diego School of Law."

2. See, e.g., *Eternal Sunshine of the Spotless Mind* (Focus Features, 2004); *Paycheck* (Paramount, 2003); *Men in Black* (Sony Pictures, 1997); *Total Recall* (Artisan Entertainment, 1990); see also Steven Johnson, The Science of Eternal Sunshine, *Slate*, Mar. 22, 2004, http://slate.msn.com/id/2097502.

3. See Part I.C. of the full article, Adam J. Kolber, Therapeutic Forgetting: the Legal and Ethical Implications of Memory Dampening. *Vanderbilt Law Review*, 59, 1561–1626.

4. See *Physicians' Desk Reference* 3423 (60th ed., 2006).

5. See Robin Marantz Henig, *The Quest to Forget*, *New York Times Magazine*, Apr. 4, 2004, at 32, 34–36.

6. For example, researchers have sought to alleviate older traumatic memories by having subjects recall those memories after ingesting propranolol. Marilynn Marchione, *A Pill to Fade Traumatic Memories? Doctors Are Working on It*, Jan. 14, 2006, http://www.signonsandiego.com/news/science/20060114-0917-traumapill.html.

7. The FDA has indicated that "once a [drug] product has been approved for marketing, a physician may prescribe it for uses or in treatment regimens of

patient populations that are not included in approved labeling." Citizen Petition Regarding the Food and Drug Administration's Policy on Promotion of Unapproved Uses of Approved Drugs and Devices; Request for Comments, 59 Fed. Reg. 59820, 59821 (Nov. 18, 1994) (quoting 12 FDA Drug Bull. 4–5 [1982]) (alteration in original); *see also* Planned Parenthood Cincinnati Region v. Taft, 444 F.3d 502, 505 (6th Cir. 2006) ("Absent state regulation, once a drug has been approved by the FDA, doctors may prescribe it for indications and in dosages other than those expressly approved by the FDA.").

8. See Editorial, "Care for the Traumatized," *Boston Herald,* Jan. 2, 2006, at 16 [hereinafter *Care for the Traumatized*] ("It seems the Department of Veterans' Affairs and its overseers in Congress worry that disability benefits for veterans diagnosed with post-traumatic stress disorder (PTSD) are becoming a budget buster.").

9. The Council's hearings on memory dampening have been divided into three parts, although the hearings took place on only two separate days. Citations are to the "printer-friendly" versions available online. See *Remembering and Forgetting: Physiological and Pharmacological Aspects: Hearings Before the President's Council on Bioethics* (Oct. 17, 2002), http://bioethics.gov/transcripts/oct02/session3.html [hereinafter Hearings, Part 1]; *Remembering and Forgetting: Psychological Aspects: Hearings Before the President's Council on Bioethics* (Oct. 17, 2002), http://bioethics.gov/transcripts/oct02/session4.html [hereinafter Hearings, Part 2]; *Beyond Therapy: Better Memories?: Hearings Before the President's Council on Bioethics* (Mar. 6, 2003), http://bioethics.gov/transcripts/march03/session4.html [hereinafter Hearings, Part 3].

10. Gilbert Meilaender, *Why Remember?, 135 First Things 20* (2003), *available at* http://www.firstthings.com/ftissues/ft0308/articles/meilaender.html, at 21–22.

11. *See* George A. Bonanno, Loss, Trauma, and Human Resilience, 59 *Am. Psychologist* 20, 25–26 (2004).

12. This example was raised by James McGaugh at the Council's hearing. *See* Hearings, Part 2, *supra* note 17, at 23–24; *see also* James N. Butcher & Chris Hatcher, The Neglected Entity in Air Disaster Planning, 43 *Am. Psychologist* 724, 728 (1988) (describing the incident).

13. Meilaender, *supra* note 11, at 22.

14. The Council acknowledges that if "bitter memories are so painful and intrusive as to ruin the possibility for normal experience of much of life and the world," the "impulse" to dampen those memories is "fully understandable." *Beyond Therapy*, *supra* note 1, at 230. The Council quickly retreats, however, adding: "And yet, there may be a great cost to acting compassionately for those who suffer bad memories, if we do so by compromising the truthfulness of how they remember." *Id.*

15. *Id.* at 291; *id.* at 208 (asking, "What qualities of character may become less necessary and, with diminished use, atrophy or become extinct, as we increasingly depend on drugs to cope with misfortune?").

16. On the relationship between memory and identity, see Derek Parfit, *Reasons and Persons* 199–345 (1984). *See also Personal Identity* (John Perry ed., 1975)

(collecting essays); Rebecca Dresser, Personal Identity and Punishment, 70 *B.U. L. Rev.* 395 (1990) (applying Parfit's reductionist approach to personal identity to theories of criminal punishment).

17. *Cf.* Agnieszka Jaworska, Respecting the Margins of Agency: Alzheimer's Patients and the Capacity to Value, 28 *Phil. & Pub. Aff.* 105 (1999) (arguing that we should respect the autonomy interests of those Alzheimer's patients who retain a capacity to value even after they have lost a coherent life narrative).

18. Parfit, *supra* note 15, at 208.

19. John Locke, Of Identity and Diversity, in *Personal Identity*, *supra* note 200, at 33, 48 ("[I]n the great day, wherein the secrets of all hearts shall be laid open, it may be reasonable to think, no one shall be made to answer for what he knows nothing of...."); see also Parfit, *supra* note 15, at 205 ("Locke claimed that someone cannot have committed some crime unless he now remembers doing so.").

20. United States v. Denny-Shaffer, 2 F.3d 999, 1016 (10th Cir. 1993) (ordering retrial with an insanity instruction where the defendant presented sufficient evidence that her dominant personality was not in control during the offense and was not aware that another personality was controlling her physical actions). See generally Elyn R. Saks, Multiple Personality Disorder and Criminal Responsibility, 10 *S. Cal. Interdisc. L.J.* 185, 190 (2001); Walter Sinnott-Armstrong & Stephen Behnke, Criminal Law and Multiple Personality Disorder: The Vexing Problems of Personhood and Responsibility, 10 S. *Cal. Interdisc. L.J.* 277 (2001).

21. Ford v. Wainwright, 477 U.S. 399, 399, 410 (1986).

22. Such a view is far from explicit, however, in the Court's decision in *Ford v. Wainwright*, which notes that there is no "[u]nanimity of rationale" behind the rule. *Id.* at 408. Among the reasons on offer, the Court deemed it "abhorrent...to exact in penance the life of one whose mental illness prevents him from *comprehending* the reasons for the penalty or its implications." *Id.* at 417 (emphasis added). Though part of the requisite "comprehension" refers to the inmate's understanding of the criminal justice system, it might also refer to the inmate's ability to understand that it is *he* who is deemed to have committed the crime for which he is to be executed. A sane but amnesic murderer is still capable of comprehending the reasons for his execution, at least at a detached cognitive level. Yet, given that the amnesic will not identify himself with his crime in the ordinary manner and may perhaps be incapable of fully appreciating the relationship between his crime and his punishment, those sympathetic to Locke's view might find that capital punishment is also inappropriate for those who genuinely cannot remember the crimes for which they are to be executed.

23. *Beyond Therapy*, *supra* note 1, at 211–12.

24. *Id.* at 212.

25. *Id.* at 217 n.

26. See David Wasserman, Making Memory Lose Its Sting, 24 *Phil. & Pub. Pol'y Q.* 12 (Fall 2004), at 14.

27. Peter Kramer quotes a patient who, after starting the SSRI antidepressant Prozac, said she felt "as if I had been in a drugged state all those years and now I am clearheaded." Peter Kramer, *Listening to Prozac* 8 (1993). Eight months after beginning Prozac, the same patient stopped the treatment and said she felt like "I am not myself." *Id.* at 18.

28. See *Beyond Therapy*, *supra* note 1, at 213 ("[B]y disconnecting our mood and memory from what we do and experience, the new drugs could jeopardize the fitness and truthfulness of how we live and what we feel....").

29. *Id.* at 233–34.

30. *Id.* at 234.

31. *Id.*

32. Robert Nozick's famous "experience machine" thought experiment is often taken to show that we want our lives to be closely connected to reality. See Robert Nozick, *Anarchy, State, & Utopia* 42–45 (1974). Nozick asked us to imagine that:

Superduper neuropsychologists could stimulate your brain so that you would think and feel you were writing a great novel, or making a friend, or reading an interesting book. All the time you would be floating in a tank, with electrodes attached to your brain. Should you plug into this machine for life, preprogramming your life's experiences?...Of course, while in the tank you won't know that you're there; you'll think it's all actually happening.

Id. at 42–43. According to Nozick, we would not choose to spend our lives connected to such a machine because we value not just particular experiences but particular *genuine* experiences. *Id.* at 43–45. At best, however, Nozick's example only shows that we value *some* connection to the real world, not that we are opposed to having *any* illusory beliefs or perceptions (for example, the drug-induced, trauma-relieving perception that one has not witnessed some atrocity that, in fact, one has).

Furthermore, even Nozick's limited conclusion that we value some connection to the real world is not robustly demonstrated by the thought experiment. The thought experiment would be more convincing if those already connected to an experience machine would also choose to disconnect from it in order to lead more genuine but substantially less enjoyable lives than they do while connected. Consistent with all available evidence, we might be connected to experience machines right now, yet I question whether we would choose to disconnect from the simulacra of our current lives, if given the choice. As I argue elsewhere, the fact that we are more willing to remain connected to an experience machine than to connect in the first place suggests that our initial intuitions about the experience machine may not be entirely trustworthy. See Adam Kolber, Mental Statism and the Experience Machine, 3 *Bard J. Soc. Sci.* 10 (Winter 1994/1995).

33. Adam Elga, On Overrating Oneself...and Knowing It, 123 *Phil. Studies* 115, 117 (2005); see also Jonathon D. Brown, Evaluations of Self and Others: Self-Enhancement Biases in Social Judgments, 4 Soc. *Cognition* 353 (1986); Darrin R. Lehman & Shelley E. Taylor, Date with an Earthquake: Coping with a Probable, Unpredictable Disaster, 13 *Personality & Soc. Psychol. Bull.* 546 (1987); Shelley Taylor & Jonathon Brown, Illusion and Well-Being: A Social Psy-

chological Perspective on Mental Health, 103 *Psychol. Bull.* 193 (1988); Shelley Taylor & Jonathon Brown, Positive Illusions and Well-Being Revisited: Separating Fact from Fiction, 116 *Psychol. Bull.* 21 (1994); *supra* notes 193–94 and accompanying text.

34. *Beyond Therapy*, *supra* note 1, at 286–87. Leon Kass (the former chairman of the Council) and Francis Fukuyama (a member of the Council until recently) have each written extensively about the importance of preserving human dignity in the face of challenges to it from allegedly dehumanizing new technologies. *See* Francis Fukuyama, *Our Posthuman Future: Consequences of the Biotechnology Revolution* (2002); Leon Kass, *Life, Liberty and the Defense of Dignity* (2002).

35. *Beyond Therapy*, *supra* note 97, at 233.

36. *Id.* at 289.

37. *Id.*

38. *Id.*

39. *Id.* at 291 n.* ("By his very nature, man is the animal constantly looking for ways to better his life through artful means and devices; man is the animal with what Rousseau called 'perfectibility.'").

40. See *supra* note 10.

41. See Marcus G. Singer, On Duties to Oneself, 69 *Ethics* 202, 202–203 (1959) ("[A] duty to oneself, then, would be a duty from which one could release oneself at will, and this is self-contradictory. A 'duty' from which one could release oneself at will is not, in any literal sense, a duty at all."). But cf. Daniel Kading, Are There Really "No Duties to Oneself"?, 70 *Ethics* 155 (1960) (raising some objections to Singer's position).

42. Kent Greenawalt, Legal Enforcement of Morality, 85 *J. Crim. L. & Criminology* 710, 718 (1995).

43. 497 U.S. 261 (1990).

44. *Id.* at 356 (Stevens, J., dissenting); *see also id.* at 343–44 (Stevens, J., dissenting) (stating that the most famous declarations of Nathan Hale and Patrick Henry "bespeak a passion for life that forever preserves their own lives in the memories of their countrymen").

45. Meilaender, *supra* note 9 at 22.

46. *Id.*

47. See generally Avishai Margalit, *The Ethics of Memory* (2002).

48. Before pursuing such cases, prosecutors generally require an extreme kind of forgetfulness that evidences gross negligence. See, e.g., Kelly v. Commonwealth, 592 S.E.2d 353, 355–57 (Va. Ct. App. 2004) (affirming the manslaughter conviction of a father who left his 21-month-old daughter unattended in a hot van for approximately seven hours where there was evidence that the father had stranded children in automobiles in the past).

49. See Mark Kelman, Interpretive Construction in the Substantive Criminal Law, 33 *Stan. L. Rev.* 591, 593–94, 600–16 (1981) (describing the "arational choice between narrow and broad time frames" in the criminal law).

50. *Beyond Therapy, supra* note 1, at 230–31 (footnotes omitted).

51. *Id.* at 231.

52. *Beyond Therapy, supra* note 1, at 228.

53. *Id.* at 229.

54. *Id.* As a preliminary observation, the example may overstate the case. According to the Council, this individual can recall what happened "without pain—or without any particular emotion." *Id.* Yet, this seems like an instance of overmedication, for there may be little reason to make absolutely horrific events seem quite ordinary.

55. *Cf. Beyond Therapy, supra* note 1, at 224.

56. See *id.* at 154.

Reading 3.3

Prozac as a Way of Life[1]

Carl Elliott

When Peter Kramer coined the term *cosmetic psychopharmacology* in his 1993 book *Listening to Prozac*, he was referring to the way psychoactive drugs could be used not just to treat illnesses but to improve a person's psychic well-being. He described this as moving a person from one normal state to another. Like many other psychiatrists in the late 1980s and early 1990s, Kramer worried that Prozac was being used as a kind of psychic enhancement, or what other psychiatrists had called "mood brighteners." Is there anything wrong with using a drug to become "better than well"?

3.3.1 Cosmetic Psychopharmacology in Historical Context

The term *cosmetic psychopharmacology* may have become popular with Prozac, but the concept had begun to take root in psychiatry almost 40 years earlier. In 1955, Wallace Laboratories began marketing meprobamate under the trade name Miltown. Miltown has a fair claim to be the first psychoactive drug developed for the anxiety of ordinary life—or as the scientist behind the drug, Frank Berger, put it, for "people who get nervous and irritable for no good reason" (Shorter, 1997). Unlike Thorazine (chlorpromazine), the new psychoactive drug introduced that same year for patients who were severely psychotic, Miltown was a prescription drug for the worried well. It was a biological treatment for people to whom psychiatrists would have otherwise offered psychotherapy, or maybe even no treatment at all. Miltown was marketed not as a new "sedative," like the barbiturates, but as a "tranquilizer" (see interview with Janssen in Healy, 1998, p. 59). Anxious Americans did not want to be sedated, but who could argue with a little more tranquility?

Miltown was an immediate success. Within months, demand for Miltown exceeded that of any drug ever introduced in the United States

(Shorter, 1997, p. 316.). Like Prozac in the 1990s, Miltown also became a pop culture phenomenon. It was joked about on talk shows (Milton Berle called himself "Miltown Berle") and worried over in the press. The *Nation* named it a "mental laxative." *Newsweek* called it "emotional aspirin," while *Time* ran the headline, "Happiness by Prescription" (Smith, 1991). Psychiatrists debated whether anxiety was normal and healthy—a spur to accomplishment or creativity—or a damaging form of psychopathology for which Miltown was a legitimate treatment.

Frank Berger, the Czech physician who developed Miltown and took it to Wallace Laboratories in the early 1950s, had no such doubts himself. "Anxiety has no relation to intelligence and scholastic achievement or to the desire to achieve," Berger wrote in a celebratory 1964 issue of the *Journal of Neuropsychiatry*. Anxiety is "not a motivating force," he claimed, "but rather a symptom of disease" (Berger, 1964). Berger then went on to defend tranquilizers against charges that sound remarkably like those that would be leveled against Prozac 30 years later. He denied that the tranquilizers are habit forming. He denied that they are sedatives. He denied that they affect the personality, and he denied that they interfere with creativity or intellectual genius. He claimed that tranquilizers facilitate psychoanalysis and make therapy more efficient. He even downplayed any resemblance between tranquilizers and alcohol, saying, "A Miltown is no substitute for a martini" (Berger, 1964).

By the time Berger wrote these words in 1964, Miltown was no longer the only tranquilizer on the market. Antidepressants were available, too, but as a cultural phenomenon the antidepressants could not compare with benzodiazepines such as Librium and Valium, commonly known as "minor tranquilizers." Librium was the best-selling prescription drug of the 1960s. At the end of the decade, Valium replaced Librium in the number one spot (Shorter, n.d., p. 319). Yet even as Frank Berger was evangelizing for tranquilizers to "liberate our minds from their primitive and outdated ways," the Rolling Stones were singing about suburban housewives who could not tolerate the mind-numbing tedium of kitchen and kids without resorting to "mother's little helpers." In the same issue of the *Journal of Neuropsychiatry* in which Berger defended the tranquilizers, writer Marya Mannes pointed out that if so many anxious Americans felt compelled to medicate themselves just so they could stand their lives, we probably ought to be asking what about their lives was producing such anxiety. For Mannes, the answer was obvious: the desire for material acquisitions and the quest for female beauty. "So great is the compulsion to acquire these things," she wrote, "so deep is the fear of their loss or lack, that the legitimate anxiety—*am I being true to myself as a human being?*—is submerged in trivia and self-deception" (Mannes, 1964).

3.3.2 Listening to Prozac

It is this very question—"Am I being true to myself as a human being?"—that Kramer revived in *Listening to Prozac*. Prozac, of course, was marketed as an antidepressant, not an antianxiety drug, but Kramer was struck by the way Prozac seemed to help patients whom he would not have previously thought to be clinically depressed—people who were shy, unhappy, emotionally rigid, or socially isolated. Some of these patients improved dramatically when Kramer prescribed Prozac. Yet Prozac did not sedate or tranquilize these people. Just the opposite: often it seemed to energize them. Nor did these patients merely return to their normal, baseline state. They claimed to feel better than they had ever felt before. This phenomenon Kramer called "cosmetic psychopharmacology," and he wondered whether it was the proper business of psychiatry.

Kramer noticed especially the striking language his patients were using to describe the changes they felt on Prozac. Some of them did not report feeling like a "different person" on Prozac. Instead they said they felt like themselves. "This is who I am," one patient told Kramer. "I just feel strong. I feel resilient. I feel confident" (Kramer, 1994, p. 219). Others agreed. After she went off Prozac, one woman said to him, "I am not myself" (Kramer, 1994, p. 19). She only felt like herself, she said, while taking Prozac. Remarks like these turned the old concern about cosmetic psychopharmacology on its head. It is one thing to worry that women are using psychoactive drugs to change their personalities, to tolerate their submissive social roles, or to blunt the anxiety of self-betrayal or self-deception. It is quite another matter when these same women, experiencing similar cultural pressures, say that they can become their true selves only on medication. As Kramer (1994, p. 294) asked, "What are we to make of patients who navigate that culture more effectively—and achieve self-realization—on medication?"

In *Listening to Prozac*, Kramer was accused of exaggerating the "better than well" phenomenon, but he was far from the only psychiatrist to notice it. Arvid Carlsson of Gothenburg University in Sweden, whose work was important for the development of zimelidine, a precursor of Prozac, said, "There are people who feel so much better, who didn't have any diagnosis really. . . . I remember from the zimelidine period, that there were people whose income went up when they started to take the drug" (Healy, 1998, p. 77). Harvard's Jonathan Cole, formerly the director of the Psychopharmacology Research Center at the National Institutes for Mental Health, says, "At McLean I treated 100 or so patients before it came on the market and a handful of them really were astoundingly better. They had been sick for 10/15 years and were clearly better than they had ever been before in their lives" (Healy, 1998, p. 259). Roland Kuhn, the Swiss psychiatrist who first identified the antidepressant effects of imipramine to Ciba-Geigy in early 1956, claims that Prozac is now "mainly used by people who are not depressed but who use it as a pure stimulant" (Healy, 1998, p. 114).

Clinical trials have provided evidence that the selective serotonin reuptake inhibitors (SSRIs), like other antidepressants, can effectively treat clinical depression, a potentially life-threatening condition (Geddes, Freemantle, Mason, Eccles, & Boynton, 2001; Spigest & Martensson, 1999). (That evidence has also been questioned recently, with some published studies suggesting that the SSRIs are little better than placebo for depression: Hypericum Depression Trial Study Group, 2002; Kirsch, Moore, Scoboria, & Nicholls, 2002[2]). Yet in the decade since *Listening to Prozac* was published, the term *antidepressant* has come to seem like a very limited way to describe the SSRIs. Soon after Prozac (fluoxetine) was introduced to the American market, it was joined by Paxil (paroxetine), Luvox (fluvoxetine), Zoloft (sertraline), Effexor (venlefaxine), and Celexa (citalopram), all drugs similar in structure to Prozac. Today clinicians use the SSRIs not just for depression but also for social phobia, panic disorder, obsessive-compulsive disorder, body dysmorphic disorder, eating disorders, post-traumatic stress disorder, the impulse control disorders, Tourette's syndrome, and sexual compulsions, among many other disorders. From sex to food to body image and self-presentation, it is a rare part of ordinary American life that has not been subjected to a clinical trial involving an SSRI. GlaxoSmithKline even markets Paxil for "generalized anxiety disorder," a medical indication that takes the drugs back full circle to the days of Miltown.

One of the most striking aspects of the Prozac phenomenon is how the drug has moved out of the doctor's office and into the culture as a whole. Prozac may have begun as a brand name for fluoxetine, but it has become as recognizable and ubiquitous a brand name as Kleenex or Pampers. And like Kleenex or Pampers, Prozac has come to stand not just for a particular market item but also for a type of technology: in this case, antidepressant drugs. Today when people say "Prozac," they may well be talking about any number of drugs whose brand names have never caught on in quite the same way.

Whatever the merits of the SSRIs, they have been among the most heavily promoted drugs of the past decade. The manufacturers of antidepressants have taken full advantage of the relaxation of U.S. Food and Drug Administration (FDA) restrictions on prescription drug advertising in 1997. In 2000, Paxil was the fourth most heavily promoted prescription drug in America, with $91.8 million in direct-to-consumer spending. Eli Lilly spent $37.7 million that same year advertising fluoxetine—$23.3 million as Prozac and $14.4 million as Sarafem. To put these figures in context: GlaxoSmithKline spent more money advertising Paxil than Nike spent advertising its top shoes (National Institute for Health Care Management, 2000). Direct-to-consumer advertising clearly works. From 1999 to 2000, antidepressants saw a 20.9% increase in sales to a figure of $10.4 billion, maintaining their position as the best-selling category of drugs in the United States. In 2000, Prozac was America's fourth most prescribed drug; Zoloft was number seven, and Paxil was number eight (National Institute for Health Care Management, 2000, pp. 17, 19).

Of course, the SSRIs could not have achieved such spectacular success if they did not work for some patients. Yet an equally important reason behind the success of psychoactive drugs in general, and the SSRIs in particular, is the elasticity of psychiatric diagnosis. Categories of "mental disorder" are in constant flux, and they often expand dramatically once a new treatment is marketed (Horwitz, 2002). For example, social anxiety disorder—the fear of being embarrassed or humiliated in public—was considered a rare disorder until physicians began treating it with Nardil (phenelzine) in the mid-1980s and then, later, with SSRIs such as Paxil (Baldwin, Bobes, Stein, Scharwachter, & Faure, 1999; Liebowitz, Fyer, Gorman, Campeas, & Levin, 1986; Liebowitz et al., 1988, 1990; Stein et al., 1998). Today, social phobia is often described as the third most common mental disorder in the United States (Kessler et al., 1994; Lamberg, 1998). Similar stories can be told for obsessive-compulsive disorder and panic disorder (the latter known among clinicians in the mid-1980s as the "Upjohn illness," after the makers of Xanax) (Shorter, 1997, p. 320). As David Healy has pointed out, the key to selling psychoactive drugs is to sell mental disorders (Healy, 1997).

But to sell a mental disorder, you must first capture it and make it your own. A drug manufacturer is not allowed to promote a product for a specific disorder until that product has FDA approval. As a result, SSRI manufacturers jockey aggressively among themselves to claim new pieces of the mental disorder market. Whereas the FDA has approved all six SSRIs on the market for depression, Paxil was until recently the only drug approved for social anxiety disorder. (In 2003 it was joined by Zoloft and Effexor.) All of the SSRIs except Celexa and Effexor have been approved for obsessive-compulsive disorder, but only Effexor and Paxil have been approved for generalized anxiety disorder. Zoloft and Paxil have claimed panic disorder and post-traumatic stress disorder, but only Prozac has been approved for bulimia. Eli Lilly's patent on Prozac expired in 2001, but Lilly has begun marketing the same drug under a different name, Sarafem, as a treatment for "premenstrual dysphoric disorder."

Conventional wisdom attributes the spectacular success of the SSRIs to their relative absence of side effects. For instance, monoamine oxidase inhibitors, an alternative type of antidepressant, can be dangerous without strict dietary restrictions, and people taking the longer-established tricyclic antidepressants often complain of drowsiness, dizziness, dry mouth, or constipation. Prozac and the other SSRIs initially appeared much less burdensome. Another significant reason for the success of the SSRIs lies in their ease of use. "One pill a day forever," says Jonathan Cole. "Fluoxetine at one pill a day is the ideal primary care physician's drug" (Healy, 1998, p. 259). Today, in fact, it is no longer even one pill a day. Prozac Weekly is a once-a-week version of Prozac that Lilly has marketed using coupons in newspapers and magazines (O'Connell & Zimmerman, 2002). The one-pill strategy has clearly worked, whether the one pill is taken daily or weekly. It has been estimated that as much as 70% of the SSRIs is prescribed not by psychiatrists but by primary care physicians (Glenmullen, 2000; Grinfeld, 1998; Pollock, 1995).

3.3.3 Treatment, Enhancement, and Identity

In bioethics, the conventional response to the phenomenon that Kramer called cosmetic psychopharmacology has been to classify it as an enhancement technology. The distinction between enhancement and treatment had gained currency during the ethical debate over gene therapy in the late 1980s and early 1990s. Many people were eager to press a research agenda into the therapeutic uses of genetic technology for conditions such as adenosine deaminase deficiency or cystic fibrosis but worried about the use of such technologies for eugenic purposes. Since then, bioethicists have used the term *enhancement technology* as shorthand for all sorts of technologies whose uses go beyond the strictly medical, from synthetic growth hormone for short boys to Botox injections for aging women. The unstated assumption behind the term has been that there is a morally important distinction between enhancement and treatment. Treating illness, it has been argued, is an essential part of medical practice. Doctors have an obligation to treat sick people. Enhancements, in contrast, are seen as extras—ethically acceptable, perhaps, but not something that a doctor has any particular obligation to provide or that a liberal society has an obligation to fund.

Yet the distinction between treatment and enhancement turns out to be much more elusive than it first appears, especially in psychiatry. Where is the line between psychopathology and social deviance, perversion, or eccentricity? When does shyness turn into social phobia, or melancholy into depression? The problem is complicated still further by the fact that so little is known about the causes or pathophysiology of mental disorders, or even about how chemical treatments for these mental disorders work. Philosophers have traditionally argued that illness is a departure from species-typical human functioning, but that definition offers us little guidance when the subject turns to the human mind and human behavior. What kind of behavior is typical of *Homo sapiens*?

It might be better to ask: What should we make of the social place that the SSRIs have come to occupy? Every culture has its own socially prescribed psychoactive substances, from peyote, kava, and betel nuts to alcohol, caffeine, and nicotine. But with the SSRIs, the gate to the drug is guarded by doctors, and the passport for access is the diagnosis of a mental disorder. Unlike alcohol, which is dispensed in bars and liquor stores, or caffeine, which is dispensed at Starbucks and Unitarian churches, SSRIs are dispensed at doctor's offices and pharmacies. It is the social place occupied by the SSRIs that has produced the ambivalence that many of us feel about their popularity. Unlike bartenders and espresso baristas, doctors have not generally thought of their job as making well people feel better than well. But that might change.

One way to begin thinking about the cultural significance of the SSRIs is to look at how late modern life pulls us in two different, often contradictory moral directions. On one hand, we have inherited a moral tradition that has come to place considerable value on the notion of authenticity (Taylor, 1989; Trilling, 1982). Concepts such as moral integrity or self-betrayal, sincerity or duplicity, being true to yourself or selling your birthright for a mess of pottage—none of this would make any sense without the idea that we all have individual selves, that these selves have unity and integrity over time and circumstance, and that (with some qualifications) we ought to be morally committed to maintaining that unity. Even the notion of self-fulfillment, controversial though it may be, has the concept of an authentic self at its core: self-fulfillment cannot be achieved without a true self to be fulfilled. An ethic of authenticity teaches us that in order to live a meaningful life we must live a fulfilled life, and fulfillment means discovering and ultimately pursuing the values, ideals, and talents that are unique to us as individuals.

Yet this moral vocabulary has been built against a social background that encourages us to adopt a flexible, adaptable identity (Gergen, 1990; Lifton, 1999). Contemporary life seems designed to fracture the unified self. The market requires the modern worker to be extraordinarily adaptable, able to develop new skills very quickly, willing to work on short-term contracts, and capable of "selling himself" to new employers when a position is terminated. Work life is sharply divided from leisure life, each with its own distinct customs, languages, and rituals. The Internet allows users to cultivate online personalities that can be vastly different from their real-world identities. The mass media reinforces the significance of public self-presentation at the expense of the inner, private self. With the aid of medical technology we can alter our face, body, personality, and even our sex. All of this uncertainty makes the notion of a unified, lasting, authentic self seem quaint at best and, at worst, stifling and oppressive.

The moral debate over Prozac contains some of this same tension and ambivalence about authenticity. Here is a drug that, at least according to some accounts, can help some of us become the people we want to be. By the standards of psychiatry, it allows us to "function" better. By our own moral standards, it gives us a better shot at self-fulfillment. Yet what if success is accompanied by dramatic changes—in our personality and behavior, in the way others perceive us, and in the way we make our way in the world? When we make these changes, what do we give up? This worry is by no means unique to Prozac or even to enhancement technologies more generally. It runs through much of American history: a tension between the values of self-improvement and personal achievement, on one hand, and, on the other, the values of stability, loyalty to your roots, and remembering where you came from. It should be no surprise that the language we use to describe how we feel on Prozac reflects a similar tension. We explain that Prozac has allowed us to become who we really are, even as it makes us feel different than we have ever felt before.

Does Prozac really change the self? This question is implicit in the title of Kramer's *Listening to Prozac.* Kramer suggests that by listening carefully to people who are taking a particular drug, by paying close attention to the changes in how they understand themselves, we may well come to think about the self in a very different way. If personality can change so dramatically on a drug, suggests Kramer, it is hard to avoid concluding that personality is largely a matter of biology. But such a conclusion would be premature. Prozac does not necessarily tell us anything about biology or human nature or the real nature of mental disorders. It simply tells us that some people interpret their lives and selves differently once they begin taking this drug. In the same way that a drinker may come to understand himself or herself as an alcoholic only after joining a support group, so a shy person may come to understand himself or herself as suffering from social anxiety disorder only after he or she starts taking Paxil. The drug and the disorder provide a new vocabulary with which to describe oneself. They give an individual a new way of understanding his or her history.

As the SSRIs have become more and more widely prescribed, and thus more and more a part of the popular culture, they have also helped to create new categories into which people can place themselves and understand their lives: depression, panic disorder, obsessive-compulsive disorder, and social anxiety disorder, among others. These drugs and diagnostic categories help people take what was previously a vague and inchoate set of psychic troubles and shape them into a recognizable narrative. Before taking Prozac, I may have simply thought of myself as melancholic or alienated. I may have considered myself introverted, self-conscious, and lonely. Or I may have simply found myself bewildered by the way my life had unfolded until that point. But Prozac can give me a new narrative. After I have taken Prozac, I understand that I have been suffering from a hidden clinical depression. Thus the drug gives me a new social identity.

By giving us a new way of understanding our lives, these narratives can help us make sense of events that may have previously been baffling or incoherent. In fact, if medication can correct the disorder, the mental disorder narrative may even offer the promise of a hopeful ending. But this narrative also carries a price. When we understand our problems as symptomatic of a mental disorder, we also change our moral status. Whereas having a disorder can relieve us of the responsibility for the illness (if I have a disorder, it is not my fault), it also places new responsibilities on us. Specifically, it implies the obligation to seek psychological help. Shyness may be part of my personality, but social anxiety disorder is a potentially remediable psychiatric illness.

The mental disorder narrative need not stand up to philosophical scrutiny. All that is necessary is for the narrative to make sense to people in psychological distress. With direct-to-consumer advertising, it is enough that I see myself in the advertisements—that I feel that tingle of recognition when I see a young mother with red eyes and tear-stained cheeks, that I feel the same sense of nauseating panic that the man in the ad feels when he has to give a business presentation, or that I identify with the middle-aged homemaker who feels anxious and worried but cannot explain why.

Maybe the best way to begin to understand the cultural phenomenon that the antidepressants have become is to think about the story that drugs tell: I am the person I am, with the problems that I have, because I have this particular mental disorder. It is a story that provides me with a sympathetic listener (my doctor or therapist), a community of like-minded sufferers (my support group), and a coherent narrative (told on television) both for myself and for those to whom I must explain myself. Increasingly, this is a story of biology and the brain, in which biological psychiatrists, pharmaceutical companies, and patient support groups all agree that disorders that respond to the SSRIs must have biological roots. Eventually this story may even create a new set of criteria for what counts as an acceptable identity, one of which will be response to treatment. If I am to be admitted to the community of social phobia, anxiety, or clinical depression, my identity papers must certify that I have responded to an antidepressant.

Notes

1. Editor's note: This reading is an abridged version of a chapter from the 2004 book *Prozac as a Way of Life*, edited by Carl Elliott and Tod Chambers, published by the University of North Carolina press, and is used with permission of publisher and author. The omitted text refers to a meeting held at the Hastings Center. Section headings were added by the editor.

2. Editor's note: In the years since Elliott's article was written, the efficacy of SSRIs has been examined in several large studies. A recent meta-analysis by Fournier, DeRubeis, and colleagues found that SSRIs are clearly helpful in severe depression, but have little or no effect on symptoms in mild and moderate depression relative to placebo. For more details see Fournier, J. C., DeRubeis, R. J., Hollon, S. D., Dimidjian, S., Amsterdam, J. D., Shelton, R. C., & Fawcett, J. (2010). Antidepressant drug effects and depression severity: A patient-level meta-analysis. *Journal of the American Medical Association, 303*, 47–53.

References

Baldwin, D., Bobes, J., Stein, D. J., Scharwachter, I., & Faure, M. (1999). Paroxetine in social phobia/social anxiety disorder: randomised, double-blind, placebo-controlled study. *British Journal of Psychiatry, 175*, 120–126.

Berger, F. (1964). The tranquilizer decade. *Journal of Neuropsychiatry, 5*, 403–410.

Geddes, J. R., Freemantle, N., Mason, N., Eccles, M. P., & Boynton, J. (2001). *Selective serotonin reuptake inhibitors for depression.* Retrieved August 29, 2001, from http://www.cochrane.org/reviews/.

Gergen, K. (1990). *The saturated self.* New York: Basic Books.

Glenmullen, J. (2000). *Prozac backlash.* New York: Touchstone.

Grinfeld, M. J. (1998). Protecting prozac. *California Lawyer, 79*, 36–40.

Healy, D. (1997). *The antidepressant era.* Cambridge, MA: Harvard University Press.

Healy, D. (Ed.). (1998). *The psychopharmacologists II: Interviews by David Healy.* London: Altman.

Horwitz, A. V. (2002). *Creating mental illness.* Chicago: University of Chicago Press.

Hypericum Depression Trial Study Group. (2002). Effect of Hypericum perforatum (St. John's wort) in major depressive disorder: A randomized controlled trial. *Journal of the American Medical Association, 287*, 1807–1814.

Kessler, R. C., McGonagle, K. A., Zhao, S., Nelson, C. B., Hughes, M., Eshleman, S., Wittchen, H. U., & Kendler, K. S. (1994). Lifetime and 12-month prevalence of DSM-III-R psychiatric disorders in the United States: Results from the National Comorbidity Survey. *Archives of General Psychiatry, 51*, 8–19.

Kirsch, I., Moore, T. J., Scoboria, A., & Nicholls, S. S. (2002). The emperor's new drugs: An analysis of antidepressant medication data submitted to the U.S. Food and Drug Administration. *Prevention and Treatment, 5*, 23.

Kramer, P. D. (1994). *Listening to prozac.* London: Fourth Estate.

Lamberg, L. (1998). Social phobia: Not just another name for shyness. *Journal of the American Medical Association, 280*, 685–686.

Liebowitz, M. R., Fyer, A. J., Gorman, J. M., Campeas, R., & Levin, A. (1986). Phenelzine in social phobia. *Journal of Clinical Psychopharmacology, 6*, 93–98.

Liebowitz, M. R., Gorman, J. M., Fyer, A. J., Campeas, R., Levin, A. P., Sandberg, D., Hollander, E., Papp, L., & Goetz, D. (1988). Pharmacotherapy of social phobia: An interim report of a placebo-controlled comparison of phenelzine and atenolol. *Journal of Clinical Psychiatry, 49*, 252–257.

Liebowitz, M. R., Schneier, F., Campeas, R., Gorman, J., Fyer, A., Hollander, E., Hatterer, J., & Papp, L. (1990). Phenelzine and atenolol in social phobia. *Psychopharmacology Bulletin, 26*, 123–125.

Lifton, R. J. (1999). *The Protean self.* Chicago: University of Chicago Press.

Mannes, M. (1964). The roots of anxiety in the modern woman. *Journal of Neuropsychiatry, 5*, 412.

National Institute for Health Care Management. (2000). *Prescription drugs and mass media drugs advertising*. Retrieved from www.nihcm.org.

O'Connell, V., & Zimmerman, R. (2002, January 14). Drug pitches resonate with edgy public. *Wall Street Journal*, p. 26.

Pollock, E. J. (1995, December 1). Managed care's focus on psychiatric drugs alarms many doctors. *Wall Street Journal*, p. 11.

Shorter, E. (1997). *A history of psychiatry: From the era of the asylum to the age of Prozac*. New York: John Wiley & Sons.

Smith, M. (1991). *Small comfort: A history of minor tranquilizers*. New York: Praeger.

Spigest, O., & Martensson, B. (1999). Fortnightly review: Drug treatment of depression. *British Medical Journal, 318*, 1188–1191.

Stein, M. B., Liebowitz, M. R., Lydiard, R. B., Pitts, C. D., Bushnell, W., & Gergel, I. (1998). Paroxetine treatment of generalized social phobia (social anxiety disorder): A randomized controlled trial. *Journal of the American Medical Association, 280*, 708–713.

Taylor, C. (1989). *Sources of the self*. Cambridge, MA: Harvard University Press.

Trilling, L. (1982). *Sincerity and authenticity*. Cambridge, MA: Harvard University Press.

Reading 3.4

The Valorization of Sadness: Alienation and the Melancholic Temperament[1]

Peter D. Kramer

At the heart of *Listening to Prozac* is a thought experiment: Imagine that we have a medication that can move a person from a normal psychological state to another normal psychological state that is more desired or better socially rewarded (Kramer, 1993). What are the moral consequences of that potential, the one I called cosmetic psychopharmacology?

The question would be overgeneral except that it occurs in the context of a discussion of psychic consequences of technologies. People now experience the self in the light of psychotherapeutic medications as lately they experienced it through psychoanalysis. In the thought experiment, the medication we are to imagine is rather like Prozac, and the less desired state is something like melancholy, when that term refers to a personality style rather than an illness. Melancholics are well described in literature that stretches back for centuries. They are pessimistic, self-doubting, moralistic, and obsessive. They have low energy but use that energy productively. They are creative in the arts. They are prone to depression, especially in response to social disappointments.

I have come to believe that much of the discussion of cosmetic psychopharmacology is not about pharmacology at all—that is to say, not about the technology. Rather, cosmetic pharmacology is a stand-in for worries over threats to melancholy. That psychotherapy caused less worry may speak to our lack of confidence in its efficacy.

3.4.1 Temperament, Creativity, and Cultural Values

We do, as a culture, value melancholy. Some months ago, I attended an exhibition of the paintings of "the young Picasso." Seeing the early canvases, I thought, "Here is a marvelous technician." I turned a corner to confront the works of the Blue Period, Picasso's response to the suicide of his friend Carlos Casagemas. Instantly I thought (as I believe the curator intended), "How profound." That pairing—melancholic/deep—is a central trope of the culture. Or to allude to another recent museum exhibition: for years the rap on Pierre Bonnard was that his paintings were too cheerful to be important. Here is the corresponding trope: happy/superficial.

Surely the central tenet of literary criticism is Franz Kafka's: "I think we ought to read only the kind of books that wound and stab us.... We need the books that affect us like a disaster, that grieve us deeply, like the death of someone we loved more than ourselves, like being banished into forests far from every one, like a suicide" (Kafka, 1904). This need may even be pragmatic. In his poetry (I am thinking of "Terence, this is stupid stuff"), A. E. Housman argues that painful literature immunizes us against the pain of life's disappointments.

And here I want to lay down two linked challenges that are intentionally provocative. The first is to say that the literary aesthetic makes most sense in relation to a particular temperament (the melancholic, in which one feels great pain in response to loss) in a particular culture (one lacking technologies to prevent or diminish that pain). What if Mithradates had had an antidote, so that he did not require prophylactic arsenic and strychnine? Might poetry appropriate to the antidepressant era be more like beer drinking? And might that new art still prove authentic to the way of the world?

The second challenge is yet more provocative; call it intentionally hyperbolic: to say that there is no neutral venue for this debate over alienation or cosmesis because our sensibility has been largely formed by melancholics. Much of philosophy is written, and much art has been created, by melancholics or the outright depressed as a response to their substantial vulnerabilities. To put the matter only slightly less provocatively (and to return to the first challenge), much of philosophy is directed at depression as a threatening element of the human condition.

As Martha Nussbaum's *The Therapy of Desire* demonstrates in detail, classical moral philosophy is a means for coping with extremes of affect that follow upon loss (Nussbaum, 1994). The ancient Greeks' recommendations for the good life, in the writings of the Cynics and Stoics and Epicureans and Aristotelians, amount to ways to buffer the vicissitudes of attachment. If loss were less painful, the good life might be characterized not by *ataraxia* but by gusto. The connection between philosophy and melancholy continues in the medieval writings on *akadie* and then in the Enlightenment, through Montaigne and through Pascal, who writes, "Man is so unhappy that he would be bored even if he had no cause for boredom, by the very nature of his temperament" (Ferguson, 1995). In a study of Kierkegaard, Harvie Ferguson writes, "Modern philosophy, particularly in Descartes, Kant, and Hegel, presupposed as a permanent condition the melancholy of modern life" (Ferguson, 1995, 32; Kafka, 1904, p. 16). Even those, like Kierkegaard, who chide melancholics do so from such a decided melancholic position that their writing reinforces the notion that melancholy is profundity.

As for literature, studies indicate that an astonishing percentage, perhaps a vast majority, of serious writers are depressives. Researchers have speculated on the cause of that connection—Does depression put one in touch with important issues of deterioration and loss? But no one has asked what it means for us as a culture or even as a species that our unacknowledged legislators suffer from mood disorders, or something like.

3.4.2 Rethinking Moral and Aesthetic Values

If there is no inherent moral distinction between melancholy and sanguinity, then we will need to worry about the association between creativity and mood. What if there is a consistent bias in the intellectual assessment of the good life or the wise perspective on life, an inherent bias against sanguinity hidden (and apparent) in philosophy and art?

An argument of this sort is worrisome—more worrisome than the conundrum we began with. Yet can we in good faith ignore the question of who sets the values? I have been, in effect, proposing still another thought experiment: Imagine a medication that diminishes the extremes of emotional response to loss, imparting the resilience already enjoyed by those with an even, sunny disposition. What would be the central philosophical questions in a culture where the use of this medication is widespread?

Aesthetic values do change in the light of changing views of health and illness. Elsewhere I have asked why we are no longer charmed by suicidal melancholics such as Goethe's Werther or Chateaubriand's René or Chekov's Ivanov (Kramer, 1993, p. 297; Kramer, 1997). Because we see major depression and affectively driven personality disorders as medically pathologic, what once exemplified authenticity now looks like immaturity or illness—as if the Romantic writers had made a category error.

A final thought experiment: Imagine that the association between melancholy and literary talent is based on a random commonality of cause: the genes for both cluster, say, side by side on a chromosome. Let us further imagine a culture in which melancholy, now clearly separate from creativity, is treated pharmacologically on a routine basis. In this culture, it is the melancholics manqués who write—melancholics rendered sanguine—so that the received notions of beauty and intimacy and nobility of character relate to bravado, decisiveness, and connections to social groups, not in the manner of false cheerleading but authentically, from the creative wellsprings of the optimistic.

What would be the notion of authenticity under such conditions? Perhaps in such a culture "strong evaluation" would find psychic resilience superior to alienation. Even today, many a melancholic looks at Panurge or Tom Jones with admiration—how marvelous to face the world with appetite! The notion of a sanguine culture horrifies those of us resonant with an aesthetics of melancholy, but morally, is such a culture inferior, assuming its art corresponds with the psychic reality? Is there a principled basis for linking melancholy to authenticity? Is there a moral hierarchy of temperaments?

I have offered an extreme version of an argument that might be more palatable in subtler form. I hope I have been convincing, or at least troubling, in one regard: the assertion that there is no privileged place to stand, no way to get outside the problem of authenticity as regards temperament.

Elliott (e.g., 1998) asks whether we do not lose sight of something essential about ourselves when we see alienation and guilt as symptoms to be treated rather than as clues to our condition as human beings. The answer is in part empirical, in part contingent (on the social conditions of human life, a culture's technological resources, and such), and altogether aesthetic. If extremes of alienation are shown to arise from neuropathology, and if aspects of that pathology respond to treatment, our notion of the essential will change. And it may be that what remains of the experience and the concept of alienation will be yet more morally admirable—alienation stripped of compulsion, alienation independent of genetic happenstance, alienation that arises from free choice.

I want to end by saying that, like Elliott, in my private aesthetic I value depression and alienation, see them as postures that have salience for the culture and inherent beauty. But the role of philosophy is to question preferences. The case for and against alienation seems to me at this moment wide open. It has become easy, in the light of the debate over Prozac, to imagine material circumstances that might cause us to reassess which aspects of alienation fall into which category. The challenge of Prozac is precisely that it puts in question our tastes and values.

Notes

1. Editor's note: This reading was excerpted from a longer chapter that appeared in the 2004 book *Prozac as a Way of Life*, edited by Carl Elliott and Tod Chambers, published by the University of North Carolina Press, and is used with permission of publisher and author. The omitted text mainly concerns the nature of alienation as clinical symptom, literary perspective, and moral stance, and was a response to Carl Elliott's (1998) essay, "The Tyranny of Happiness." Section headings were added by the editor.

References

Elliott, C. (1998). The tyranny of happiness: Ethics and cosmetic psychopharmacology. In E. Parens (Ed.), *Enhancing human traits: Ethical and social implications*. Washington, DC: Georgetown University Press, pp. 177–188.

Ferguson, H. (1995). *Melancholy and the critique of modernity: Søren Kierkegaard's religious psychology*. London: Routledge.

Kafka, F. (1904; 1977). Letter to Oskar Pollak, January 27, 1904. In: R. Winston & C. Winston (Eds.), *Letters to friends, family, and editors*. New York: Schocken Books.

Kramer, P. D. (1993). *Listening to Prozac*. New York: Viking.

Kramer, P. D. (1997, December 21). Stage view: What Ivanov needs is an antidepressant. *New York Times*, section 2, p. 5.

Nussbaum, M. C. (1994). *The therapy of desire: Theory and practice in Hellenistic ethics*. Princeton, NJ: Princeton University Press.

4

Brain Reading

The ability to read minds, like the ability to fly or become invisible, is an age-old subject of fantasy. Of course, there is a sense in which we read minds every day, every time we decode the facial expressions, tone of voice, and body language of others or infer what they are feeling by putting ourselves in their situations. This everyday mind reading is fallible, however, and leaves many aspects of other people's beliefs, attitudes, and intentions entirely out of reach for us. "Real" mind reading would give us access to this internal mental content directly, without relying on behavior or context for clues. Could brain imaging accomplish this?

In principle the answer would seem to be "yes." This answer is based on our current understanding of the relation between mind and brain. Most contemporary philosophers of mind believe that thought just *is* brain activity, whether they hold an "identity theory" of mind–brain relations or a "functionalist" theory (e.g., Block & Fodor, 1972; Lewis, 1980). Even those who do not equate mental states with brain states, but instead speak of the "supervenience" of mental states on brain states, nevertheless see the relationship as involving a necessary correlation between these states (Davidson, 1970; Kim, 2005; see Clark, 2000, for a review). According to these views, there is no *a priori* reason why a person's brain states could not be decoded to reveal their mental states.

In practice, however, brain imaging delivers information about some aspects of brain activity and not others, and indeed current methods offer very limited and crude information. In addition, we still have much to learn about the relationships between those aspects of brain activity we can visualize and psychological states. In sum, the potential of brain imaging for mind reading is uncertain but cannot be dismissed *a priori*.

As for the current state of the art, at least some researchers and corporations have claimed the ability to ascertain psychological information about individuals with brain imaging. Whatever the current capabilities are, they will surely improve in the future. From an ethical point of view this raises an unprecedented privacy issue; namely, the threat to mental privacy. In addition, the scientific aura of brain imaging and the appealingly concrete nature of the images themselves lead to another risk: that society will adopt certain brain imaging methods prematurely or use them uncritically. This would be to the detriment of people whose mental states and traits are misidentified, as well as to society as a whole if guilty or dangerous people are misclassified.

Some of the more ethically freighted uses of imaging include lie detection, prediction of dangerousness, and measurement of unconscious attitudes toward other racial groups. The readings in this section review some of actual and potential uses of brain imaging and analyze the most pressing ethical, legal, and societal concerns. We begin with an explanation of how the different methods of brain imaging work, including some of the recent analytic techniques that have expanded the "brain reading" capabilities of brain imaging far beyond what seemed possible just 5 years ago. After this introduction to the technical foundations of brain imaging, we will highlight some of the key concerns about the uses of "brain reading" raised in the readings.

Methods of Brain Imaging

The great granddaddy of functional neuroimaging is electroencephalography, or EEG, invented in the 1920s when a German researcher discovered that the electrical activity of the brain is detectable outside the head. Scalp-recorded EEG reveals only a small portion of the brain's electrical activity; among its restrictions is that only synchronized activity among large populations of neurons with similar orientations is detectable. Most current applications of this approach use event-related potentials (ERPs), which are the components of the EEG that consistently occur in conjunction with an "event" such as the presentation of a stimulus. The brain fingerprinting method discussed by Wolpe and colleagues uses ERPs.

A very different approach to studying human brain function began in the 1960s, based on the measurement of cerebral blood flow rather than electrical activity. By having patients inhale radioactive xenon gas, which entered the bloodstream through the lungs, the amount of blood flow to the brain could be gauged according to the amount of radioactivity detected outside the head. The breakthrough that transformed this relatively crude method for estimating cerebral blood flow into an imaging modality was computer-assisted tomography (CAT), which came to be known as computed tomography (CT). CT was initially developed in the 1970s for use with x-rays. By passing x-rays through the head from various directions and similarly positioning detectors all around the head, it was possible to computationally reconstruct the three-dimensional structure inside. The adaptation of CT for functional neuroimaging involved applying the same computational strategy to the radioactivity emanating from the head rather than passing through it. This enabled the three-dimensional distribution of radioactively labeled blood within the head to be represented. Because more active brain regions require more blood to support their metabolic needs, this method essentially yielded a three-dimensional image of brain activity.

The two types of radioactive substances used in functional neuroimaging are single photon emitters and positron emitters, and hence the two general types of radioactive neuroimaging are called single photon emission computed tomography (SPECT) and positron emission tomography (PET). By radioactively labeling an inhaled gas or an intravenously administered glucose-like molecule or water, the most metabolically active regions of the brain can be identified. Activity specific to single types of neurotransmitter receptors or transporters (see chapter 2) can be measured and localized by radioactively labeling ligand molecules with affinities specific for these neurochemical targets of interest. These methods are primarily used in research on psychiatric disorders. For example, much of what we know about the effects of drug abuse on the dopamine system, discussed in chapter 5 and in the reading by Steven Hyman, has come from PET imaging with radioligands. The use of these methods for clinical research should be distinguished from their use for diagnosis. Despite the claims of some for-profit companies offering SPECT-based diagnosis for a wide range of psychological problems, these methods are not generally appropriate for psychiatric diagnosis outside of certain forms of dementia (Foster et al., 2007).

Most cognitive neuroimaging research today uses functional magnetic resonance imaging (fMRI). Whereas the tracers used in SPECT and PET are exogenous radioactive materials introduced into the blood, fMRI in effect uses the blood itself as a tracer. The noninvasive nature of fMRI is probably its biggest advantage. Its negligible risks make the risk:benefit ratio of just about any study favorable, undergraduates can be induced to volunteer for $10 to $20 per hour, and it makes possible the scanning of children as well as repeated scans for longitudinal studies of learning, development, intervention effects, and so forth. The most common fMRI technique, blood oxygen level dependent (BOLD) fMRI, uses the different magnetic properties of oxygenated and deoxygenated hemoglobin to mark regional differences in brain activity. Because regional brain activity changes the amount of oxygenated versus deoxygenated hemoglobin in the vicinity, differences in the magnetic susceptibility of blood can be used to measure brain activity.

The temporal and spatial resolution of fMRI is limited by hemodynamics rather than by the physics of the method. Temporally, the delivery of blood in response to increased neuronal activity is not instantaneous; blood flow changes in a time frame of seconds. Spatially, these changes extend into nearby tissue, just as a gardener trying to water a specific plant will water the surrounding plants. In practice, fMRI has a spatial resolution of about 2–3 mm and a temporal resolution of about 2 s (Aguirre, 2005), which is adequate to distinguish among at least some psychologically meaningful differences in brain activity.

In sum, a number of different methods have been devised for monitoring human brain function "in vivo," based on fundamentally different physical principles. EEG is based on the measurement of the electrical potential fields generated by large populations of neurons. SPECT and PET are based on the measurement of radioactive decay products from tracers in the blood. fMRI methods are based on the measurement of the magnetic behavior of blood, specifically, in the case of BOLD, the difference in susceptibility between deoxygenated and oxygenated hemoglobin. In each case the signal being measured is different and in no case is it a direct measure of all functionally relevant brain activity. Inferences and assumptions are required for the use of any method of visualizing brain function. Brain images might be captioned in magazine articles as "snapshots of the brain at work," but the images in fact bear a much less direct relationship to brain function than that. Nevertheless, "indirect" is not the same thing as "inaccurate" or "misleading." Properly applied and interpreted, brain imaging is a powerful method that has already taught us much.

From Brain Imaging to "Brain Reading"

The earliest uses of functional brain imaging were clinical, but by the 1980s these methods had been adapted for the basic science goal of understanding how the mind works and how it is implemented in the brain. The collaboration of cognitive psychologist Michael Posner with neuroradiologist Marcus Raichle led to the first uses of functional brain imaging as a tool for cognitive neuroscience. Rather than imaging the brain activity associated with the performance of a task, which might involve the use of visual processes to perceive an object, knowledge to understand what the object is, phonological retrieval to access the name the object in memory, and motor processes to utter the name aloud, they devised a way to image the brain activity associated with a single component operation, such as visual or phonological processing alone (see, e.g., Posner, Petersen, Fox, & Raichle, 1988). By the 1990s, this approach had been refined and became widely used with fMRI. From the point of view of neuroethics, a key development in the 1990s was the extension of functional neuroimaging from the study of cognition (memory, language, attention, and so on) to the study of affective processes (emotion, mood, motivation).

In the first decade of the twenty-first century, the advances have shifted toward disaggregating data to enable inferences about individuals and events. Up until this point, studies used groups of subjects, or if they used individual subjects, such subjects were assumed to be typical, and differences between subjects were not the focus of studies. Also up until this point, the many stimuli or events of a scanning session were not studied individually but were combined for purposes of analysis. The ability to draw conclusions about individual subjects, and about what such subjects might be thinking on individual trials, raises yet more neuroethical questions.

As the readings to follow make clear, individual differences in intelligence and personality have long been a subject of study in psychology because of their relevance to education and employment. In recent years, neural correlates of these traits have been found in neuroimaging studies. Compared with the strength of correlation between specific genes and these psychological traits, patterns of brain activation are considerably more predictive (although in some cases the strengths of correlation have been inflated by in effect undercorrecting for multiple statistical comparisons (see Vul et al., 2009). In some cases, brain imaging correlates of psychological traits are capable of delivering useful information about individuals. For example, my colleagues and I have found that a number of published studies have revealed sufficiently strong correlations between aspects of brain activation and traits of interest that one could substantially narrow the range of possible values for an individual on the trait by scanning the person (Farah, Smith, Gawuga, Lindsell, & Foster, 2009).

Psychological states as well as traits have measurable correlates in functional brain imaging, including socially relevant states such as the state of lying or otherwise deceiving. For example, Langleben and colleagues (2005) have found that regions of the brain associated with cognitive effort tend to be more active when people lie about the identity of the playing card they are viewing compared with when they truthfully report the card's identity. This may be an indication of the effort required to hold in mind two incompatible concepts, namely the concept of what the card is and the concept of the deceptive response. Two companies, No Lie MRI and Cephos, are currently selling lie detection services based on versions of this technique. A different technique, based on event-related potentials, has been used to discriminate stimuli that are familiar to the viewer from stimuli that are not. This technique has also been commercialized under the name *brain fingerprinting* and can in principle be used to probe for "guilty knowledge"; that is, knowledge of objects, locations, or people that only the perpetrator of a crime would be expected to have.

The most recent development to contribute to the "mind reading" potential of brain imaging is an approach to data analysis called *pattern classification*, which has enabled researchers to extract surprisingly specific information from a brain image, such as which one of 1000 pictures someone is viewing (Kay et al., 2008). Pattern classification analysis differs from conventional analyses of brain imaging data in two main ways. The first way involves the use of finer-grained images. Whereas in conventional image analysis the small-scale "salt and pepper" variation in the image is treated as noise and smoothed away, in pattern classification it is treated as potentially information-bearing. Second, whereas in conventional image analysis, the relation of brain activity to a psychological state is considered one location at a time—in effect asking a series of questions of the form, "When in this psychological state, is this location active? How about this location..." for the tens of thousands of locations typically represented in fMRI, pattern analysis asks a different kind of question, to wit, "What pattern of activation, across voxels, is associated with this psychological state as opposed to that one?" Such patterns, or combinations of activations across disparate small regions, may be reliably associated with certain psychological states and can thereby serve as indicators of people's psychological state (Haynes & Rees, 2006).

Neuroethics of Brain Imaging

The Readings

The readings that follow present a closer look at the ways in which brain imaging reveals information about specific kinds of psychological traits and states, from personality traits to the states of intentional deception, as well as the ways in which the public interprets brain images. They also present various ethical challenges raised by these phenomena, and some discuss the pros and cons of different policy solutions.

Canli and Amin discuss the neuroethical issues raised by our growing ability to image the neural correlates of emotion and personality. They identify three kinds of methodological challenge and discuss the way each affects the social, legal, and ethical impact of brain imaging. Whereas most of the ethically weighty uses of imaging concern these affective dimensions of the human mind, *Gray and Thompson* discuss the imaging of a contentious cognitive dimension, namely intelligence. Their article reviews a small but growing body of imaging research on intelligence, which is beginning to cohere but still contains some gaps and inconsistencies. They point out that the very idea of a biological basis for intelligence is associated, in the minds of some, with eugenics and racism and review the additional complications this adds to the task of responsible research conduct in this area. *Wolpe, Foster, and Langleben* present a framework for evaluating technologies for the detection of deception, including brain-based lie detection, and emphasize the current state of uncertainty regarding these methods. They also enumerate the many ways in which inappropriate use of these methods could lead to a variety of social, legal, and ethical problems. *Racine, Bar-Ilan, and Illes* describe their research on media portrayals of fMRI. They report that the public seems to believe (or at least is frequently told!) that imaging evidence provides special validation of the reality of psychological phenomena and is an especially compelling basis for making policy. *Phelps and Thomas* tackle some of the basic assumptions that may lie behind these misconceptions. They take, as their test case for working through these issues, recent research on the neuroscience of racial group identity. In this context they examine the roles that neuroscience and behavioral evidence play in the evaluation of psychological claims, and challenge the idea that neuroscience evidence is more informative than behavioral evidence. Finally, *Tovino* (2007) considers possible policy responses to the expanding application of neuroimaging outside the research laboratory, drawing on precedents that include phrenology and x-rays to identify risks to be managed and means of managing them.

Selected Cross-cutting Issues

In overview, there are two general types of ethical issue raised by brain imaging, which bring to mind the old lament "damned if you do, damned if you don't." On the one hand, any brain imaging method that delivers precise and accurate information about mental content will raise significant new threats to privacy. On the other hand, regardless of their accuracy, these methods have a way of inspiring confidence. If they are in fact not capable of delivering the kinds of information people expect of them, then serious harm could be done in contexts such as legal proceedings, employment, or security screening.

Let us begin with the "damned if you do" category of worry. In principle, brain imaging could pose an unprecedented challenge to privacy by breaching the privacy of the human mind. *Wolpe et al.* intriguingly state that such methods "would force a reexamination of the very idea of privacy." Even the most familiar and commonsensical notion of privacy is endangered in a world where personal thoughts, feelings, and traits can be ascertained without the knowledge of the subject. Consider, for example, the research summarized by *Phelps and Thomas* showing a correlation between unconscious racial attitudes and differential amygdala activation to same-race and other-race faces (Phelps et al., 2000). On the one hand, behavioral measures exist of these unconscious attitudes. Indeed, the imaging measures are validated by their correlation with the behavioral measures. On the other hand, the imaging protocol used by Phelps and colleagues has no obvious relevance to race. One need only say "step into my scanner and have a look at some faces," rather than administer a task involving explicit decisions about the race of faces and the positive or negative valence of other things.

Gray and Thompson point out another way in which brain reading could be problematic, even if it is accurate. Biological measures of group differences have a social impact that goes beyond the effects of individual measures. They warn that, accurate or inaccurate, brain imaging studies of group differences have the potential to be inflammatory or stigmatizing.

All of the readings point out that brain images convey the impression of accuracy and objectivity, even when the impression is unwarranted. One source of this misunderstanding is the failure to appreciate the statistical nature of brain images and their dependence on the specific tasks and comparisons (*Canli & Amin; Wolpe et al.*) and samples of subjects (*Gray & Thompson*) used. An additional misconception is that brain differences observed in imaging studies are genetic and immutable (*Gray & Thompson; Phelps & Thomas*) and that localizing a psychologic function in the brain shows that the function is somehow more "real" (*Phelps & Thomas; Racine et al.*). Furthermore, it is not only the man on the street who is prone to uncritical acceptance of claims supported by brain images: Science writers (*Racine et al.*), judges and security agencies (*Tovino; Wolpe et al.*), and even behavioral scientists and neuroscientists (*Phelps & Thomas*) are susceptible.

Of course, this is not to say that brain images are always misleading, nor to deny that they are uniquely well suited to testing and supporting certain claims. For example, *Canli and Amin* point out that imaging measures can be more selective and hence sensitive than behavioral measures, and *Phelps and Thomas* remind us that imaging is a valuable tool for the study of brain–behavior relations at a "circuits" level. In addition, we need not demand perfection of imaging methods under all circumstances; regulation of brain imaging for match-making purposes could be considerably more lax than for national security (*Tovino*).

Questions for Discussion

1. Like brain enhancement technologies, brain imaging technologies were originally developed for medical use but are increasingly applied outside the clinic. Each reading mentions possible applications of brain imaging for nonmedical purposes. Which such applications are mentioned in the readings and how many are there? Can you think of other applications that are currently being explored by researchers or entrepreneurs or new applications that could be explored?
2. To what extent are some of the so-called neuroethical issues discussed in these readings not ethical (or social or legal) at all but rather scientific? Which problems just boil down to scientific incompetence or poor scientific quality control? In contrast, which problems, if any, are inherently ethical, social or legal; that is, cannot be resolved just by getting the science right?
3. Some of the readings describe specific policies to address some of the problems they discuss; others do not mention policy. Identify the policy proposals in the readings and come up with criteria for evaluating them; for example, ease of initial implementation, enforceability, and effectiveness. Identify a problem for which no policy solution was proposed, and develop one of your own. Finally, evaluate it by the same criteria.

References

Aguirre, G. K. (2005). Functional imaging in cognitive neuroscience I: Basic principles. In M. J. Farah & T. E. Feinberg (Eds.), *Patient-based approaches to cognitive neuroscience* (2nd ed.). Cambridge, MA: MIT Press, pp. 35–46.

Block, N., & Fodor, J. (1972). What psychological states are not. *Philosophical Review*, *81*, 159–181.

Clark, A. (2000). *Mindware: An introduction to the philosophy of cognitive science*. New York: Oxford University Press.

Davidson, D. (1970). Mental events. In *Essays on actions and events*. Oxford: Clarendon Press, pp. 207–224.

Farah, M. J., Smith, M. E., Gawuga, C., Lindsell, D., & Foster, D. (2009). Brain imaging and brain privacy: A realistic concern? *Journal of Cognitive Neuroscience*, *21*, 119–127.

Foster, N. L., Heidebrink, J. L., Clark, C. M., Jagust, W. J., Arnold, S. E., Barbas, N. R., et al. (2007). FDG-PET improves accuracy in distinguishing frontotemporal dementia and Alzheimer's disease. *Brain: A Journal of Neurology*, *130*, 2616–2635.

Haynes, J. D., & Rees, G. (2006). Decoding mental states from brain activity in humans. *Nature Neuroscience*, *7*, 523–534.

Kay, K. N., Naselaris, T., Prenger, R. J., & Gallant, J. L. (2008). Identifying natural images from human brain activity, *Nature*, *452*, 352–355.

Kim, J. (2005). *Physicalism, or something near enough*. Princeton, NJ: Princeton University Press.

Langleben, D. D., Loughead, J. W., Bilker, W. B., Ruparel, K., Childress, A. R., Busch, S. I., Gur, R. C. (2005). Telling truth from lie in individual subjects with fast event-related fMRI. *Human Brain Mapping*, *26*, 262–272.

Lewis, D. (1980). Mad pain and Martian pain. In: N. Block (Ed.), *Readings in the philosophy of psychology* (Vol. I, pp. 216–222). Boston, MA: Harvard University Press.

Phelps, E. A., O'Connor, K. J., Cunningham, W. A., Funayama, E. S., Gatenby, J. C., Gore, J. C., & Banaji, M. R. (2000). Performance on indirect measures of race evaluation predicts amygdala activation. *Journal of Cognitive Neuroscience*, *12*, 729–738.

Posner, M. I., Petersen, S.E., Fox, P.T., & Raichle, M.E. (1988). Localization of cognitive functions in the human brain. *Science, 240*, 1627–1631.

Vul, E., Harris, C., Winkielman, P., & Pashler, H. (2009). Puzzlingly high correlations in fMRI studies of emotion, personality, and social cognition. *Perspectives on Psychological Science*, *4*, 274–290.

Reading 4.1

Neuroimaging of Emotion and Personality: Ethical Considerations[1]

Turhan Canli and Zenab Amin

Common concerns in biomedical ethics, such as privacy, confidentiality, and agency, are equally relevant to neuroimaging of emotion and personality. However, these issues have been contemplated elsewhere (Glannon, 2001). Here, we want to focus on ethical considerations as they apply uniquely to neuroimaging. This is important because the greatest concern we have is that exaggerated and naive expectations about the power of neuroimaging can lead to unethical use of this technology. Therefore, this section will discuss the limitations of the neuroimaging approach with respect to three applications: (1) structural and functional brain characterization; (2) association between brain state and psychological or behavioral measures; and (3) prediction of future psychopathology. Each of these applications has its own set of methodological limitations that relate to ethical concerns.

4.1.1 Structural and Functional Brain Characterization

An applicant for a life insurance policy may be asked to submit to a structural brain scan to determine the presence of contraindications. A defense attorney may want to demonstrate that her client is incapable to stand trial on account of some organic brain damage. These are the kinds of scenarios where neuroimaging may be used to determine whether a person's brain has some form of pathology.

Although some structural changes (e.g., discrete tumors or lesions) can be unambiguously identified, other structural abnormalities can be much more subtle. In such cases, one would like to compare a scan image from the individual in question with a normative template. This can be difficult for three reasons. First of all, there are few templates of "normal" brains against which one could compare an individual sample. Second, it is unclear which brain measure should be used as a normative standard. Should it be size? If so, should one use overall volume or rather specific measures of certain brain regions or nuclei? Should it be the ratio of white matter to gray matter? Neurons to glia? Or should it be the size of the ventricles? There may be no single measure that can serve as a gold standard. Third, the decision whether a given sample is "normal" or "pathologic" is statistical, not absolute. This is because a normative template would consist of the average multiple brains, so that any given brain structure is an averaged representation whose shape will vary along a normal distribution in each dimension. To compare a test sample against the normative template means to determine whether its measures are significantly different from the normal distribution. This leads to the fourth and final point. The decision whether a given brain structure falls within the normative range will likely vary from one brain structure to another, so that any given brain may qualify as "normal" by one measure but not another.

Statistical uncertainty is an even bigger factor when one attempts to characterize brain function in terms of activation patterns. When one sees an image of a brain's activation pattern, it is difficult not to be struck by its visual persuasiveness. However, the image of an activation pattern from a poorly designed study is visually indistinguishable from one based on an exemplary study. It takes a skilled practitioner to appreciate the difference. Therefore, one great danger lies in the abuse of neuroimaging data for presentations to untrained audiences, such as courtroom juries. What can be easily forgotten when looking at these images is that they represent statistical inferences, rather than absolute truths.

When researchers identify a region as being "active," they mean to say that the activation in that region during one condition (e.g., seeing a sad face) is *significantly greater than* during a control condition (e.g., seeing a neutral face). What constitutes a "significantly greater" activation is, in a way, in the eye of the beholder. Quantitatively, that decision will be made through statistical analysis: activation in one condition will be called "significantly greater" than another if a certain threshold of statistical certainty is crossed. Thus, lowering the threshold will create more regions that are statistically significant, whereas raising the threshold will reduce the number of significant regions. The choice of the threshold is largely determined by convention among researchers, rather than an absolute standard. Reporting a brain activation pattern is therefore primarily a *statistical interpretation* of a very complex data set, and may be interpreted differently by different researchers.

Another factor that greatly affects the appearance of a brain activation pattern is the choice of test and control conditions. From the preceding paragraph, it is evident that the activation pattern is based on a comparison between a test and a control condition. However, there may be more than one control condition and different control conditions may produce different activation patterns. This was demonstrated in a functional magnetic resonance imaging (fMRI) study by Stark and Squire (Stark & Squire, 2001). The investigators asked subjects to respond to various visual or auditory stimuli, do a simple motor task, or do nothing (the so-called rest condition). When "rest" was used as a baseline condition, only the right parahippocampal cortex was significantly activated, and only in one cognitive task (looking at novel pictures). However, when a simple decision task was used as the baseline, significant activation was registered bilaterally in both the parahippocampal cortex and the hippocampal region in three different tasks. Thus, regions that did not appear to be involved in a cognitive task when rest was used as a baseline were significantly activated when an alternative baseline was used. The study also illustrates that even a "rest" baseline condition is associated with some degree of brain activation.

4.1.2 Association of Brain State and Psychological or Behavioral Measures

Correlational studies to date have related brain activation to emotional memory, race evaluation, and personality traits. Future applications could be relevant in the context of eyewitness testimony, trials based on recovered traumatic memories, assessment of defendants in hate crime trials, or simply screening of job applicants for undesirable personality traits. It needs to be stressed, however, that this approach identifies an association between two variables, not a causal link between them. It is possible that both measures are caused by another factor that has not yet been identified.

This does not mean that the correlational approach is flawed or scientifically invalid. It is a useful approach that allows one to search for relationships between brain activation and potentially very many other variables. To the extent that there is an *a priori* reason to search for such relations, it is entirely appropriate to report such data. In the absence of an *a priori* hypothesis, it is an efficient method to develop research hypotheses. Furthermore, neuroimaging is a much more powerful tool to detect subtle relationships between two variables than traditional behavioral methods. For example, a study of 15 subjects yielded a highly significant correlation between amygdala activation to happy faces and subjects' extraversion scores (see fig. 2 in Canli, Sivers, Gotlib, &

Gabrieli, 2002). At the behavioral level of analysis, most studies would require dozens, perhaps hundreds of subjects to attain similar statistical reliability. Why the difference in effect size? One explanation is that observed behavior constitutes *the sum of all neural activation patterns*, whereas brain imaging can focus on a *single* brain structure, and may therefore pick up relationships that are too subtle to be noticed at the behavioral level of analysis.

Correlational studies can be strengthened if they address discriminant and convergent validity. Discriminant validity means that measures of two unrelated concepts should also be unrelated to each other. Convergent validity means that measures of two related concepts should also be related to each other. For example, in the extraversion study cited above, discriminant validity was established by showing that amygdala activation to happy faces did not significantly correlate with constructs that were unrelated to extraversion (e.g., neuroticism or openness to experience). Convergent validity was not tested in this study. A demonstration of convergent validity would require that amygdala activation to happy faces also correlates with a construct related to extraversion (e.g., gregariousness).

4.1.3 Prediction of Future Behavior or Pathology

The greatest concern about the ethical use of neuroimaging of emotion and personality is in regard to its power to predict future psychopathology. The importance of this concern is illustrated by proposed legislation in Britain's Mental Health Act (Health, 1999). This act would allow for the detention of individuals who have not yet committed a crime, but are deemed a potential threat to public safety. Such individuals would be diagnosed with "Dangerous Severe Personality Disorder," or DSPD, a term without defined or sanctioned legal or medical status (Buchanan & Leese, 2001). Setting aside the thorny question of whether detaining such individuals is morally defensible, a recent analysis focused on the empirical question "to what extent is it possible to identify those who will behave violently in the future?" (Buchanan & Leese, 2001). The authors estimated that six individuals would have to be detained for 1 year to prevent one violent act due to DSPD during that year. Advocates of the proposed legislation may want to argue that neuroimaging methods could produce a more accurate prediction of future violent behavior.

The proposed Mental Health Act legislation is about the public's right to safety versus the individual's right to freedom. With respect to neuroimaging methodologies, these contrasting interests translate into statistical decisions about so-called type I ("false positive") and type II ("false negative" or "miss") errors. A type I error would be committed if a person were considered potentially dangerous when, in fact, he or she is not. A type II error would be committed if a person were considered harmless when, in fact, he or she is potentially dangerous. Because these two errors are related, one cannot reduce the likelihood of committing one error without increasing the likelihood of committing the other. As discussed earlier, the detection of both structural and functional brain abnormalities is based on statistical inferences. Therefore, the determination whether one individual's brain is "normal" or "pathologic" depends on a statistical threshold that balances type I (e.g., incorrectly calling someone dangerous) and type II (e.g., incorrectly calling someone harmless) errors. The tolerance for one type of error or the other depends on the cost of being wrong, which itself is a subjective assessment.

Whereas the cost of preventive incarceration of potentially dangerous persons who have not yet committed any crimes may seem high to some, the cost of detaining a convicted felon beyond his sentence to protect the public may appear lower. Indeed, this procedure is legal in some instances. On January 22, 2002, the United States Supreme Court ruled that states can confine violent sexual offenders beyond their prison term if it can be shown that they have a mental or personality disorder that makes it difficult for them to control dangerous behavior (Vicini, 2002). This ruling represents a lower burden of proof, relative to the Kansas Supreme Court Ruling, which would have required an *absolute* lack of control.

It is very likely that future imaging studies will be used to determine a felon's ability to control his behavior, because they may be more sensitive in picking up dysfunction in impulse inhibition than a behavioral test. A particularly relevant study investigated neural systems engaged in self-control of sexual arousal in non-criminal male subjects (Beauregard, Levesque, & Bourgouin, 2001). Sexual arousal in response to erotic films produced activation in limbic and paralimbic regions, but attempted inhibition of sexual arousal was restricted to activation of the right superior frontal gyrus and anterior cingulate. Both prosecutors and defense attorneys of sexual offenders are likely to adapt these functional imaging methodologies in the future to provide evidence for control of sexual impulses. The limitations of the neuroimaging approach already discussed make it clear that the evaluation of such data must be conducted with great sophistication to not be abused by either party.

In a similar vein, a neuroimaging approach might be applied to monitor the mental health treatment of convicted psychopaths. Psychopaths are four times as likely to commit a violent act on release, relative to offenders diagnosed with a non-psychopathic form of antisocial personality disorder (APD) (Abbott, 2001; Hemphill, Hare, & Wong, 1998). Although measures have been developed to identify psychopathic individuals, clinical interventions are often unsuccessful. Indeed, persons with high PCL-R (revised Psychopathy Checklist) scores who appear to respond well to behavioral therapy are more likely to re-offend than those with lower scores who appear to respond less favorably to the intervention (Abbott, 2001; Seto & Barbaree, 1999), perhaps because they are particularly skilled in manipulating the impression made upon the therapist. Thus, some kind of objective verification of treatment progress would be desirable. At the present time, forensic neuroimaging cannot deliver such an evaluation, but the goal is clearly defined, albeit within the limitations discussed in this section.

Another ethical concern about using neuroimaging to predict future psychopathology relates to issues of confidentiality. The concern is minimal for convicted pedophiles because many states have equivalents to Megan's law mandating public access to information regarding the whereabouts of such criminals. The issue would be thornier if brain scans could be used to identify potential pedophiles among non-criminal persons (for reasons already discussed, this would, at best, be a statistical prediction). Who would conduct the scan and what would precipitate such an evaluation in the first place? If the scan suggested some likelihood of pedophilic tendencies, how should such data be used? Would it warrant registration of this person with law enforcement authorities? Should it be made accessible to potential employers?

The use of prognostic scan data could be modeled after that of "notifiable" diseases, particularly with respect to psychopathy, pedophilia, and related disorders that put the public at risk. A notifiable disease is defined as "one for which regular, frequent, and timely information regarding individual cases is considered necessary for the prevention and control of the disease" (U.S. Department of Health & Human Services, 1999). The occurrence of a notifiable disease must be relayed to the National Notifiable Diseases Surveillance System (NNDSS), which is operated by the Centers for Disease Control and Prevention (CDC) in collaboration with the Council of State and Territorial Epidemiologists (CSTE). Among the 58 notifiable diseases (as of January 1, 1999) are a number of sexually transmitted diseases (STDs) such as gonorrhea, HIV infection, and syphilis. In the case of HIV infection, 33 states and 1 U.S. territory reported HIV infections by the patient's name (as of January 1, 1999). Patients identified with STDs are encouraged to seek counseling and treatment services. A parallel treatment for persons with prognosed violent tendencies could call for referral to treatment services and perhaps the establishment of a database accessible to qualified parties.

Prognostic neuroimaging could not only lead to treatment referral, but could even guide treatment strategy. For example, violent patients with amygdala damage may require a different treatment strategy than violent patients with prefrontal damage. This is because amygdala damage is more likely associated with instrumental violence and psychopathic tendencies whereas prefrontal damage is more likely associated with reactive violence that is non-psychopathic (Blair, 2001). One can speculate that some form of anger management therapy may be effective in patients who exhibit reactive violence (i.e., may have prefrontal damage), but not in patients who exhibit instrumental violence (i.e., may have amygdala damage).

These scenarios illustrate the wide range of potential applications for neuroimaging of emotion and personality. A scrupulous use of this methodology could greatly benefit individuals and society. Naive use and uninformed interpretations of the data, on the other hand, may compromise its benefits. For this reason, it is important to be mindful of the limitations of the neuroimaging approach.

4.1.4 Concluding Remarks

In light of the caveats listed above and the ethical considerations, what benefits are to be derived from neuroimaging of emotion and personality? The likely benefits apply to the prevention, diagnosis, and treatment of a variety of disorders and social ills. Neuroimaging approaches could identify vulnerability factors for psychopathology such as structural abnormalities, dysfunctional metabolism or activation patterns. Preventive interventions such as medication, cognitive therapy, or lifestyle changes could be motivated by such imaging data and progression toward psychopathology could be monitored. Treatment of affective or personality disorders could be greatly aided by neuroimaging techniques that can identify specific subpopulations and optimize patient-treatment match. The mental status of felons to-be-released to the public could perhaps be ascertained to provide continued mental health treatment to those who require it or continue incarceration of those who show no improvement. It is the prospect of such beneficial applications that motivates the continued development of neuroimaging approaches to the study of emotion and personality.

Notes

1. Editor's note: This reading is an excerpt from a 2002 article published in *Brain and Cognition*, volume 50, pages 414–431, under the title "Neuroimaging of emotion and personality: Scientific evidence and ethical considerations," and is used with permission of publisher and author. The omitted sections review the science of affective neuroimaging in healthy and psychopathic individuals.

References

Abbott, A. (2001). News feature: Into the mind of a killer. *Nature*, *410*, 296–298.

Beauregard, M., Levesque, J., & Bourgouin, P. (2001). Neural correlates of conscious self-regulation of emotion. *Journal of Neuroscience*, *21*(18), RC165.

Blair, R. J. (2001). Neurocognitive models of aggression, the antisocial personality disorders, and psychopathy. *Journal of Neurology, Neurosurgery, and Psychiatry*, *71*(6), 727–731.

Buchanan, A., & Leese, M. (2001). Detention of people with dangerous severe personality disorders: A systematic review. *Lancet*, *358*(9297), 1955–1959.

Canli, T., Sivers, H., Gotlib, I. H., & Gabrieli, J. D. E. (2002). Amygdala activation to happy faces as a function of extraversion. *Science*, *296*, 2191.

Glannon, W. (Ed.). (2001). *Contemporary readings in biomedical ethics*. Belmont, CA: Wadsworth.

Health, H. O. a. D. o. (1999). *Managing dangerous people with severe personality disorder*. London: Home Office and Department of Health.

Hemphill, J. F., Hare, R. D., & Wong, S. (1998). *Legal Criminology and Psychology*, *3*, 141–172.

Seto, M. E., & Barbaree, H. E. (1999). *Journal of Interpersonal Violence*, *14*, 1235–1248.

Stark, C. E., & Squire, L. R. (2001). When zero is not zero: The problem of ambiguous baseline conditions in fMRI. *Proceedings of the National Academy of Sciences of the United States of America*, *98*(22), 12760–12766.

U.S. Department of Health and Human Services, C. f. D.C. a. P.C. (1999). *Morbidity and mortality weekly report: MMWR summary of notifiable diseases, United States 1999* (Vol. 48, No. 53). Atlanta, GA: U.S. Department of Health and Human Services, Centers for Disease Control and Prevention (CDC).

Vicini, J. (2002, January 22). Supreme Court decides sex predator requirement. Reuters. Retrieved January 22 2002 from http://dailynews.yahoo.com/h/nm/20020122/ts/court_predator_dc_20020121.html.

Reading 4.2

Neurobiology of Intelligence: Science and Ethics[1]

Jeremy R. Gray and Paul M. Thompson

In the United States it is mildly impolite to dwell on an obvious fact—individual differences are the rule, not the exception. Furthermore, it is distinctly impolite to suggest that individual differences in ability have a biological basis (Gould, 1981/1996; Pinker, 2002). The root fear is that evidence about the brain might be misconstrued as evidence about an individual's or group's inherent quality or fitness, in the sense of an immutable social and moral value (Kagan, 1998; Ordover, 2003).

In this review, we emphasize intelligence in the sense of reasoning and novel problem-solving ability. Also called *fluid intelligence* (Gf) (Cattell, 1971), it is related to analytical intelligence (Sternberg, 1985). Intelligence in this sense is not at all controversial and is best understood at multiple levels of analysis. Empirically, Gf is the best predictor of performance on diverse tasks, so much so that Gf and general intelligence (*g*, or general cognitive ability) might not be psychometrically distinct (Carroll, 1993; Jensen, 1980). Conceptions of intelligence(s) and methods to measure them continue to evolve, but there is agreement on many key points; for example, that intelligence is not fixed, and that test bias does not explain group differences in test scores (Neisser et al., 1996). Intelligence research is more advanced and less controversial than is widely realized (Deary, 2001a; Lubinski, 2004; Neisser et al., 1996), and permits some definitive conclusions about the biological bases of intelligence to be drawn.

4.2.1 Neural Bases of Intelligence

Patients with brain damage provided early data that are still important—causal evidence that intelligent behavior depends on the integrity of specific neural structures. More than 125 years ago, the frontal lobes were implicated in abstract reasoning (Finger, 1994). Modern studies have reinforced and refined these conclusions (Duncan, Emslie, Williams, Johnson, & Freer, 1996; Kolb & Whishaw, 1996; Piercy, 1964). One notable finding of many studies is that patients with damaged frontal lobes often have normal IQs as assessed by tests that typically measure skills and knowledge (*crystallized intelligence*, Gc); for example, the Wechsler Adult Intelligence Scale (WAIS). By contrast, posterior lesions often cause substantial decreases in IQ (Kolb et al., 1996). Duncan, Burgess, and Emslie (1995) suggested that the frontal lobes are involved more in Gf and goal-directed behavior than in Gc. In addition, Gf is compromised more by damage to the frontal lobes than to posterior lobes (Duncan, Burgess, & Emslie 1995; Duncan, Emslie, Williams, Johnson, & Freer, 1996). Other studies indicate that the frontal lobes are crucial for integrating abstract relationships (Waltz et al., 1999), a key aspect of resolving many reasoning problems (but not of previously learned skills or knowledge).

Modern neuroimaging methods reveal aspects of brain function with greater spatial precision than patient studies and can do so in healthy individuals. Imaging studies provide correlational rather than causal evidence (for discussion, see Gray, Chabris, & Braver, 2003), but they have contributed considerably to our understanding of the neurobiology of intelligence.

Correlations between intelligence and total brain volume or gray matter volume have been replicated in magnetic resonance imaging (MRI) studies, to the extent that intelligence is now commonly used as a confounding variable in morphometric studies of disease (see McDaniel & Nguyen, 2002, for a meta-analysis). Using a twin study design, brain regions correlating with *g* and under genetic control have been mapped (Thompson et al., 2001). This study found that volume of frontal gray matter is predictive of *g* over and above the predictive effect of total brain volume.

Measuring brain activity while participants are performing an intelligence test, and contrasting it with activity under control conditions, reveals common regions of activation that probably support intelligent behavior. Lateral prefrontal cortex is consistently activated during many different kinds of intelligence tasks relative to control tasks (e.g., Duncan et al., 2000).

A complementary empirical approach is to examine how people, rather than tasks, differ (Deary, 2001b; Kosslyn et al., 2002). Both are needed for a full understanding (Gray, Chabris, & Braver, 2003). Here, the focus is on individual differences in brain activity and how they correlate with differences in psychometric intelligence. In the largest imaging study of individual differences in intelligence ($n = 48$; Gray et al., 2003), we tested whether Gf is mediated by neural mechanisms that support the executive control of attention during working memory. This hypothesis was based on a large body of cognitive literature emphasizing resistance to distraction or interference under these conditions (Engle, Tuholski, Laughlin, & Conway, 1999; Kane & Engle, 2002, 2003). Participants performed verbal and nonverbal working memory tasks (see Gray et al., 2003, for details of the study design and results) while their brain activity was measured with functional MRI (fMRI). Participants with higher Gf showed greater neural activity in many regions, including the frontal, parietal, and temporal lobes, dorsal anterior cingulate, and lateral cerebellum, and with the lateral prefrontal and parietal activity mediating the correlation between Gf and task performance.

Overall, intelligence in the sense of reasoning and novel problem-solving ability is consistently linked to the integrity, structure, and function of the lateral prefrontal cortex, and possibly to that of other areas. Regions within both the lateral prefrontal cortex and posterior areas are under genetic control. The lateral prefrontal cortex supports the executive control of action and attention (Miller & Cohen, 2001), but how this brain area (and other regions) contributes specifically to intelligent behavior is less well understood (as discussed in Gray et al., 2003). Several imaging studies indicate that the parietal cortex and other areas (such as the anterior cingulate cortex) might also contribute. Patient-based studies support this view, but only for Gc (Kolb & Whishaw, 1996) and not for Gf.

4.2.2 Beyond the Data: The Good, the Bad, and the Ugly

In our view—which is shared by most investigators—the data unambiguously indicate a neurobiological basis for intelligence, particularly for reasoning and novel problem-solving ability (which strongly predicts psychometric *g*). Neuroimaging and neuroanatomical data are consistent with sophisticated behavioral studies of intelligence and specific aspects of working memory. From this vantage point, the formulation of detailed neurobiological models of intelligence is inevitable.

The field is at an exciting juncture because nuanced conceptual and empirical approaches are now available, and intelligence is important for both practical (Gottfredson, 1997; Lubinski & Humphreys, 1997; Neisser et al., 1996) and theoretical (Deary, 2000; Duncan, 1995;

Duncan et al., 2000) reasons. Biological models of intelligence will help elucidate the structure of intelligence and the processes and mechanisms that underlie intelligent behavior. These mechanisms might indicate avenues for enhancing intelligence, where society deems this to be useful (Little, 2002). We have highlighted progress within neurobiology, but a great deal remains to be discovered both within neurobiology and in its relation to genetics. Biological measures have much to contribute to the study of human abilities, but psychometric and social psychological research is equally indispensable.

The empirical successes also raise ethical issues (Farah, 2002; Kulynych, 2002; Wolpert, 1999) that the science does not—and in principle cannot (Hume, 1739/1964; Loehlin, 2000; Wolpert, 1999)—resolve. The ethics of how knowledge is applied are very important and have been discussed in detail elsewhere (Farah, 2002; Illes, Kirshen, & Gabrieli, 2003; Kulynych, 2002; Wolpert, 1999). For example, there are evolving standards concerning privacy and concerns about equitable access to methods for enhancing intelligence. Public scientific literacy is vital for informed discussion of policy options. Intelligence research is relevant to social policy (and so scientific literacy is crucial), but the data in no way force any particular policy. Even so, scientific literacy alone is unlikely to resolve all of the ethical concerns regarding intelligence research.

We consider in detail a practical concern about a highly polarized research topic. Is it ever ethical to assess population-group (racial or ethnic) differences in intelligence (for example, recent projects described in Hunt, 1999)? It is easier to set aside such difficult and distasteful questions. In reviewing the neurobiological bases of intelligence it is not necessary, on scientific grounds, to consider race. Most of the variance in intelligence is within racial groups not between them (Loehlin, 2000), and the causes of individual differences are relatively tractable with available methods, whereas the causes of racial differences are not. Although the topic of race differences is only a minor area within the field of intelligence research, it has had a disproportionately large (and strongly negative) impact on the public perception of intelligence research (Gottfredson, 1994; Hunt, 1999; Miele, 2002). Science is generally perceived as a noble and honorable pursuit, yet "The field of intelligence itself is widely suspect" (Gottfredson, 1994). Given the history of misuse of intelligence research (Gould, 1981/1996; Ordover, 2003), a statement about biology and intelligence that ignores the question of race can be mistaken as being complicit with a racist agenda. To a nonspecialist, the field of intelligence research has become stereotyped as elitist and socially divisive. We disavow—and hope to weaken—these unfortunate and unnecessary associations.

Further, we offer the opinion that research on race and intelligence is unethical if it lacks the consent of the target group. Intelligence and race are rarely addressed in neuroethics, which has emphasized individual-level issues (but see Hunt, 1999; Little, 2002; Wolpert, 1999; for a discussion of the term *race*, see Feldman et al., 2003). The issue of race is not unique to biological investigations of intelligence, but it is more visceral in a biological context (in part because heritability can be misunderstood to imply both that group differences must be genetic and that intelligence is a fixed rather than a context-sensitive ability—both of these interpretations are incorrect). Many scientists find the question of group differences in intelligence to be distasteful to contemplate, let alone investigate—we are among them. But it is probably more harmful to simply censor all such work because this would set a terrible precedent of allowing an extrascientific agenda to constrain objective inquiry (Hunt, 1999; Pinker, 2002; Wolpert, 1999). The existence of knowledge is not an ethical problem in itself. Freedom of inquiry is rightly defended on the basis that scientific knowledge is inherently neither good nor bad (Hume, 1739/1964; Hunt, 1999; Wolpert, 1999). At the same time, it is also firmly established that ethical safeguards must constrain the conduct of science.

Informed consent is a bedrock principle of research with human participants. In the arena of potential race differences, however, the imperative to investigate seems to have been placed above the imperative to obtain consent. For example, some have argued that it is unethical not to investigate the world as we find it, including the possibility of group differences in intelligence, with no apparent consideration of consent (Hunt, 1999; Loehlin, 1992; Miele, 2002). Yet one person's feeling of an obligation to explore leads to the identity and psychic space of a great many people being rudely probed without consent or recourse.

In light of such unresolved ethical issues, many neuroscientists have been reluctant to investigate individual or group differences in intelligence. Few scientists investigate race differences in intelligence; those who do are overwhelmingly white. Under the status quo, target groups will continue to feel alienated and attacked, unimpressed by the need for freedom of inquiry when other important freedoms are lacking. The credibility of intelligence research is suffering. The quality of the science will be affected in turn if there is a (mistaken) perception that most scientists who study intelligence are tacitly racist or tolerate racism among their colleagues.

The key dilemma is how to preserve freedom of scientific inquiry while upholding the highest standards of ethical conduct; neither can be compromised. We are not seeking to stimulate research on potential race differences in intelligence. Nor can we advocate censorship. In health care, patients are given the final say over testing and treatment to be performed for their own benefit (Wolpert, 1999). Can such principles of informed consent and self-determination be generalized to a group level? Probably not perfectly, yet there are doubtless advantages to be gained from making the effort. In our view, a study of race differences in intelligence that does not meet the following criteria is ethically dubious. We consider the following to be points for discussion, not prohibitions: (i) All participants contributing to any group comparisons should be fully debriefed about the study's aims and predictions, and given a chance to withdraw from the study and have their data destroyed (or excluded from racial comparisons in data sets in the public domain or other databases). (ii) Target groups should actively support the study, including financially. If the experimental aims are dubious, such support would be difficult to secure. Appropriate representatives should endorse the design, conduct, interpretation, and dissemination of the study and its results. An advisory group could include experts in the science and ethics, as well as advocates for the interests of the target group. (iii) The procedures must eliminate all known confounds, including asking participants to indicate their race before they take a test. This simple act induces stereotype threat (Steele & Aronson, 1995), which impairs test performance by diverting working memory resources (Schmader & Johns, 2004)—threat-related emotion (anxiety) can modulate the activity of the lateral prefrontal cortex (Bishop, Duncan, Brett, & Lawrence, 2004; Gray, Braver, & Raichle, 2002). (iv) Groups should be matched using pairwise matched controls (including matching for age, parental education, health and nutritional history, and familiarity and fluency with testing procedures). Appropriate sampling of the true populations must be ensured. Samples must be large enough to allow inferences to the population—samples of convenience should not be used. (v) Descriptions of the results should use noninflammatory language; for example, in terms of percentage of the variance explained (Loehlin, 1992). There is much more variation within groups than between them, which should be emphasized (Loehlin, 2000). It should be noted that the findings are necessarily correlational and potentially confounded by environmental effects.

Such standards are difficult to achieve, but are worthwhile to pursue. If there is an inclusive consensus that each specific study has a legitimate motivation (scientific inquiry), then the process remains both inherently open and protected from being hijacked by extrascientific agendas.

Another way to preserve freedom of inquiry and reduce the potential for mischief (or worse) might be to require any studies of race and intelligence to be conducted by adversarial collaboration (Mellers, Hertwig, & Kahneman, 2001). This framework offers important advantages over exchanges in the peer-reviewed literature, because investigators on both sides agree beforehand on the precise research question and the methods for testing it. They then commit themselves before data is gathered to specific interpretations of the possible outcomes, and a neutral third party conducts the work. However, adversarial collaboration speaks less directly to the issues of consent and self-determination.

Research on human intelligence has recently advanced at multiple levels of analysis—social, cognitive, psychometric, neural, and genetic. By bridging these levels (Deary, 2001b; Kosslyn et al., 2002) and including measures from neurobiology, the field is now taking early steps toward a credible mechanistic understanding of individual variation. Intelligence research has implications for conceptions of human nature, so all ethical issues must be addressed proactively. Concerted efforts to encourage the highest standards for conducting research can only help to bolster public confidence in the legitimacy and value of research on human mental abilities.

Notes

1. Editor's note: This reading was excerpted from a substantially longer article that appeared in 2004 in *Nature Reviews Neuroscience*, volume 5, pages 471–482, and is used with permission. The omitted sections included reviews of the literature from behavioral genetics and brain imaging, and the excerpt has been abridged with the permission of the authors. The work described in the original article was supported in part by research grants from the National Institute of Mental Health to J.R.G., and from the National Institute for Biomedical Imaging and Bioengineering and the National Center for Research Resources to P.M.T. The authors thank R.J. Sternberg and W.R. Gray for their comments on a draft.

References

Bishop, S., Duncan, J., Brett, M., & Lawrence, A. D. (2004). Prefrontal cortical function and anxiety: Controlling attention to threat-related stimuli. *Nature Neuroscience,* 7, 184–188.

Carroll, J. (1993). *Human cognitive abilities: A survey of factor–analytic studies.* Cambridge, UK: Cambridge University Press.

Cattell, R. B. (1971). *Abilities: Their structure, growth, and action*. Boston: Houghton Mifflin.

Deary, I. J. (2000). *Looking down on human intelligence*. New York: Oxford University Press.

Deary, I. J. (2001). Human intelligence differences: A recent history. *Trends in Cognitive Science, 5*, 127–130.

Deary, I. J. (2001). Human intelligence differences: Toward a combined experimental-differential approach. *Trends in Cognitive Science. 5*, 164–170.

Duncan, J., Burgess, P., & Emslie, H. (1995). Fluid intelligence after frontal lobe lesions. *Neuropsychologia, 33*, 261–268.

Duncan, J., Emslie, H., Williams, P., Johnson, R., & Freer, C. (1996). Intelligence and the frontal lobe: The organization of goal-directed behavior. *Cognitive Psychology, 30*, 257–303.

Duncan, J., Seitz, R. J, Kolodny, J., Bor, D., Herzog, H., Ahmed, A., Newell, F. N., & Emslie, H. (2000). A neural basis for general intelligence. *Science, 289*, 457–460.

Engle, R. W., Tuholski, S. W., Laughlin, J. E., & Conway, A. R. A. (1999). Working memory, short-term memory, and general fluid intelligence: A latent-variable approach. *Journal of Experimental Psychology: General, 128*, 309–331 (1999).

Farah, M. J. (2002). Emerging ethical issues in neuroscience. *Nature Neuroscience, 5*, 1123–1129.

Feldman, M. W., Lewontin, R. C., & King, M. C. (2003). Race: a genetic melting pot. *Nature, 424*(6947), 374.

Finger, S. (1994). *Origins of neuroscience*. New York: Oxford University Press.

Gottfredson, L. (1994). Egalitarian fiction and collective fraud. *Society, 31*, 53–59.

Gottfredson, L. S. (1997). Why *g* matters: the complexity of everyday life. *Intelligence, 24*, 79–132.

Gould, S. J. (1981/1996). *The mismeasure of man*. New York: Norton.

Gray, J. R., Braver, T. S., & Raichle, M. E. (2002). Integration of emotion and cognition in the lateral prefrontal cortex. *Proceedings of the National Academy of Sciences of the United States of America, 99*, 4115–4120.

Gray, J. R., Chabris, C. F., & Braver, T. S. (2003). Neural mechanisms of general fluid intelligence. *Nature Neuroscience, 6*, 316–322.

Hume, D. (1739/1964). *A treatise of human nature*. Oxford: Clarendon.

Hunt, M. (1999). *The new know-nothings: The political foes of the scientific study of human nature*. Piscataway, NJ: Transaction.

Illes, J., Kirschen, M. P., & Gabrieli, J. D. (2003). From neuroimaging to neuroethics. *Nature Neuroscience, 6*, 205.

Jensen, A. R. (1980). *Bias in mental testing*. New York: Free Press.

Kagan, J. (1998). *Three seductive ideas*. Cambridge, MA: Harvard University Press.

Kane, M. J., & Engle, R. W. (2002). The role of prefrontal cortex in working-memory capacity, executive attention and general fluid intelligence: An individual-differences perspective. *Psychonomic Bulletin & Review*, *9*, 637–671.

Kane, M. J., & Engle, R. W. (2003). Working-memory capacity and the control of attention: The contributions of goal neglect, response competition, and task set to Stroop interference. *Journal of Experimental Psychology: General, 132*, 47–70.

Kolb, I., & Whishaw, B. (1996). *Fundamental of human neuropsychology*. New York: W. H. Freeman.

Kosslyn, S. M., Cacioppo JT, Davidson RJ, Hugdahl K, Lovallo WR, & Spiegel D, et al. (2002). Bridging psychology and biology: The analysis of individuate in groups. *American Psychologist, 57*, 341–351.

Kulynych, J. (2002). Legal and ethical issues in neuroimaging research: Human subjects protection, medical privacy, and the public communication of research results. *Brain Cognition, 50*, 345–357.

Little, P. (2002). *Genetic destinies*. New York: Oxford University Press.

Loehlin, J. C. (1992). Should we do research on race differences in intelligence? *Intelligence, 16*, 1–4.

Loehlin, J. C. (2000). Group differences in intelligence. In R. J. Sternberg (Ed.), *Handbook of intelligence*. New York: Cambridge University Press, pp. 176–193.

Lubinski, D. (2004). Introduction to the special section on cognitive abilities: 100 years after Spearman's (1904) 'general intelligence,' objectively determined and measured. *Journal of Personality and Social Psychology, 86*, 96–111.

Lubinski, D., & Humphreys, L. G. (1997). Incorporating general intelligence into epidemiology and the social sciences. *Intelligence, 24*, 159–201.

McDaniel, M. A., & Nguyen, N. T. (2002). A meta-analysis of the relationship between MRI-assessed brain volume and intelligence. *Proceedings International Society for Intelligence Research*.

Mellers, B., Hertwig, R., & Kahneman, D. (2001). Do frequency representations eliminate conjunction effects? An exercise in adversarial collaboration. *Psychological Science, 12*, 269–275.

Miele, F. (2002). *Intelligence, race, and genetics: Conversations with Arthur R. Jensen*. Boulder, CO: Westview.

Miller, E. K., & Cohen, J. D. (2001). An integrative theory of prefrontal cortex function. *Annual Review of Neuroscience, 21*, 167–202.

Neisser, U., et al. (1996). Intelligence: Knowns and unknowns. *American Psychologist, 51*, 77–101.

Ordover, N. (2003). *American eugenics: Race, queer anatomy, and the science of nationalism*. Minneapolis, MN: University of Minnesota Press.

Peters, M. (1995). Does brain size matter? A reply to Rushton and Ankney. *Canadian Journal of Experimental Psychology, 49*, 570–576.

Piercy, M. (1964). The effects of cerebral lesions on intellectual function: A review of current research trends. *British Journal of Psychiatry, 110*, 310–352.

Pinker, S. (2002). *The Blank Slate*. New York: Viking.

Schmader, T., & Johns, M. (2004). Converging evidence that stereotype threat reduces working memory capacity. *Journal of Personality and Social Psychology, 85*, 440–452.

Steele, C. M., & Aronson, J. (1995). Stereotype threat and the intellectual test performance of African Americans. *Journal of Personality and Social Psychology, 69*(5), 797–811.

Sternberg, R. J. (1985). *Beyond IQ: A triarchic theory of human intelligence*. Cambridge, UK: Cambridge University Press.

Thompson, P. M., Cannon, T. D., Narr, K. L., van Erp, T., Poutanen, V. P., Huttunen, M., et al. (2001). Genetic influences on brain structure. *Nature Neuroscience, 4*, 1253–1258.

Waltz, J. A., Knowlton, B. J., Holyoak, K. J., Boone, K. B., Mishkin, F. S., et al. (1999). A system for relational reasoning in human prefrontal cortex. *Psychological Science, 10,* 119–125.

Wolpert, L. (1999). Is science dangerous? *Nature 398*, 281–282.

Reading 4.3

Emerging Neurotechnologies for Lie Detection: Promises and Perils[1]

Paul Root Wolpe, Kenneth R. Foster, and Daniel D. Langleben

Rapid advances in diagnostic medical imaging over the past decade have revolutionized neuroscience. Scientists are gaining a new understanding of brain function and structure and uncovering exciting and challenging insights into the nature of human behavior. Advances in magnetic resonance imaging, electroencephalography (EEG), and other modern techniques can, for the first time, reliably measure changes in brain activity associated with thoughts, feelings, and behaviors, in principle allowing researchers to link brain activity patterns directly to the cognitive or affective processes or states they produce (e.g., Canli & Amin, 2002; Fischer, Wik, & Fredrikson, 1997; Sugiura et al., 2000).

Although most of this work is still in the basic research stage, its potential social, legal, and ethical implications are significant (see, e.g., Farah, 2002; Foster, Wolpe, & Caplan, 2003; Illes, Kirschen, & Gabrieli, 2003; Wolpe, 2002, 2004). For the first time, using modern neuroscience techniques, a third party can, in principle, bypass the peripheral nervous system—the usual way in which we communicate information—and gain direct access to the seat of a person's thoughts, feelings, intention, or knowledge (Berns, Cohen, & Mintun, 1997). Given the current state of the art in neuroscience research, speculations about any impending ability to "read thoughts" of unsuspecting citizens are not realistic, and free-form mind-reading in the style described in recent films such as *Minority Report* remains science fiction (see Ross, 2003). Nevertheless, there has been real, if limited, progress in finding brain correlates of certain simple memories, emotions, and behaviors, and potential applications in the social arena are foreseeable (Donaldson, 2004).

One application of these techniques has been the attempt to develop reliable brain-imaging lie-detection technology. In the United States, defense-related agencies have dedicated significant funds to the development of new lie-detection strategies for eventual use in criminal and terrorist investigations. A number of universities and private companies are trying to develop lie-detection technologies, using functional magnetic resonance imaging (fMRI), electroencephalography (EEG), near-infrared light, and other strategies to directly access brain function.

The ethical issues that would arise from a reliable (or thought-to-be-reliable) brain-imaging deception technology are complex. Using these technologies in court rooms and for security screening purposes, for example, raises many of the same difficult ethical and legal issues already present in the debate over conventional polygraphy. However, some of the ethical issues that such technologies would present are novel. For the first time, we would need to define the parameters of a person's right to "cognitive liberty," the limits of the state's right to peer into an individual's thought processes with or without his or her consent, and the proper use of such information in civil, forensic, and security settings. Clearly, a comprehensive and probing debate concerning the limits and proper use of brain-imaging technologies is needed and timely. Our goal in this essay is to inform that debate through a description of the technical limitations in neuroscience research on detecting deception, and to raise concerns about their premature and inappropriate use.

4.3.1 New Methods, Old Paradigms

A lie-detection system such as the polygraph, or any system aimed at determining physiologic correlates of behavior, consists of two components. One is the set of physiologic parameters being measured, and the other (no less important but often overlooked) is the paradigm or model used to produce the target behavior (such as deception) in a standardized fashion. Conventional polygraphy measures the subject's physiologic responses by monitoring chest expansion, pulse, blood pressure, and electrical conductance of the skin. The physiologic data measured in polygraphy signify the activity of the autonomic nervous system, and so may reflect not only arousal during deception but anxiety in general, no matter the cause.

To overcome this limitation, a number of recent studies have attempted to employ more direct measurements of brain activity to indicate deception and the presence of concealed information. Some of these studies, for example recent applications of fMRI for lie detection, have attracted attention because of the novelty of the physiologic parameters being measured (see Spence et al., 2004). What is less recognized, however, is that many of these studies have used, sometimes unknowingly, variants of decades-old paradigms to produce the target behavior.

In order to test any means of lie detection, a standardized protocol to generate the behavior must be developed. There are two prototype paradigms that have been used to generate instances of truth-telling and deception to be subjected to measurement. The first is the comparison question test (known in the polygraph literature as the control question test, CQT), which forms the basis of conventional polygraphy. The CQT requires a subject to respond to a series of yes–no questions of one of three kinds. "Relevants" are intended to produce a presumed lie and would, in a standard polygraph test, be related to the matter under investigation (e.g., "Did you kill your wife?"). Comparison or control questions are designed to induce a strong response in all subjects (e.g., "Did you ever steal something?"). Finally, there are irrelevant questions to establish a baseline ("Are you sitting in a chair?"). A consistently stronger physiologic response in a subject to the relevants than to the control questions is taken as evidence of deception.

In contrast, the second paradigm, the guilty knowledge test (GKT), seeks to determine the salience ("attentional value") of information to a subject by comparing his or her responses to "relevant" and "neutral" questions. For example, in a crime investigation involving a stolen red car, a sequence of questions could be: "Was the car yellow? Was the car red? Was the car green?" The questions are chosen so that subjects with knowledge of the crime (but not other individuals) would have an amplified physiologic response to the relevant question—that the car was red—which is dubbed "guilty knowledge" (Ben-Shakhar & Elaad, 2003; Lykken, 1991).

Whereas the CQT involves measurement of physiologic or psychophysical responses to classify a response as a lie, the GKT uses such responses to indicate the presence of concealed knowledge. The tester then uses this information to make inferences about the truth. Thus, the GKT does not detect deception directly and indeed, in the polygraphic literature, the term *lie-detector* is reserved for the CQT-based applications (Lykken, 1991). In fact, the GKT need not rely on verbal responses from the subject at all; physiologic responses to simply hearing the relevant question can suffice. This has given rise to the claims that GKT "directly" probes the information stored in a person's brain (Farwell & Smith, 2001).

The debate about the relative advantages of the CQT versus GKT as research and applied paradigms has been raging for decades. The main criticism against the CQT has been the inability to standardize the selection of the control questions (though the choice of the neutral items in the GKT could also affect the results). From a neuropsychological perspective, both the CQT and GKT are "forced-choice" protocols that seek to detect differences in psychological salience between question by examining the physiologic responses of the subject to target and baseline conditions. Though investigators generally agree that the GKT is methodologically more robust than the CQT (Rosenfeld et al., 1988; Stern, 2002), it has been less popular with forensic practitioners in the field because the test requires reliable and specific crime-related information known only to the investigators and the perpetrator, which is often difficult to obtain.

In recent years, investigators have used the GKT (or variants) to explore the usefulness of a variety of neuroscience techniques for detecting deception. One group, for example, has used infrared photography to detect changes in temperature patterns (and thus blood flow) near the eye, and proposed it for "deception detection on the fly" such as screening airline passengers (Pavlidis & Levine, 2002; Pavlidis, Eberhardt, & Levine, 2002). Another group has applied the GKT using scattering of near-infrared light (NIR) using sensors placed in contact with the scalp that detect infrared light shone through the skull and reflected off the blood vessels of the cortex (Chance & Kang, 2002). Another set of studies has employed a variety of GKT-like forced-choice paradigms with fMRI (e.g., Langleben et al., 2002, 2004; Spence et al., 2001). All of these examples, however, are laboratory based and are in early stages of research.

One technique, however, has been applied in an actual forensic situation and has drawn considerable media attention. Dubbed "brain fingerprinting" by its developer, Lawrence Farwell (Farwell & Donchin, 1991; Farwell & Smith, 2001), it involves application of the GKT while using EEG as a measurement tool. The signals picked up by the EEG, known as event-related potentials (ERPs), can be measured on the scalp 300–500 ms after the subject is exposed to a stimulus; their precise origin is unknown, but they are associated with novelty and salience of incoming stimuli. Through this technique, Farwell claims to be able to tell whether a stimulus is familiar or unfamiliar to the subject (e.g., whether or not a suspect's response indicates familiarity with a picture of a crime scene). "Brain fingerprinting" is thus not really a deception- or lie-detection technology. It is also not new; the use of the GKT coupled with ERP was reported as long ago as 1988 (Rosenfeld et al., 1988). Farwell's "brain fingerprinting," in fact, is a proprietary version of the technology that has been developed commercially by Farwell and is being actively promoted by his firm Brain Fingerprinting Laboratories, Inc. (http://www.brainwavescience.com) for forensic, medical, advertising, and security applications.

4.3.2 Reliability Concerns

Polygraph testing in civil and judicial settings has been subject to ongoing concerns about accuracy of measurement, reliability of the questioning paradigm used, and the relevance of the test to the field situations in which it is used (Stern, 2002). Neurotechnological means of lie detection suffer from many of the same weaknesses as conventional polygraphy. Whereas monitoring brain activity directly, rather than monitoring peripheral responses such as skin conductance, may improve the measurement component of a lie-detection system, there is no assurance that changing the measurement component alone will result in improved overall performance for any particular application.

A simple example, using concepts familiar in medical testing, shows the difficulty of the problem. In a meta-analysis of a number of GKT studies used with polygraph, Ben-Shakhar and Elaad found an effect size (the ratio of the difference in the mean responses in "knowledge present" versus "knowledge absent" subjects to the standard deviations in responses) ranging from 1.1–1.3 to 2.09 standard deviations, with the higher effect size being found in studies involving mock crime tests (Ben-Shakhar & Elaad, 2003). In terms used to characterize medical tests, this corresponds with a sensitivity and specificity ranging from 0.7 to 0.85. A similar sensitivity and specificity of 0.8–0.82 was found in a separate review of the GKT in laboratory experiments (MacLaren, 2001). Ben-Shakhar and Elaad conclude: "when properly administered, the GKT may turn out to be one of the most valid applications of psychological principles.... This raises a question regarding the limited usage of the GKT in criminal investigations in North America" (Ben-Shakhar & Elaad, 2003).

Measures of accuracy determined under laboratory conditions, however, may not be relevant to the performance of a test under field conditions. Moreover, what counts as high accuracy by the standards of a laboratory scientist may not be adequate when used to characterize test performance in forensic and civil populations. The probability of a true-positive test result depends not only on the specificity and the sensitivity

4.3.2.1 External Validity

External validity refers to the ability of a test to yield information about the things it claims to test. For example, many laboratory studies of deception employ protocols in which participants are instructed by the investigators to lie and are then monitored by the same investigators. As, by definition, deception is an interactive process that requires an unknowing target (victim), such a study, though scientifically useful, could not be considered a valid indication of the ability of the test to detect deception in a situation when only the test subject knows when, or even whether, he will be lying. In short, lying can be a complex, situation-dependent activity, with a variety of degrees and levels of prevarication, and the ability to detect simple deceptions in laboratory settings may not translate into a usable technology in less controlled situations.

Another issue is the relevance of a study to predict the performance of a test with a specific population or individual: For example, the first three studies on lie detection with fMRI were performed in young, healthy controls (Langleben et al., 2002; Lee et al., 2002; Spence et al., 2001). The baseline brain activity, and thus fMRI signals, of subjects varies with age, health status, and multitude of other variables (including the use of prescription or illicit drugs, depression, or the presence of a personality disorder). Clearly, the results of these studies cannot be generalized to the "real world" populations of criminal and terrorist suspects.

4.3.2.2 Internal Validity

The internal validity of a test (also called *reproducibility*) depends on the success of a method in controlling possible confounding variables. Factors relevant to internal validity include both how the test is designed and how data is collected and analyzed.

The reproducibility of a test can be affected by a number of factors, including the scenario used in the test (e.g., what is the test about: a crime, espionage, or hidden playing cards?), the level of risk that the test carries to the subject (e.g., whether the test is being applied to real-life crime suspects or to college students role-playing in a simulated crime scenario or asked to lie about playing cards), the paradigm used by the test (e.g., GKT or CQT), or to specific design features of the test (e.g., frequency of presentation, order, duration, semantic significance, and graphic properties of the stimuli).

To give a concrete example of such concerns, the state of Iowa objected to Lawrence Farwell's use of "brain fingerprinting" on Terry

Harrington, in *Harrington v. State of Iowa* (a post-conviction relief action undertaken 23 years after the crime). In his testing, Farwell claimed to show that Harrington had no memory of the crime scene, using Harrington's familiarity response to probes that included "across street," "parked cars," "weeds and grass," "drainage ditch," "by trees," and "straight ahead." The state argued, however, that familiarity or lack of familiarity with probes of such a general nature was neither a robust nor specific enough measure to prove his innocence, particularly given the long period since the crime had occurred.

This case has been cited by Farwell and others as setting a precedent for use of "brain fingerprinting" in court. However, whereas the district judge in the post-conviction relief hearing (a non-jury proceeding) heard Farwell's evidence, he denied Harrington's petition on other grounds and indicated that Farwell's evidence would not have affected the results of the proceedings. An appeal to a higher court reversed the district court's decision, on grounds unrelated to Farwell's testing (the recantation of a witness), and ordered a new trial for Harrington; the local prosecutors declined to pursue the case and Harrington was freed. Thus, despite the claims in the media and on Farwell's website implying its success in the Harrington case, "brain fingerprinting" in fact had been heard by a judge only in a non-jury proceeding, and was judged irrelevant to the outcome of the case. To our knowledge, the technology has not been admitted to any court proceedings since that case.

To create a test that truly measures verisimilitude or salience, the relation between the measured signal and the physiologic chain of events coupling a behavior with the signal must be fully characterized. In studies using fMRI (specifically, using a technique called blood oxygenated level dependent fMRI, or BOLD fMRI), the local change in the concentration of oxygenated hemoglobin in the brain is used as an indicator of neuronal activity. Although local blood flow in the brain is related to neural activity, the relationship remains incompletely understood (Heeger, Huk, Geisler, & Albrecht, 2000; Miezin, Maccotta, Ollinger, Petersen, & Buckner, 2000; Mintun et al., 2001; Vafaee & Gjedde, 2004). "Brain fingerprinting" suffers from an even more basic problem: Though EEG has been around for quite a while, the specific techniques used in brain fingerprinting rely on a proprietary (and nondisclosed) method of analysis, and therefore cannot be validated independently.

New truth-detecting technologies should not be used for socially important applications until their capabilities and limitations are adequately understood—not that neuroscience cannot yield reliable technologies for determining truth-telling for legal or security applications. There are fundamental differences between deception and truth-telling at the neurologic level, and neuroscience may provide the tools to detect these differences with sufficient reliability—or they may not. The requirements for "sufficient reliability" will clearly depend on the social purposes for which the technologies will be applied, and an adequate evaluation of new truth-telling technologies has not even begun. Whatever its other problems, considerable effort has been spent over the years to standardize polygraph testing (Kleiner, 2002). Similar work would have to be done before any new technique is ready for routine use for real-world applications.

4.3.3 Countermeasures

Effective measures to thwart conventional (CQT) polygraphy have long been known. Most attempt to increase the response of a subject to the comparison (control) questions using physical (e.g., biting the tongue or pressing the toes to the floor) or mental (e.g., counting backward by 7) techniques (Honts, Raskin, & Kircher, 1994).

Countermeasures against the GKT when used with polygraphy have also been demonstrated. There is no reason to doubt that countermeasures against the GKT could be used with other brain-measurement techniques as well. Additionally, Rosenfeld (2004) has reported that "tests of deception detection based on P300 amplitude as a recognition index may be readily defeated with simple countermeasures that can be easily learned." As brain fingerprinting is based on the P300 [a component of the ERP], this suggests that countermeasures against brain fingerprinting are also available. Recently, Langleben et al. (2004) provided preliminary data suggesting that similar countermeasures could reduce the robustness of the GKT-fMRI technology as well. Thus, until conclusively proven otherwise, brain imaging should be expected to be no less sensitive to countermeasures than the polygraph.

4.3.4 The Hype

Despite the caveats that many investigators themselves have raised about the various methods, there is an obvious attraction of new techniques for detection of deception in a society that is newly concerned with internal security and foreign threats. It is not surprising, therefore, that the media have spread an overly optimistic perception that these methods will soon become useful for practical application. "Truth and Justice, by the Blip of a Brain Wave" was the headline in one *New York Times* article (Feder, 2001), and the *San Francisco Chronicle* simply announced "Fib Detector" (Hall, 2001).

A television news broadcast in October 2003 on the "cognoscope," a helmet-mounted instrument using NIR scattering to detect changes in brain blood flow, showed an enthusiastic student saying that the technique "works," followed by a fictional scenario showing airline passengers being screened by beams of light. Such scenarios go far beyond the claims of the investigators themselves; indeed, neither the accuracy of the method for lie detection nor the ability of fNIR [functional near infrared light] to measure changes in brain blood flow without direct skin contact have been conclusively demonstrated. Press coverage of the studies by the University of Pennsylvania group investigating use of fMRI using the GKT (Langleben et al., 2002) often include speculation about the imminent usefulness of the technology in civil or forensic settings, a claim not made by the investigators and not justified by the state of current research.

Farwell's brain fingerprinting technique has been the most aggressively promoted of all neurotechnology for detecting deception. On his company's website (http://www.brainwavescience.com), Farwell is shown in a white lab coat, surrounded by testimonials from a U.S. senator, media clips, and praise of the technique for applications including forensic investigation, counterterrorism efforts, early detection of Alzheimer's disease, studies of efficiency of advertising campaigns, and security testing. Indeed, brain fingerprinting is on the verge of more widespread use. Several countries have purchased equipment for "brain fingerprinting," and India is beginning to use the method for forensic investigations (The Statesman, 2003). In May 2004, the DaVinci Institute, a Colorado "futurist think tank" (http://www.davinciinstitute.com), announced funding for a task force to develop a curriculum to train 1000 "brain fingerprinting" technicians by September 2005.

Media reports have been bolstered by excessive claims made for these methods. Farwell has been quoted as claiming "100% accuracy" for "brain fingerprinting" and the ability to detect "scientifically" if certain information is "stored" in the brain (BBC, 2004). In his testimony in *Harrington v. State of Iowa*, Farwell compared the P300 phenomenon to the sound made by a computer when it replaces a computer file with an updated version of the same file (*Harrington v. State of Iowa*). Our understanding of the workings of human memory is insufficient to support the implications of such an analogy (Squire, Stark, & Clark, 2004), which suggests an erroneous model of both human memory and the P300 wave generation (Bledowski et al., 2004). Moreover, the proprietary "brain fingerprinting" technology has been the subject of few peer-reviewed publications, and those that exist are by Dr. Farwell and his colleagues, covering less than 50 subjects altogether and raising obvious concerns about conflict of interest. (On his website, Farwell claims that "nearly 200 scientific tests" prove the accuracy of "brain fingerprinting." This appears to refer to tests conducted over time on 200 individual subjects, not to 200 independent studies. Most of the data is not published in peer-reviewed literature.) Thus, the true accuracy, validity, and relevance of this method to any real-world applications must be deemed unknown by any modern scientific standard.

Polygraphy, despite its considerable limitations, is commonly used not only for testing criminal suspects, but also for civil purposes such as screening employees or applicants to sensitive positions. The widespread use of polygraphy, even in the face of critical reports such as the one by the National Academy of Sciences (Stern, 2002) and an earlier report by the U.S. Office of Technology Assessment (1990), shows how strongly lie detection technologies are desired. Alternatives are welcomed and implemented even though they suffer from the same, or new, limitations.

4.3.5 Ethical Concerns

Traditional polygraphy has long been the topic of ethical debate. Questions have been raised concerning its validity, reliability, misuse of results, testing biases, coercion of examinees, and even possible harm due to comparison questions in the CQT (Furedy, 1995; Kokish, 2003). Many of these concerns are also relevant to brain-imaging technologies.

In addition, the current state of development of brain imaging and the existence of societal and political demand for improvements in the methods of lie-detection raise some other ethical concerns worthy of consideration, specifically (1) premature adoption; (2) misapplication through misunderstanding of the technology; (3) privacy concerns; (4) collateral information; and (5) forensic use.

4.3.5.1 Premature Adoption

Much of the funding for development of new methods to detect deception and concealed information comes from federal (U.S.) defense-related security agencies, who are looking for practical products from the research in the shortest time possible. The competition over funding and the need to attract new sources of investment have led researchers to promote the technology in the media as well as to federal agencies. Clearly there are benefits to being an early player in the marketplace. However, such competition to win potentially lucrative government contracts for these products can lead to premature translation of new technologies into practice before they are established scientifically.

Conventional polygraphy was introduced when the standards of scientific research and publications were significantly less rigorous than today. In fact, polygraph testing was shielded for many years from independent scrutiny, as were many other forensic technologies (Risinger & Sacks, 2003), due in part to lack of interest by the mainstream scientific community. Current standards of practice in conventional polygraphy are therefore strikingly behind those used in commercial psychological testing, in evaluating medical devices and therapies, or in research that is acceptable to most peer-reviewed science journals. This regrettable situation should not be allowed to develop with new technologies coming into existence.

Some investigators have promoted these technologies with claims that can be taken out of context. Pavlidis and Levine (2002), for example, suggest the use of their thermographic technique in airports or borders and comment: "The machine's recommendation will serve as an additional data point to the traveler's on-line record." Given the reaction of U.S. security agencies to even weak evidence of terrorist activity in specific individuals, one wonders whether agencies will pay heed to the second part of Pavlidis and Levine's recommendation: to give such evidence a weight that is "commensurate with how well the machine proves itself in actual practice" (Pavlidis & Levine, 2002).

4.3.5.2 Misapplication through Misunderstanding of the Technology

None of the new imaging technologies actually detect "lies." Techniques such as fMRI, P300 electrophysiology, or "brain fingerprinting" detect physiologic changes, such as blood flow or increased electrical activity in the regions of the brain that might be activated by the act of deception per se, or by the visual or psychological salience of a particular test item to the individual being tested. Separation of a deception-related signal from the host of potentially confounding signals is a complicated matter and depends on the careful construction of the deception task rather than the measurement technology. Sophisticated application of the technology and interpretation of results will therefore be crucial to the successful translation of these technologies to settings outside the laboratory. The technical limitations can be easily overlooked in civil settings. If employers, for example, started screening employees using these methods, they might find it easier to simply eliminate individuals with ambiguous results rather than understand the confounding factors that can lead to ambiguous results even in an innocent person.

Currently, compiling and interpreting brain-imaging data requires highly specialized skills in neuropsychology, physics, and statistics. Unlike polygraphy, which yields an irregular multichannel tracing that is uninterpretable by the uninitiated, the graphic appearance of processed functional brain images may give a false sense of security to anyone lacking relevant experience. Yet, those images are not the raw data itself, but pictorial renderings of statistical maps of brain activity that have been thresholded for display at an arbitrary level of significance and projected over a brain template that may not even belong to the person being imaged. Individuals with experience in generating, processing, analyzing, and interpreting functional brain imaging data are currently available only at major research institutions, and there are currently no training or professional standards for their skills.

Who will be allowed to use the technology and in what settings? Will private firms begin offering deception detection to banks looking for honest employees, parents trying to determine whether their children are really using drugs, and boy scout troops looking to weed out child molesters? The potential for misuse might require a careful system of licensing practitioners, should the technology develop to the point where it is used widely for consumer applications (Rosen & Gur, 2002). However, this will require a more open process than licensing practitioners by the company that produces the equipment, as is currently the case with "brain fingerprinting." The safest approach may be to continue applying the privacy and safety standards of medical information use to any data acquired using medical technology regardless of indication.

4.3.5.3 Privacy Concerns

Does a person have an alienable right to keep his or her subjective thoughts private? If technology develops to the point where, for example, remote fNIR could be used covertly to monitor a person's frontal lobe patterns during questioning, would it be mandatory in all cases to reveal that one is being probed? Would a reliable lie detector, if one can be developed, find its way into airports and courtrooms, stores and offices, the Olympics, the schools? Reliable, safe lie detectors (and other potential uses of imaging not discussed in this paper) would force a re-examination of the very idea of privacy, which up until now could not reliably penetrate the individual's cranium. A number of organizations have already begun to advocate for the right to cognitive freedom.

4.3.5.4 Collateral Information

Brain imaging data that has been acquired for research purposes in the United States is subject to strict ethical and legal standards provided by the Declaration of Helsinki and the federal regulations, however, there is no guarantee that similar standards could be maintained in civil, forensic, or security settings. MRI images usually cover more of the brain than the discrete area of concern. Therefore, imaging for a nonmedical indication could reveal medically relevant information. It is easy to foresee a lawsuit by a person who was given a brain scan in the course of pre-employment screening in which an early-stage brain tumor was clearly visible on the scan, yet the candidate was not informed (see, e.g., Illes et al., 2004; Katzman, Dagher, & Petronas, 1999). In addition, researchers are discovering that brain scans may reveal a great deal of information about us. Data indicates that brain scans could potentially reveal rudimentary information about personality traits, mental illness, sexual preferences, or predisposition to drug addiction (Andreasen, 1997; Hamann, Herman, Nolan, & Wallen, 2004; Kiehl et al., 2001; Lindsey, Gatley, & Volkow, 2003). If disclosed without proper consent, such information could lead to unanticipated insurance, employment, or legal problems for the individual being tested. Most of this research, so far, has been conducted by comparing groups, not individuals, and consequently its potential for identifying such information in individuals is unknown (Farah, 2002). Still, some traits are distinguishable on an individual level, and as research continues, more such traits are likely to be discovered. The ability to store brain-scan images indefinitely suggests a scenario that we are already facing in genetics: Genetic information that was inconsequential when originally stored in tissue samples becomes increasingly revealing as our knowledge of genetics grows more sophisticated.

4.3.5.5 Forensic Use

Results of polygraph examinations are not admissible in most U.S. courts (or in courts in most other countries) because of well-justified concerns about the reliability of the results. Is "brain fingerprinting" a more reliable technology? Nobody really knows, and the appropriate studies have not been done. As the state of Iowa complained in its brief against brain fingerprinting in *Harrington*, the most critical problem with admission of "brain fingerprinting" evidence is the lack of any track record establishing its reliability.

High-technology tools such as brain scans can give a persuasive scientific gloss to what in reality are subjective interpretations of the data. The implied certainty and authority of science can be prejudicial to juries, and when it is accompanied by images to reinforce expert testimony, it can be particularly persuasive. This concern has been raised about the use of computer-generated visual displays in the courtroom in general (Borelli, 1996). Brain scan images might influence juries even when the images add no reliable or additional information to the case. In addition, if such scans gain currency in judicial settings, subjects may face intense pressures to undergo such testing to "prove" guilt or innocence, and their refusal to undergo such testing might be used against them in subsequent proceedings.

4.3.6 Conclusion

Neuroscience research has begun to establish brain correlates of specific cognitive processes. In a real though very limited sense, we have begun to probe the subjective contents of the mind. Brain-imaging technology has created the potential for powerful new ways to understand the workings of the human brain, as well as concerns about misusing that potential. Limitations of the existing methods to detect lies and verify truth and changing priorities of the federal defense agencies have led to attempts to apply these research advances for forensic and defense purposes. Though promising, it remains unknown whether those early research findings will ever lead to a better lie-detection methodology. Whereas media and research attention has been focused on the impressive images medical-imaging technology can produce, the limitations of the existing forms of questioning formats and deception paradigms (CQT, GKT, etc.) that include sensitivity to countermeasures and the choice of appropriate questions remain unchanged.

Premature commercialization will bias and stifle the extensive basic research that still remains to be done, damage the long-term applied potential of these powerful techniques, and lead to their misuse before they are ready to serve the needs of society. Society must be ready to come to a decision about the value of cognitive privacy before these technologies become widespread. Scientists, ethicists, and other advocates must take an active role in the discussion of the threat to civil liberties that their research might make possible. The discussion about the implications of reliable, as well as involuntary, lie-detection technologies should begin in scientific, legal, and civil forums in anticipation of the further development of these promising and challenging technologies.

Notes

1. Editor's note: This reading originally appeared in 2005 in the *American Journal of Bioethics*, volume 5, pages 39–49, and is used with permission.

References

Andreasen, N. C. (1997). Linking mind and brain in the study of mental illnesses: A project for a scientific psychopathology. *Science, 275*(5306), 1586–1593.

BBC News World Edition. (2004, February 17). Brain fingerprints under scrutiny. Retrieved from: http://news.bbc.co.uk/2/hi/science/nature/3495433.stm.

Ben-Shakhar, G., & E. Elaad. (2003). The validity of psychophysiological detection of information with the Guilty Knowledge Test: A meta-analytic review. *Journal of Applied Psychology, 88*(1), 131–151.

Berns, G. S., Cohen, J.D., & Mintun, M. A. (1997). Brain regions responsive to novelty in the absence of awareness. *Science, 276*(5316), 1272–1275.

Bledowski, C., Prvulovic, D., Hoechstetter, K., Scherg, M., Wibral, M., Goebel, R., & Linden, D. E. (2004). Localizing P300 generators in visual target and distractor processing: A combined event-related potential and functional magnetic resonance imaging study. *Journal of Neuroscience, 24*(42), 9353–9360.

Borelli, M. (1996). The computer as advocate: An approach to computer-generated displays in the courtroom. *Indiana Law Journal, 71*, 439.

Canli, T., & Amin, Z. (2002). Neuroimaging of emotion and personality: Scientific evidence and ethical considerations. *Brain and Cognition, 50*, 414–431.

Chance, B., & Kang, K. A. (2002). Vision statement: Interacting brain. In M. C. Roco & W. S. Bainbridge (Eds.), *Converging technologies for improving human performance* (pp. 199–201). Washington, DC: Technology Administration. Available at: http://www.technology.gov/reports/2002/NBIC/Part3.pdf.

Donaldson, D. I. (2004). Parsing brain activity with fMRI and mixed designs: What kind of a state is neuroimaging in? *Trends in Neurosciences 27*(8), 442–444.

Farah, M. J. (2002). Emerging ethical issues in neuroscience. *Nature Neuroscience, 5*, 1123–1129.

Farwell, L. A., & Donchin, E. (1991). The truth will out: Interrogative polygraphy ("lie detection") with event-related brain potentials. *Psychophysiology, 28*(5), 531–547.

Farwell, L. A., & Smith, S. S. (2001). Using brain MER-MER testing to detect knowledge despite efforts to conceal. *Journal of Forensic Sciences, 46*(1), 135–143.

Feder, B. J. (2001, October 9). Truth and justice, by the blip of a brain wave. *New York Times*, section F, p. 3.

Fischer, H., Wik, G., & Fredrikson, M. (1997). Extraverion, neuroticism, and brain function: A PET study of personality. *Personality and Individual Differences, 23*, 345–352.

Foster, K. R., Wolpe, P. R., & Caplan, A. (2003). Bioethics and the brain. *IEEE Spectrum, June*, 34–39.

Furedy, J. J. (1995). The 'control' question 'test' (CQT) polygrapher's dilemma: Logico-ethical considerations for psychophysiological practitioners and researchers. *International Journal of Psychophysiology, 20*(3), 199–207.

Hall, C. T. (2001, November 26). Fib detector. *San Francisco Chronicle*, p. A10.

Hamann, S., Herman, R. A., Nolan, C. L., & Wallen, K. (2004). Men and women differ in amygdala response to visual sexual stimuli. *Nature Neuroscience, 7*(4), 411–416.

Harrington v. State of Iowa. (2003). Supreme Court of Iowa 659 N.W.2d 509; 2003 Iowa Sup. LEXIS 35.

Heeger, D. J., Huk, A. C., Geisler, W. S., & Albrecht, D. G. (2000). Spikes versus BOLD: What does neuroimaging tell us about neuronal activity? *Nature Neuroscience, 3*(7), 631–633.

Honts, C. R., Raskin, D. C., and Kircher, J. C. (1994). Mental and physical countermeasures reduce the accuracy of polygraph tests. *Journal of Applied Psychology, 79*(2), 252–259.

Illes, J., Kirschen, M. P., & Gabrieli, J. D. (2003). From neuroimaging to neuroethics. *Nature Neuroscience, 6*(3), 205.

Illes, J., Rosen, A. C., Huang, L., Goldstein, R. A., Raffin, T. A., Swan, G., & Atlas, S. W. (2004). Ethical consideration of incidental findings on adult brain MRI in research. *Neurology, 62*(6), 888–890.

Katzman, G. L., Dagher, A. P., & Patronas, N. J. (1999). Incidental findings on brain magnetic resonance imaging from 1000 asymptomatic volunteers. *JAMA, 282*(1), 36–39.

Kiehl, K. A., Smith, A. M., Hare, R. D., Mendrek, A., Forster, B. B., Brink, J., & Liddle, P. F. (2001). Limbic abnormalities in affective processing by criminal psychopaths as revealed by functional magnetic resonance imaging. *Biological Psychiatry, 50*(9), 677–684.

Kleiner, M. (2002). *Handbook of polygraph testing.* San Diego, CA: Academic Press.

Kokish, R. (2003). The current role of post-conviction sex offender polygraph testing in sex offender treatment. *Journal of Child Sex Abuse, 12*(3–4), 175–194.

Langleben, D. D., Schroeder, L., Maldjian, J. A., Gur, R. C., McDonald, S., Ragland, J. D., et al. (2002). Brain activity during simulated deception: An event-related functional magnetic resonance study. *NeuroImage, 15*(3), 727–732.

Langleben, D. D., Loughead, J. W., Bilker, W., Phend, N., Buseh, S., Chilress, A. R., et al. (2004). Imaging deception with fMRI: The effects of salience and ecological relevance. Program No. 372.12. San Diego, CA: Society for Neuroscience. Available at: http://www.sfn.org.

Lee, T. M., Liu, H. L., Tan, L. H., Chan, C. C., Mahankali, S., Feng, C. M., et al. (2002). Lie detection by functional magnetic resonance imaging. *Human Brain Mapping, 15*(3), 157–164.

Lindsey, K. P., Gatley, S. J., & Volkow, N. D. (2003). Neuroimaging in drug abuse. *Current Psychiatry Reports, 5*(5), 355–361.

Lykken, D. T. (1991). Why (some) Americans believe in the lie detector while others believe in the guilty knowledge test. *Integrative Physiological and Behavioral Science, 26*(3), 214–222.

MacLaren, V. V. (2001). A quantitative review of the guilty knowledge test. *Journal of Applied Psychology, 86*, 674–683.

Miezin, F. M., Maccotta, L., Ollinger, J. M., Petersen, S. E., & Buckner, R. L. (2000). Characterizing the hemodynamic response: Effects of presentation rate, sampling procedure, and the possibility of ordering brain activity based on relative timing. *NeuroImage, 11*, 735–759.

Mintun, M. A., Lundstrom, B. N., Snyder, A. Z., Vlassenko, A. G., Shulman, G. L., & Raichle, M. E. (2001). Blood flow and oxygen delivery to human brain during functional activity: Theoretical modeling and experimental data. *Proceedings of the National Academy of Sciences of the United States of America 98*(12), 6859–6864.

Pavlidis, I., Eberhardt, N. L., & Levine, J. A. (2002). Seeing through the face of deception. *Nature, 415*(6867), 35.

Pavlidis, I., & Levine, J. (2002). Thermal image analysis for polygraph testing. *IEEE Engineering in Medicine and Biology Magazine, 21*(6), 56–64.

Risinger, D. M., & Saks, M. J. (2003). A house with no foundation. *Issues in Science and Technology, 20*(1), 35–39.

Rosen, A. C., & Gur, R. C. (2002). Ethical considerations for neuropsychologists as functional magnetic imagers. *Brain and Cognition, 50*(3), 469–481.

Rosenfeld, J. P. (2004). Simple, effective countermeasures to P300-based tests of detection of concealed information. *Psychophysiology, 41*(2), 205–219.

Rosenfeld, J. P., Cantwell, B., Nasman, V. T., Wojdac, V., Ivanov, S., & Mazzeri, L. (1988). A modified, event-related potential-based guilty knowledge test. *International Journal of Neuroscience, 42*(1–2), 157–161.

Ross, P. (2003). Mind readers. *Scientific American, September*, 74–77.

Spence, S. A., Farrow, T. F., Herford, A. E., Wilkinson, I. D., Zheng, Y., & Woodruff, P. W. (2001). Behavioural and functional anatomical correlates of deception in humans. *Neuroreport, 12*(13), 2849–2853.

Spence, S. A., Hunter, M. D., Farrow, T. F., Green, R. D., Leung, D. H., Hughes, C. J., & Ganesan, V. (2004). A cognitive neurobiological account of deception: Evidence from functional neuroimaging. *Philosophical Transactions of the Royal Society of London. Series B: Biological Sciences, 359*, 1755–1762.

Squire, L. R., Stark, C. E., & Clark, R. E. (2004). The medial temporal lobe. *Annual Review of Neuroscience, 27*, 279–306.

Stern, P. C. (2002). *The polygraph and lie detection: Report of The National Research Council Committee to Review the Scientific Evidence on the Polygraph*. Washington, DC: The National Academies Press.

Sugiura, M., Kawashima, R., Nakagawa, M., Okada, K., Sato, T., Goto, R., et al. (2000). Correlation between human personality and neural activity in cerebral cortex. *NeuroImage, 11*, 541–546.

The Statesman (India). (2003; May 23). "Brain fingerprinting" arrives in state.

U.S. Office of Technology Assessment. (1990). *The use of integrity tests for pre-employment screening*. Washington, DC: Author.

Vafaee, M. S., & Gjedde, A. (2004). Spatially dissociated flow-metabolism coupling in brain activation. *NeuroImage, 21*(2), 507–515.

Wolpe, P. R. (2002). Treatment, enhancement, and the ethics of neurotherapeutics. *Brain and Cognition, 50*, 387–395.

Wolpe, P. R. (2004). Neuroethics. In S. G. Post (Ed.), *Encyclopedia of bioethics*, 3rd edition. New York: Macmillan Reference USA.

Reading 4.4

fMRI in the Public Eye[1]

Eric Racine, Ofek Bar-Ilan, and Judy Illes

Functional neuroimaging techniques, such as functional magnetic resonance imaging (fMRI) and positron emission tomography (PET), have evolved as key research approaches to studying both disease processes and the basic physiology of cognitive phenomena in contemporary neuroscience. In the clinical domain, they carry hope for guiding neurosurgical mapping, monitoring drug development, and providing new approaches to disease diagnosis and management at early, possibly even presymptomatic stages. However, issues relating to these capabilities, such as technical readiness and the possibility of disease screening in advance of effective therapeutic intervention, raise substantial ethical challenges for investigators, health care providers, and patients alike. In basic neuroscience, increasing numbers of non-health-related fMRI studies that touch on our personal values and beliefs have also forced us to expand our ethical perspectives (Illes, Kirschen, & Gabrieli, 2003). The wide dissemination of this research, growing applications of the technology, and continuously improving resolution have not escaped the attention of the neuroscience and neuroethics communities, the media, or the broader public (Butler, 1998; Editorial, 1998; Editorial, 2004; Jaffe, 2004; President's Council on Bioethics, 2003). However, are the boundaries of what this technology can and cannot achieve being effectively communicated to the public? Are its limitations understood? Are the applications of the technology viewed as useful and meaningful? Are some studies more conducive to misinterpretation than others? What are the associated risks to society? From a scientific perspective, important methodological and technical assumptions guide fMRI research. However, from the public's point of view, once research results are publicized, especially when they concern personality, self-identity, and other social constructs, they are bound to interact with lay conceptions of these phenomena.

4.4.1 Print Media Coverage of fMRI

To understand this complex interaction between neuroscience and society, we focused on the coverage of fMRI—as one model of frontier neurotechnology—in the print media. We investigated how both neuroscience and the media shape the social understanding of fundamental aspects of our reality and how this, in turn, points to issues of scientific communication and public involvement in neuroscience. To this end, we frame our perspective according to three trends that we have observed in press coverage of fMRI—"neuro-realism," "neuro-essentialism," and "neuro-policy"—and explore how neuroethics can attend to the related ethical, legal, and social issues by promoting multidirectional communication in neuroscience.

We carried out a press content analysis (Neuendorf, 2002) of samples of print media coverage of fMRI. Using this method, we were able to capture salient messages about the research as they are conveyed to readers. We did not study content from the point of view of scientific accuracy, but rather as a phenomenon of communication that could affect public perceptions. In this respect, the dynamics of news production, including the original interaction between journalists and scientists, are not directly addressed, but this information is also unavailable to readers.

We conducted a key-word search on fMRI using the LexisNexis Academic database of General News (major newspapers), General News (magazines and journals), Medical News (medical and health news), University News, and Legal News. We retrieved 132 nonredundant articles for our sample, returned from a search beginning in January 1991 and ending in June 2004. Seventy-nine articles were from general sources such as the *New York Times* and the *Washington Post*, and 53 were from specialized sources such as *Pain & Central Nervous System Week* and *New Scientist*. After an initial pilot of a coding scheme and testing for reproducibility, two of us (E.R. and O.B.-I.) coded the articles for tone, presence of ethical issues, and type of research reported.

We found descriptions of clinical research in 35% of the articles, of nonclinical research (in particular, studies of higher-order cognition and emotions) in 44%, and both in 20%. We also found discussion of clinical benefits (for example, early or improved diagnosis, therapy, and monitoring of health interventions such as drug effects and neurosurgery) in 65% of the articles, but nonclinical benefits (such as technical improvements of fMRI, non-health-related early childhood interventions, and improved techniques for lie detection) in only 17%.

4.4.2 Neuro-realism, Neuro-essentialism, and Neuro-policy

To put these findings in perspective, we draw on three concepts that we have termed neuro-realism, neuro-essentialism, and neuro-policy, and use examples from the articles themselves for illustration. The interaction of these concepts, which encompasses lay perceptions of reality, subjectivity, and policymaking, combined with both hope and leaps of faith about the meaning of the data across the life span, contribute to public appreciation of the benefits and risks of functional neuroimaging. These interactions undoubtedly also influence the evolution of the science itself, as researchers are not isolated from wider social and cultural beliefs about the brain.

Our concept of neuro-realism describes how coverage of fMRI investigations can make a phenomenon uncritically real, objective, or effective in the eyes of the public. This occurs most notably when qualifications about results are not brought to the reader's attention. For example, commenting on an fMRI study of fear, one article (Anonymous, 1999b) states, "Now scientists say the feeling is not only real, but they can show what happens in the brain to cause it". Many occurrences of neuro-realism deal with the effectiveness of health-related procedures such as acupuncture. For example, "Patients have long reported that acupuncture helps relieve their pain, but scientists don't know why. Could it be an illusion? Now brain imaging technology has indicated that the perception of pain relief is accurate" (Anonymous, 1999a). Another headline: "A relatively new form of brain imaging provides visual proof that acupuncture alleviates pain" (Anonymous, 1999c). Furthermore, because fMRI investigation shows activation in reward centers when subjects ingest high-fat foods, one reads, "Fat really does bring pleasure" (Biskup, 2004). So, neuro-realism reflects the uncritical way in which an fMRI investigation can be taken as validation or invalidation of our ordinary view of the world. Neuro-realism is, therefore, grounded in the belief that fMRI enables us to capture a "visual proof" of brain activity, despite the enormous complexities of data acquisition and image processing.

The concept of neuro-essentialism reflects how fMRI research can be depicted as equating subjectivity and personal identity to the brain. In this sense, the brain is used implicitly as a shortcut for more global concepts such as the person, the individual, or the self. This is the case in many expressions where the brain is used as a grammatical subject. Headline examples of this phenomenon are: "Brain can banish unwanted memories" (Cookson, 2004), "How brain stores languages" (Blakeslee, 1997), and "Brain stores perceptions into small meaningful chunks" (Anonymous, 2001).

Other statements imply that fMRI is scrutinizing our minds. For example, "The better fMRI systems become, and the more adept scientists get at extracting information from them, the more they will be able to piece together the neural circuits that make us who we are" (Sample & Adam, 2003) and "The brain can't lie: brain scans reveal how you think and feel and even how you might behave. No wonder the CIA and big businesses are interested." Other claims insinuate that individual differences can be reduced to brain differences (Begley, 2001), "Odds are that gambling addict's brain is built differently" (Anonymous, 2003), or that neuroscience provides ultimate explanations (Bacon, 2004), "How it all starts inside your brain" (Begley, 2001). Although studies of the mind and brain are a cornerstone of cognitive neuroscience, neuro-essentialism represents a hasty reduction of identity to the brain.

Neuro-policy describes attempts to use fMRI results to promote political and personal agendas, as in the case of interest groups that uphold the investigation of social problems using fMRI. For example, the Lighted Candle Society, a Utah-based nonprofit organization that is dedicated to the enhancement of moral values, advocates the use of fMRI to prove that pornography is addictive (Bacon, 2004). Another example of neuro-policy has been reported by a neuroscientist who has received queries from "both sides of the current California debate on bilingual education" (Hall, 1998).

These examples represent press coverage that shows how neuroscience is extending to new areas of social concern and, accordingly, how neuroscientists are being tapped for advice in policy development. Undoubtedly, an element of neuro-policy can imply neuro-realism and neuro-essentialism, as the regulatory appeal of neuroscience might be increased by the beliefs that brain research shows reality and that the brain is the core of our identity.

Neuro-policy also creates practical challenges for the neuroscientist. What happens when neuroimagers receive calls from journalists to explain their (or others') provocative findings? How can the early state of certain findings be communicated, especially before peer review? What is the proper reaction to "social demand" for brain findings? Brain–machine neuroengineering research has shown that military demand for neuroscience research can certainly be controversial (Editorial, 2003; Hoag, 2003; Rudolph, 2003). These are some of the challenges faced by neuroscientists when participating in research efforts that are framed in terms of real-world policy.

To manage the consequences of our three original concepts of neuro-realism, neuro-essentialism, and neuro-policy, media reporters need to reflect on how they select and frame fMRI studies to ensure balanced and sensitive communication. In parallel, and despite the time and career pressures of interfacing with the media, the neuroscience community must respond by reflecting on its own ways of conveying the complexities of fMRI and other neurotechnologies to the press, and ensure proper interpretation of the results. Communication of the limitations of fMRI technology and critical appraisal of claims of health benefits are immediate challenges. New collaborations with bioethics and humanities scholars in the design of investigations into the lay perception of neurotechnologies, on medical, scientific, and public perceptions of neurologic and psychiatric diseases, and on patient narratives will also be instrumental in broadening the reflection and bringing new elements to the discussion. To fully realize and appreciate the transition, we will need increasingly elaborate and diverse research perspectives.

Resistance to some of these suggestions is not unexpected, but this does not dispirit the endeavor. In the end, debate among neuroscientists, life science colleagues, the media, and the public represents an exercise in critical thinking and self-reflection. It brings into focus and strengthens the pillars of science, medicine, and our pluralist society.

Notes

1. Editor's note: This reading was excerpted from an article that appeared in 2005 in *Nature Reviews Neuroscience*, volume 6, pages 159–164, and is used with permission of the publisher and authors. The omitted sections included further reflections on media portrayals of neuroscience. The authors were supported by The Greenwall Foundation, the National Institutes of Health, and the National Institute of Neurological Disorders and Stroke (J.I.) and an FQRSC postdoctoral fellowship to E.R. They extend their thanks to C. Jennings for inspiring this project and to S. W. Atlas, T. A. Raffin, P. Schraedley Desmond, and M. Gallo.

References

Anonymous (1999a, December 17). The cutting edge. *Washington Post*, p. Z05.

Anonymous (1999b, June 22). Fear of pain may be worse than pain itself. *New York Times*, p. 14.

Anonymous (1999c, December 17). Acupuncture: Brain images demonstrate pain relief. *Pain & Central Nervous System Week*, p. 13.

Anonymous (2001, July 27). Brain stores perceptions into small meaningful chunks. *The Hindu.*

Anonymous (2003, September 18). Odds are that gambling addict's brain is built differently. *Times–Picayune,* 3.

Bacon, J. (2004, May 12). Group to prove pornography is addictive. *Daily Universe.*

Begley, S. (2001, February 12). How it all starts inside your brain. *Newsweek, 137*, 40–42.

Biskup, A. (2004, April 13). Fat really does bring pleasure. *Boston Globe*, p. C3.

Blakeslee, S. (1997, July 27). How brain stores languages. *Plain Dealer*, 2J.

Butler, D. (1998). Advances in neuroscience 'may threaten human rights.' *Nature*, *391*, 316.

Cookson, C. (2004, January 9). Brain can banish unwanted memories. *Financial Times,* 11.

Editorial. (1998). Does neuroscience threaten human values? *Nature Neuroscience, 1*, 535–536.

Editorial. (2003). Silence of the neuroengineers. *Nature, 423*, 787.

Editorial. (2004). Brain scam? *Nature Neuroscience, 7*, 683.

Hall, S. S. (1998, April 5). The scientific method: test–tube moms. *New York Times,* section 6, p. 22.

Hoag, H. (2003). Neuroengineering: remote control. *Nature, 423*, 796–798.

Illes, J., Kirschen, M. P., & Gabrieli, J. D. (2003). From neuroimaging to neuroethics. *Nature Neuroscience, 6*, 205.

Jaffe, S. (2004). Fake method for research impartiality (fMRI). *Scientist, 18*, 64.

Neuendorf, K. A. (2002). *The content analysis guidebook*. Thousand Oaks, CA: Sage Publications.

President's Council on Bioethics. (2003). *Beyond therapy: Biotechnology and the pursuit of happiness*. Washington, DC: HarperCollins.

Rudolph, A. (2003). Military: Brain-machine could benefit millions. *Nature*, *424*, 369.

Sample, I., & Adam, D. (2003, November 20). The brain can't lie: Brain scans reveal how you think and feel and even how you might behave. *The Guardian*, p. 4.

Reading 4.5

Race, Behavior, and the Brain: The Role of Neuroimaging in Understanding Complex Social Behaviors[1]

Elizabeth A. Phelps and Laura A. Thomas

Over the past decade, we have seen unprecedented advances in our understanding of the human brain. Perhaps most exciting has been research with functional neuroimaging that allows us a window into the brain activity underlying complex human behaviors. Using neuroimaging techniques, recent studies have begun to explore the brain systems involved in behaviors we consider to be defining as individuals, such as moral reasoning (e.g., Greene, Sommerville, Nystrom, Darley, & Cohen, 2001), social cooperation (Rilling et al., 2002), violent tendencies (Davidson, Putnam, & Larson, 2000), responses to race groups (e.g., Phelps et al., 2000), and love (e.g., Bartels & Zeki, 2000). This is certainly good news. From a scientific perspective, these research tools provide additional avenues to help us understand these complicated and important behaviors. However, it remains unknown whether this knowledge has any significance beyond furthering basic behavioral and neural science.

The implications are clear. As we begin to uncover the neural basis of behaviors that are relevant to our social and political lives, could we use brain science to guide social and political choices? In this article we argue that at this time the answer is "no." There is a disturbing trend developing in the interpretation of brain imaging research in the general public, as well as among some scientists. This trend is rooted in the assumption that a biological understanding of a behavior is more informative or reliable than a psychological understanding of a behavior. Although brain science can inform our understanding of complex human behaviors, it cannot help us predict human behavior with any more certainty than can be derived from examining behavior itself. To explore this topic, we review what has been learned about the neural basis of social group processing, in particular social groups defined by race.

4.5.1 Direct and Indirect Measures of Race Evaluation

A major finding that has emerged from research on the neural basis of social group processing concerns biases in the evaluation of individuals on the basis of race group membership. These studies have focused on a brain region called the amygdala. The amygdala is a small almond-shaped structure in the medial temporal lobe that has been shown to be important for emotional learning and memory and for some aspects of emotional evaluation (e.g., LeDoux, 1996). Specifically, the amygdala has been shown to be important for the expression of a learned evaluation when it is assessed indirectly, such as through a physiologic response. The amygdala does not seem to be important when this same evaluation is expressed explicitly or directly. For example, fear conditioning is a paradigm by which a neutral stimulus, such as a blue square, comes to acquire aversive properties by repeatedly being paired with an aversive event, such as a mild shock to the wrist. After a few pairings of square and shock, normal subjects are able to explicitly report that the blue square predicts a shock, and they demonstrate a physiologic fear response (e.g., increased sweating) to the blue square presented alone. Patients with amygdala damage also explicitly report that the blue square predicts a shock; however, they fail to show any indirect, physiologic indication of this learned fear response (Bechara et al., 1995; LaBar, LeDoux, Spencer, & Phelps, 1995). This preferential involvement of the amygdala in the indirect (implicit) expression of an evaluation, but not direct evaluation, has also been shown with other types of emotional stimuli and measures of evaluation (Funayama, Grillon, Davis, & Phelps, 2001).

The evaluation of social groups defined by race is a topic of undeniable importance that has been studied by social psychologists for decades. One finding that has emerged from this literature is that there has been a consistent decrease over the years in the explicit report of negative attitudes toward black Americans expressed by white Americans (Biernat & Crandall, 1999; Schuman, Steeh, & Bobo, 1997). Although it was not uncommon for white Americans to express negative attitudes toward black Americans 40 years ago, recent studies have found that white Americans' explicit attitudes are significantly less biased today. However, there is robust evidence that when attitudes are assessed indirectly or implicitly, most white Americans demonstrate a negative bias toward black Americans (Banaji, 2001; Bargh & Chen, 1997; Devine, 1989; Fazio, Jackson, Dunton, & Williams, 1995; Fiske, 1998; Nosek, Banaji,

& Greenwald, 2002). There are a few possible reasons for this observed dissociation between an implicit negative bias and an explicit unbiased report. One is that this negative implicit bias is consistent with a biased explicit attitude, but subjects are simply reluctant to admit a biased attitude. However, it is also possible that even those subjects who consciously believe that there is no reason for a biased attitude are influenced by cultural stereotypes and limited experiences, in such a way that biases are expressed on implicit tests that are not amenable to conscious control and mediation. This notion of unintentional and unconscious bias was recently expressed in the Study on the Status of Women Faculty in Science at MIT, in which the dean of science, when acknowledging a pattern of sex discrimination, indicated that "it was usually totally unconscious and unknowing" (Birgeneau, 1999).

In an effort to determine whether the dissociation in the neural systems underlying implicit and explicit assessments of evaluation could be linked to the dissociation observed among white Americans in explicit and implicit measures of race bias, my lab, in collaboration with Banaji and colleagues (Phelps et al., 2000), used functional magnetic resonance imaging (fMRI) to examine the relationship between activation of the amygdala and behavioral measures of race bias. During brain imaging, white American subjects were shown pictures of unfamiliar black and white male faces. They were simply asked to indicate when a face was repeated. After imaging, subjects were given a standard explicit assessment of race attitudes (the Modern Racism Scale, or MRS; McConahay, 1986) and two indirect assessments of race bias. The first indirect test measured startle potentiation while viewing the same faces. Startle is a reflex that is potentiated in the presence of stimuli considered to be negative (Grillon, Ameli, Woods, Merikangus, & Davis, 1991; Lang, Bradley, & Cuthbert, 1990); the difference in the magnitude of the startle reflex in the presence of the black versus white faces served as the indirect measure of bias. The second indirect measure of bias was the Implicit Association Test (IAT; Greenwald, McGhee, & Schwartz, 1998). In this study, the IAT involved the presentation of a series of trials on which a black or white unfamiliar face was presented; subjects were asked to classify these faces according to race as quickly as possible. Interspersed among the face trials were trials in which a word was presented and subjects were asked to classify this word as representing something "good" (e.g., joy, love, peace) or "bad" (e.g., cancer, bomb, devil), again as quickly as possible. On half of the trials, the subjects pressed one key to indicate a "good" word or a black face and another key for a "bad" word or a white face. For the other half of the trials, these pairings were reversed. Reaction time differences in the black + good/white + bad pairing versus the black + bad/white + good pairing provided an indirect assessment of social group evaluation.

The behavioral results from this study (Phelps et al., 2000) mirrored those of previous studies (see Banaji, 2001). The white American subjects showed a relatively pro-black bias as measured by the explicit test, the MRS, while at the same time they exhibited an anti-black bias as measured indirectly by the IAT. On the startle potentiation test, they showed a nonsignificant trend toward greater startle while viewing black versus white unfamiliar faces. The imaging results did not yield a main effect for race. Although the majority of these subjects showed greater amygdala activation to the black faces than to the white faces, this effect was not observed in all of the subjects. When the variability in amygdala activation to black versus white faces across subjects was correlated with the variability in the behavioral measures, an interesting pattern emerged: Those subjects who showed greater negative bias on the indirect measures of race evaluation (IAT and startle) also showed greater amygdala activation to the black faces than to the white faces. This correlation between the indirect measures of bias and amygdala activation was not observed when the amygdala response was compared with the MRS, the explicit test of race attitudes. In other words, the amygdala response was predictive of race bias evaluation when this bias was measured indirectly, but not by explicit report.

This study is the first to link what is known about the neural systems of emotional learning and evaluation to the evaluation of social groups. It is important because it relates social evaluations to the ordinary mechanisms of everyday emotional learning and memory. Also, by showing a brain region whose activation response is correlated with indirect but not direct measures of evaluation, this study supports the idea proposed by social psychologists that attitudes toward social groups can be expressed both directly and indirectly and that these two means of expression may represent different underlying processes. However, like the study with face recognition mentioned earlier, it is unclear what we have learned about the behavior of race bias that we did not know before identifying these neural mechanisms. It may be appealing to extrapolate from these neuroimaging results that such measurements may be able to detect biases that individuals are unwilling to admit, but as this study indicates, behavioral tests are already able to detect such biases. Indeed, the only avenue we have to understand the behavioral significance of this or any other brain activation result is by linking it to behaviors we have defined. In addition, there are several other factors to consider when interpreting these results.

First, how general are these results? The subjects in the study by Phelps et al. (2000) were white Americans. Would members of other race groups perform similarly? Overall, black Americans show more variability than white Americans on measures of indirect race bias, with some showing a pro-black bias and others showing no bias or a pro-white bias (Nosek et al., 2002). If amygdala activation to same-race versus other-race faces predicts a negative bias on indirect measures of race evaluation, we might expect a less consistent amygdala response to same-race versus other-race faces in black American subjects. However, a study by Hart et al. (2000) assessed amygdala activation to black versus white faces in both black and white American subjects and observed greater amygdala activation to other-race faces than to same-race faces in both groups. In this study, the activation response was not linked to behavioral measures of race evaluation, making it difficult to know whether this link between implicit race bias and amygdala activation extends to black American subjects.

Second, does this effect extend to other stimuli that vary by race? Both the Hart et al. (2000) and Phelps et al. (2000) studies presented photographs of unfamiliar black and white male faces. A previous study (DuBois et al., 1999) reported that the amygdala shows increased activation to unfamiliar versus familiar neutral faces. Could it be that the amygdala response to same-race versus other-race faces is driven by greater familiarity with one's own race group? How might the responses observed in these studies be changed by familiarity? In an effort to address this question, Phelps et al. (2000) conducted a second study in which the faces presented belonged to familiar black and white male individuals. In this study, the white American subjects did not show any consistent evidence of stronger amygdala activation to the black versus white faces. The indirect startle test also failed to show any differentiation between race groups. Although the IAT, in which subjects are asked to classify these faces by race, did show a negative bias toward black faces, performance on this task was not correlated with an amygdala response. This second study suggests that this amygdala response can be modified by experience and familiarity.

Finally, what does an "activation" response measured with fMRI tell us about the precise role of the amygdala in this study? Brain activation is usually measured in response to a mental challenge created by the experimenter. In the Phelps et al. study, pictures of faces were presented during brain imaging and subjects were asked to respond to the identity of the individual faces, but clearly subjects also coded race group information from these stimuli. The result was a relationship observed between the differential amygdala response to the race groups and indirect measures of race bias. This relationship or correlation between the brain response and the behavioral measures does not tell us how or if the amygdala is involved in generating these behaviors; it indicates simply that there is some relationship. An activation response does not inform us as to what, exactly, a brain region does in the generation of a behavior. To determine the precise role of a given brain region in a task, we must use other techniques.

In an effort to determine whether the amygdala plays a critical role in the indirect evaluation of race groups, our lab tested a patient with bilateral damage to the amygdala (Phelps, Cannistraci, & Cunningham, 2003). This white American patient was given the same explicit measure of race bias mentioned earlier (MRS) along with the IAT to assess indirect race bias. Her performance was similar to the normal white American control subjects and consistent with previous results. This patient and the control subjects demonstrated a dissociation between the explicit measure of race attitude, on which they showed a pro-black bias, and the IAT, on which they showed a negative, anti-black bias. In other words, damage to the amygdala did not eliminate the indirect expression of race bias. These results suggest that even though the amygdala may play some role in the learning or expression of indirect race bias, the amygdala is not critical for the expression of this behavior, at least as measured by the IAT.

As is clear from the Phelps et al. (2000) and Hart et al. (2000) studies, we can begin to use brain imaging to help inform our understanding of social group evaluation and of race bias in particular. However, it is inappropriate to assume that these techniques yield results that are more informative about the behavior of race bias than studying the behavior itself. Instead, combining the psychological and neural approaches is the best way to advance our understanding of these complex human behaviors more rapidly and with more clarity than could be achieved using either approach in isolation.

4.5.2 General Conclusions About Brain Imaging and Behavior

In this brief review, we have tried to highlight how neural science can contribute to psychological science. Studies examining how the brain processes race information have provided support for psychological theories concerning the dissociation between direct and indirect measures of race group evaluation. As these studies indicate, a good understanding of the potential contributions of brain imaging can help us discover the structure and organization of a behavior. However, a poor understanding can lead us to conclusions that are inappropriate and possibly hurtful. As we develop techniques that allow us to investigate the biological basis of complex behavior, we need to be clear about what it means to say that a behavior is "in the brain."

Showing a behavior "in the brain" does not indicate that it is innate, "hardwired," or unchangeable. Every experience leads to an alteration in the brain. Some of these alterations may be long-lasting and result in learning or memory. For example, the second study by Phelps et al. (2000) showed that the amygdala response to same-race versus other-race faces can be altered by familiarity and learning. Although it is often exciting to demonstrate a neural basis for a given behavior, it should not be surprising to show that *any* behavior has an identifiable neural substrate. To take a simple example, imagine that a week ago you visited a new restaurant and enjoyed it. This experience resulted in a favorable opinion of this restaurant. Is this preference in your brain? Of course it is; there is a neural signature underlying this preference. If the next visit to this restaurant is a disappointment, you may change your opinion, and your neural representation of this restaurant will change accordingly. This rather trivial example makes an important point. We all change, learn, and grow over time. It is easy to recognize the dynamic nature of behavior. Most of us also recognize the interdependence of behavior and the brain. We all have heard of or know someone who has suffered a brain injury resulting in a change in behavior. However, it is somehow more difficult to make the additional connection and recognize the dynamic nature of the brain. Changes in behavior correspond with changes in brain activity and structure. Discovering the representation of a behavior in the brain does not discount the influence of learning in generating, maintaining, or changing this behavior.

Showing a behavior "in the brain" does not say something more important or fundamental about who we are than our behavior. Functional neuroimaging techniques pick up on signals indicating brain activity. These signals, by themselves, do not specify a behavior. Only by linking these brain signals with behavior do they have psychological meaning. For example, recent research has shown different brain activity patterns during reading in individuals with dyslexia (e.g. Habib, 2000). Discovering the brain activity related to this disorder does not change the defining characteristic of dyslexia, which is difficulty reading. We would not label someone "dyslexic" solely on the basis of his or her brain activity pattern; likewise, we should not label someone "racist" because of the pattern of his or her brain response. Assessing brain activity may aid our understanding of a behavior, but the psychological meaning of these brain signals comes from their link to behavior. Discovering the brain activation pattern linked to a behavior does not change the importance of that behavior. It is also unlikely to tell us something about ourselves that we could not conclude from the behavior itself.

Over the past decade, we have become accustomed to seeing pictures of the human brain that are colored like junior high school geography maps, indicating where certain behaviors occur. These colored brain maps are a necessary oversimplification in our initial efforts to describe the representation of behavior in the brain, but they can be misleading. There is not a one-to-one correspondence between a behavior and a brain structure. Most behaviors recruit a network of brain regions, and a given brain region may be important for a number of behaviors. It is a mistake to assume any single brain region "does" a given behavior, just as it is a mistake to assume that activity in a given brain region predicts a single behavior.

Despite these misconceptions, these maps of brain activation patterns are compelling demonstrations of our newfound ability to investigate the biology of human behavior. For neuroscientists, it is an exciting time to try to unravel the complex circuits that tie together the brain and behavior. However, we need to be reasonable in our interpretation of this research and use it to enhance, not subtract from, other means of investigation. These advances in neuroscience should not negate nor substitute for advances in the psychological understanding of behavior.

Notes

1. Editor's note: This reading was excerpted from a 2003 article published in *Political Psychology*, volume 24, pages 747–758, with the permission of the publisher and authors. The omitted section was a review of the neuroimaging literature on the "other race" effect in the perception and memory of faces. The authors' work was supported by the James S. McDonnell Foundation 21st Century Scientist Award.

References

Banaji, M. R. (2001). Implicit attitudes can be measured. In H. L. Roediger III, J. S. Nairne, I. Neath, & A. Surprenant (Eds.), *The nature of remembering:*

Essays in honor of Robert G. Crowder (pp. 117–150). Washington, DC: American Psychological Association.

Bargh, J. A., & Chen, M. (1997). Nonconscious behavioral confirmation processes: The self-fulfilling consequences of automatic stereotype activation. *Journal of Experimental Social Psychology, 33*, 541–560.

Bartels, A., & Zeki, S. (2000). The neural basis of romantic love. *NeuroReport, 11*, 3829–3834.

Bechara, A., Tranel, D., Damasio, H., Adolphs, R., Rockland, C., & Damasio, A. R. (1995). Double dissociation of conditioning and declarative knowledge relative to the amygdala and hippocampus in humans. *Science, 269*, 1115–1118.

Biernat, M., & Crandall, C. S. (1999). Racial attitudes. In J. P. Robinson, P. H. Shaver, & L. S. Wrightsman (Eds.), *Measures of political attitudes* (pp. 291–412). San Diego, CA: Academic Press.

Birgeneau, R. J. (1999, March). Introductory comments. *MIT Faculty Newsletter, 11*(4) (special edition on Study on the Status of Women Faculty in Science at MIT), p. 2.

Davidson, R. J., Putnam, K. M., & Larson, C. L. (2000). Dysfunction in the neural circuitry of emotion regulation: A possible prelude to violence. *Science, 289*, 591–594.

Devine, P. G. (1989). Stereotypes and prejudice: Their automatic and controlled components. *Journal of Personality and Social Psychology, 56*, 680–690.

DuBois, S., Rossion, B., Schiltz, C., Bodart, J. M., Michel, C., Bruyer, R., & Crommelinck, M. (1999). Effect of familiarity on the processing of human faces. *NeuroImage, 9*, 258–289.

Fazio, R. H., Jackson, J. R., Dunton, B. C., & Williams, C. J. (1995). Variability in automatic activation as an unobtrusive measure of racial altitudes: A bona fide pipeline? *Journal of Personality and Social Psychology, 69*, 1013–1027.

Fiske, S. T. (1998). Stereotyping, prejudice, and discrimination. In D. T. Gilbert, S. T. Fiske, & G. Lindzey (Eds.), *The handbook of social psychology* (vol. 2, pp. 357–411). New York: Oxford University Press.

Funayama, E. S., Grillon, C. S., Davis, M., & Phelps, E. A. (2001). A double dissociation of affective modulation of startle eyeblink in humans: Effects of unilateral temporal lobectomy. *Journal of Cognitive Neuroscience, 13*, 721–729.

Greene, J. D., Sommerville, R. B., Nystrom, L. E., Darley, J. M., & Cohen, J. D. (2001). An fMRI investigation of emotional engagement in moral judgment. *Science, 293*, 2105–2108.

Greenwald, A. G., McGhee, J. L., & Schwartz, J. L. (1998). Measuring individual differences in social cognition: The Implicit Association Test. *Journal of Personality and Social Psychology, 74*, 1464–1480.

Grillon, C., Ameli, R., Woods, S. W., Merikangus, K., & Davis, M. (1991). Fear-potentiated startle in humans: Effects of anticipatory anxiety on the acoustic blink reflex. *Psychophysiology, 28*, 588–595.

Habib, M. (2000). The neurological basis of developmental dyslexia: An overview and working hypothesis. *Brain, 12*, 2373–2399.

Hart, A. J., Whalen, P. J., Shin, L. M., McInerney, S. C., Fischer, H., & Rauch, S. L. (2000). Differential response in the human amygdala to racial outgroup vs. ingroup face stimuli. *NeuroReport, 11*, 2351–2355.

LaBar, K. S., LeDoux, J. E., Spencer, D. D., & Phelps, E. A. (1995). Impaired fear conditioning following unilateral temporal lobectomy in humans. *Journal of Neuroscience, 15*, 6846–6855.

Lang, P. J., Bradley, M. M., & Cuthbert, B. N. (1990). Emotion, attention, and the startle reflex. *Psychological Review, 97*, 377–395.

LeDoux, J. E. (1996). *The emotional brain: The mysterious underpinnings of emotional life*. New York: Simon & Schuster.

McConahay, J. P. (1986). Modern racism, ambivalence, and the Modern Racism Scale. In J. F. Dovidio & S. L. Gaertner (Eds.), *Prejudice, discrimination, and racism* (pp. 91–125). Orlando, FL: Academic Press.

Nosek, B. A., Banaji, M. R., & Greenwald, A. G. (2002). Harvesting implicit group attitudes and beliefs from a demonstration web site. *Group Dynamics, 6*, 101–115.

Phelps, E. A., Cannistraci, C. J., & Cunningham, W. A. (2003). Intact performance on an indirect measure of race bias following amygdala damage. *Neuropsychologia, 41*, 203–208.

Phelps, E. A., O'Connor, K. J., Cunningham, W. A., Funayama, E. S., Gatenby, J. C., Gore, J. C., & Banaji, M. R. (2000). Performance on indirect measures of race evaluation predicts amygdala activation. *Journal of Cognitive Neuroscience, 12*, 729–738.

Rilling, J. K., Gutman, D. A., Zeh, T. R., Pagnoni, G., Berns, G. S., & Kilts, C. D. (2002). A neural basis for social cognition. *Neuron, 35*, 395–405.

Schuman, H., Steeh, C., & Bobo, L. (1997). *Racial attitudes in America: Trends and interpretations*. Cambridge, MA: Harvard University Press.

Reading 4.6

Regulating Neuroimaging[1]

Stacey A. Tovino

The question I address here is whether our experiences with phrenology, x-ray, positron emission tomography (PET), and single photon emission computed tomography (SPECT) can assist us in thinking about the appropriateness of other legal protections for individuals whose brains are scanned using functional neuroimaging technology.

4.6.1 A Complete Prohibition on Functional Neuroimaging?

After phrenology's demise, the City of Lincoln, Nebraska, passed an ordinance making it unlawful for an individual to "exercise, carry on, advertise, or engage" in the business of phrenology.[2] Several other jurisdictions passed similar prohibitions against the practice of phrenology, character reading, and mind reading.[3] Perhaps, then, we should consider a complete prohibition on the practice of functional neuroimaging. Given the proven value of functional magnetic resonance imaging (fMRI) in pre-neurosurgical brain mapping, its emerging value in the treatment of depression and dozens of other physical and mental health conditions, and its continuing contributions to neurology, psychiatry, and other areas of medicine and science,[4] this option should receive no further consideration. Phrenology was determined to be a pseudoscience in all its applications, thus warranting a blanket prohibition by local governments. Functional MRI, however, has both proven and potential clinical and scientific applications. It has the potential to benefit many individuals who have been diagnosed with brain tumors, other brain abnormalities, acquired and traumatic brain injuries, mental illness, and many other physical and mental health conditions. At the very least, clinical and research uses of fMRI must be continued.

4.6.2 A Limited Prohibition on Functional Neuroimaging?

In the first year after the discovery of x-ray, remember that a New Jersey assemblyman reportedly introduced a bill to the New Jersey Legislature that would prohibit the use of x-ray glasses in theaters and other public places.[5] This legal response suggests a second option, which would be to prohibit the use of fMRI in nonclinical and nonresearch contexts. For example, we could prohibit the advertising, marketing, or other offering of fMRI scanning services for nonclinical or nonresearch uses. Or, we could prohibit the use of fMRI for certain purposes, such as lie detection; or by just certain persons or organizations, such as employers, educators, health and life insurers, governments, lawyers, and judges.

This option has the benefit of allowing physicians and scientists to continue to use fMRI to benefit current and future patients. To the extent that fMRI is not capable, or not yet capable, of accurately identifying deception and other behaviors, conditions, and characteristics, this option also has the benefit of preventing individuals and third parties from wasting money on, relying on, or using inaccurate functional neuroimaging tests to the detriment of individual citizens.

One possible risk of this option is that it could drive commercial fMRI services underground, perhaps increasing the chance that less-than-honest individuals will provide such services illegally, thus lowering the standard of care in the provision of these services. A second, more important, issue relates to the desirability, or the necessity, of establishing limited prohibitions on functional neuroimaging. Some authors, including myself, have suggested that now may be the time to craft limited prohibitions on the use of functional neuroimaging technology for certain nonclinical and nonresearch uses.[6] Other authors questioned the necessity, and worried about the cost and administrative burden, of additional regulation. Still others suggested that we were lending undue credence to neuroimaging technology by talking about its legal implications and considering potential methods of regulation.

My viewpoint is shaped in large part by fMRI's perceived, rather than its actual, capabilities.[7] Even though fMRI may never be capable of accurately reading an individual's mind, I am concerned that the intense media hype[8] surrounding functional neuroimaging technology may cause employers, insurers, criminal justice officials, governmental agencies, and other third parties to believe that fMRI is capable of doing so.[9] An fMRI that accurately reveals an individual's thoughts is one thing. An fMRI that is incorrectly interpreted to reveal a condition, thought, characteristic, or behavior that does not exist, and that is used to an individual's detriment in an employment, criminal justice, or insurance capacity, is another.[10] Functional MRI, like other sophisticated technologies, possesses an illusory accuracy and objectivity[11] that I think is dangerous in the hands of employers, insurers, jurors, lawyers, judges, and government officials who lack the scientific and statistical training necessary to understand published fMRI studies and interpret fMRI test results.[12] Yet, these are the individuals to whom commercial fMRI services currently are being marketed.[13] For these reasons, I believe that protections against the use of functional neuroimaging technology outside the clinical and research contexts may be desirable.

In light of the varying viewpoints, I hope that those who continue this dialogue will examine the following questions. First, which uses of functional neuroimaging technology (e.g., efforts to detect lies, racial and social evaluation, pedophilia, sexual preferences, mental health conditions, etc.) concern us the most? For example, do we think it is simply too dangerous—ethically, legally, and socially—to use fMRI to attempt to identify deception or racial preferences outside of the research context at this point and time? On the other hand, is it safe and acceptable to allow individuals to purchase brain scans for "fun" purposes, such as dating? Second, which organizations (employers, health and life insurers, government agencies, criminal justice officials, educators, lawyers and judges, individual citizens, etc.) are we most worried about using functional neuroimaging technology or obtaining functional neuroimaging information? For example, is it too dangerous—ethically, legally, and socially—to allow an employer to obtain functional neuroimaging test results about a job applicant? On the other hand, is it acceptable for a judge to use a functional neuroimaging test result to exculpate a criminal defendant? Thinking through these questions may help further the discussion regarding the contexts, if any, in which functional neuroimaging regulation may be needed.

4.6.3 Taxing and Licensure of Functional Neuroimaging Services?

Rather than prohibiting phrenology, some jurisdictions taxed or licensed individuals who offered phrenological services to the public.[14] This legal response suggests a third option, which is to permit but tax, license, or otherwise regulate the commercial offering of fMRI in an attempt to protect the public's health and safety. The benefit to the public of licensing or otherwise regulating the offering of medical and other similar services is textbook health law, although such regulation can be criticized as costly, anticompetitive, and administratively burdensome.[15] In light of the safety issues raised by magnetic resonance imaging (MRI), perhaps licensure, regulation, or even the imposition of minimum insurance coverage limits should be considered. In her article, Jennifer Kulynych examines several safety issues raised by MRI, including the issue whether MRI scanner operators are adequately trained and whether MRI screening procedures are sufficiently detailed and redundant to minimize the risk of physical injury to individuals.[16] The Food and Drug Administration has found that lapses in screening and safety procedures in clinical uses of MRI have caused patient injury and death, and Kulynych suggests that safety procedures may be even less standardized (and the risks of adverse events may be greater) in the research setting.[17] The question here is whether the commercial provision of fMRI services is or will be performed by credentialed persons and subject to the same safety procedures as scanning performed in the clinical setting.[18] If not, requiring trained radiology technicians, minimum safety and screening procedures, and minimum insurance coverage as part of a licensure process or through other regulation may be desirable.

4.6.4 Consumer Law and Truth-in-Advertising

After the fall of phrenology, a national television programming code made programming material relating to phrenology "unacceptable if it encourage[d] people to regard [phrenology] as providing commonly accepted appraisals of life."[19] This legal response suggests a fourth option, which would be to adopt a specific law requiring anyone who offers fMRI services in any context to offer and advertise the services truthfully. A variation of this option is to ensure that current federal and state regulatory agencies are aware of commercial and other uses of fMRI and will enforce truth-in-advertising rules with respect to such uses. The Federal Trade Commission Act,[20] state deceptive trade practices acts,[21] and state consumer laws[22] already require some advertisers to be truthful and nondeceptive and advertisers to have evidence backing their claims. The truth-in-advertising principles that underlie these laws certainly could be applied or extended to apply to fMRI.

One company offering fMRI services to the public states on its website that fMRI is the "first and only direct measurement of truth verification and lie detection in human history."[23] This statement presumably is meant to distinguish polygraph, which measures a response of the peripheral nervous system, from fMRI, which involves the central nervous system. But these statements do raise additional questions. For example, is it fair to state that fMRI is a direct measurement of truth verification given that fMRI uses blood-oxygenation-level dependent (BOLD) signal as a proxy for neuronal activity and usually is referred to as an indirect measure of neuronal activity?[24] Or, is it good enough that BOLD signal has been found to be a "close approximation," or a "faithful signal," of neuronal activity?[25] Or, would these descriptions be considered nonmaterial because they likely would not affect a reasonable consumer's decision to purchase an fMRI test? Or, does the complexity of the science behind fMRI give these companies some legal grace in describing their tests to the public?

One company offering fMRI services to the public states that its fMRI tests are "fully automated" and "[o]bserver independent (objective)."[26] A second company states that its fMRI testing is "Non-subjective—humans do not ask the questions or examine the scans."[27] If scientists and radiology technicians do not ask any test questions or otherwise examine or interpret the fMRI scans, then fMRI testing is more objective than I previously thought. But the concept of objective fMRI testing runs counter to the subjective traits attributed to fMRI in both the popular and scientific literature. In the past 2 years, observers have referred to fMRI as an "interpretive practice," noting that, "Sometimes, the difference between seeing higher activity in the parietal lobe compared to the occipital lobe is akin to deciding whether Van Gogh or Matisse is the more colorful artist"[28] and that, "What constitutes a 'significantly greater' activation is, in a way, in the eye of the beholder."[29] So, is fMRI testing an objective or subjective activity, or is it both? Does it depend on how the fMRI test is designed? To clarify the legal question, is it truthful, fair, nondeceptive, and nonmisleading to state that an fMRI test is objective and fully automated? Or, does the complexity of fMRI again require legal grace?

The accuracy of fMRI testing also is featured prominently in these web materials. According to one representation, "Current accuracy is over 90% and is estimated to be 99% once product development is complete."[30] A second company states that its product is "Accurate—currently 90% accuracy in clinical testing."[31] Although there is no suggestion that these statements are untruthful, deceptive, or not backed by evidence—indeed, both companies cite and link to particular scientific studies supporting their claims[32]—one concern is that these statements will cause nonscientifically trained parties to think that "over 90%" means that fMRI is capable of identifying all instances of deception.

4.6.5 Conclusion

At first glance, phrenology, x-ray, PET, SPECT, and fMRI are an odd collection of both junk and real sciences, dramatically different methods of imaging body structure and mapping brain function. All of these developments were introduced in the name of science but quickly moved into the commercial, employment, government, and judicial contexts. The legal responses to these transitions included, but certainly were not limited to, absolute practice prohibitions; limited practice prohibitions; taxing, licensure, and regulation; and the application of consumer law and truth-in-advertising principles. These legal responses can help us think about appropriate responses to advances in functional neuroimaging.

I certainly do not think that functional neuroimaging should be prohibited in the clinical or research contexts. I do think, however, that there may be a role for nonclinical and nonresearch practice prohibitions that are time-limited, such as prohibitions against using fMRI to detect deception until using fMRI to detect deception has been determined to be highly effective. There also may be a role for the licensure or regulation of the commercial offering of fMRI services (due to safety concerns), and the application of truth-in-advertising principles (due to intense media speculation regarding and public interest in neuroimaging technology). I hope that the desirability and appropriateness of these legal responses continue to be examined as the field of neuroethics develops.

Judicial opinions involving phrenology, x-ray, PET, and SPECT also revealed several themes. These themes include the general duty of the law to keep up with advances in medicine and science, the more specific duty of the law to adopt technologies that will assist the jury in seeking the truth, uneasiness about the illusory objectivity of body imaging and brain mapping (including concern that body images and brain scans can be inaccurate and misleading to jurors, employers, and other nonscientists), and the difficulty of balancing advances in science and medicine against the risks associated with junk science and charlatans. As scientists continue to develop new methods of body imaging and brain mapping, these themes undoubtedly will reappear, and the law will continue to balance individual interests, including interests in confidentiality, privacy, and identity, against society's desire for greater transparency of the body and the brain.

Notes

1. Editor's note: This reading was excerpted from a substantially longer article that appeared in 2007 in the *American Journal of Law and Medicine*, volume 33, pages 193–228 under the title "Imaging body structure and mapping brain function: A historical approach," and is used with permission of the publisher and author. The omitted sections provide a historical context for current discussions of the use and regulation and of fMRI by recounting the development of phrenology and x-rays and society's attempts to control their uses. The author is grateful to Bill Winslade, Cheryl Ellis Vaini, Judy Illes, Ron Carson, Melvyn Schreiber, and Adam Kolber for their comments on earlier versions of the article and Sarah Vallely for her research assistance.

2. Lincoln, Nebraska Municipal Ordinances § 9.40.030 (1997), cited in Argello v. City of Lincoln, 143 F.3d 1152, 1152 (1998). See also Azusa Municipal Code § 8.52.060 ("No person shall practice or profess to practice or engage in the business or art of...phrenology...or any similar business or art, who either solicits or receives a gift or fee or other consideration for such practice, or where admission is charged for such practice."), cited in Spiritual Psychic Science Church v. City of Azusa, 39 Cal. 3d 501, 506 (1985).

3. See Stacey A. Tovino (2007), Functional Neuroimaging Information: A Case for Neuro Exceptionalism? 34 Florida State University Law Review 415, Part I(C).

4. See Tovino, *supra* note 3, at Part II(A) (discussing some of the clinical uses of fMRI).

5. See, e.g., Kevles, *supra* note 150, at 27 & n.14; Goodman, *supra* note 165, at 1043. *But see* Howell, *supra* note 165, at 142 & n.39 ("There is no record of the bill's passage; in fact, there is reason to doubt whether the bill was actually ever introduced.")

6. See, e.g., Tovino, *supra* note 3 at Part VI (arguing that generic privacy protections, including privacy protections applicable to functional neuroimaging information, are needed at least in the employment and insurance contexts); Henry T. Greely & Judy Illes, Neuroscience-Based Lie Detection: The Urgent Need for Regulation, 33 *Am. J. L. & Med.* 377, 413–418 (2007) (arguing that the federal government or state governments should ban any nonresearch use of new methods of lie detection, including fMRI-based lie detection, unless or until the method has been proven safe and effective to the satisfaction of a regulatory agency and has been vetted through the peer-reviewed scientific literature).

7. See Tovino, *supra* note 3, at Part VI(A).

8. See *id.* at Part VI(B) (examining the media hype surrounding fMRI).

9. See *id.* at Part VI(A).

10. See Steve Olson, Brain Scans Raise Privacy Concerns, 307 *Science* 1550, 1550 (2005).

11. Martha J. Farah, Emerging Ethical Issues in Neuroscience, 5 *Nature Rev. Neuroscience* 1127, 1127 (2002).

12. Greely H, *supra* note 253, at 118–20.

13. See No Lie MRI, http://www.noliemri.com/ (accessed March 1, 2007); Cephos Corp., http://www.cephoscorp.com/ (accessed March 1, 2007); Malcolm Ritter, Brain Scans as Lie Detectors: Ready for Court Use?, *Live Sci.*, Jan. 29, 2006, http://www.livescience.com/humanbiology/060129_brain_lie.html (accessed March 1, 2007).

14. FLA. STAT. § 205.41 (1941) ("Every . . . phrenologist . . . shall pay a license tax of one hundred dollars; provided, that this section shall not be construed to require members of any recognized christian denomination who pray for the sick to obtain a license."), cited in Curley v. State, 153 Fla. 773, 776–77 (1943); Henry County, Virginia, Code art. III, ch. 5, § 5–10 (1983), cited in Adams v. Board of Supervisors, 569 F. Supp. 20, 21 (1983); GA. CODE ANN. § 36-1-15 (2006) ("The county governing authority may by proper ordinance . . . regulate, or tax the practice of fortunetelling, phrenology, astrology, clairvoyance, palmistry, or other kindred practices, businesses, or professions where a charge is made or a donation accepted for the services and where the practice is carried on outside the corporate limits of the municipality.").

15. See, e.g., Ralph Reisner, Christopher Slobogin, & Arti Rai, *Law and the Mental Health System: Civil and Criminal Aspects* 74–94 (4th ed., 2004) (examining the state's interest in ensuring the quality of professional services offered to the public); Mark A. Hall, Mary Anne Bobinski & David Orentlicher, *Health Care Law and Ethics* 809–821 (6th ed., 2003) (discussing the public health benefits and the anticompetitive effects of the licensing and regulation of health care providers).

16. Jennifer Kulynych, The Regulation of MRI Neuroimaging Research: Disentangling the Gordian Knot, 33 *Am. J.L. & Med.* 295, 2007, at 311–312.

17. *Id.*

18. *Id.*

19. National Association of Broadcasters, The Television Code §§ IV(12) and IX(10) (19th ed. 1976), cited in Gemini Enterprises, Inc. v. WFMY Television Corp., 470 F. Supp. 559, 562 (1979).

20. *See,* e.g., 15 U.S.C. § 41 et seq. (2006) (Federal Trade Commission Act); FTC Policy Statement on Deception (Oct. 14, 1983), available at http://www.ftc.gov/bcp/policystmt/ad-decept.htm (accessed Oct. 21, 2006); FTC Policy on Unfairness (Dec. 17, 1980), available at http://www.ftc.gov/bcp/policystmt/ad-unfair.htm (accessed Oct. 21, 2006).

21. *See,* e.g., Minnesota Unlawful Trade Practices Act, MINN. STAT. § 325D.09–.16 (2006); Texas Deceptive Trade Practices-Consumer Protection Act, TEX. BUS. & COM. CODE § 17.41–17.63 (2006).

22. See, e.g., Minnesota False Statement in Advertising Act, MINN. STAT. § 325F.67 (2006); Minnesota Prevention of Consumer Fraud Act, MINN. STAT. § 325F.69, subd.1 (2006).

23. See No Lie MRI, http://www.noliemri.com/ (accessed March 1, 2007).

24. National Institutes of Health, National Institute of Mental Health, FMRI Signal Found "Faithful" to Neuronal Activity (2001), available at http://www.nimh.nih.gov/press/fmrisignal.cfm (accessed March 1, 2007).

25. *Id.*

26. No Lie MRI, http://www.noliemri.com/products/Overview.htm (accessed March 1, 2007).

27. Cephos Corporation, http://www.cephoscorp.com/ (accessed March 1, 2007).

28. Sam Jaffee, Fake Method for Research Impartiality, 18 *Scientist* 64 (2004).

29. D.I. Donaldson, Parsing Brain Activity with fMRI and Mixed Designs: What Kind of State is Neuroimaging In?, 27 *Trends in Neurosciences* 442, 442 (2004).

30. No Lie MRI, http://www.noliemri.com/products/Overview.htm (accessed March 1, 2007).

31. Cephos Corporation, http://www.cephoscorp.com/ (accessed March 1, 2007).

32. No Lie MRI, Publications, http://www.noliemri.com/pressNPubs/Publications.htm (accessed Oct. 21, 2006) (listing and linking to published, peer-reviewed fMRI studies); Cephos Corporation, http://www.cephoscorp.com/fmri_deception.htm (accessed Oct. 21, 2006) (same).

5

Neuroscience and Justice

As neuroscience provides increasingly detailed explanations of human behavior, it becomes increasingly relevant to understanding human *mis*-behavior. We know that certain kinds of brain injury can make upstanding, law-abiding citizens dangerous and antisocial, but is all dangerous and antisocial behavior the result of brain dysfunction? And if it is, then how can we hold anyone responsible for their misdeeds when "their brain made them do it?" The answers to these questions depend in part on what one means by "dysfunction" and "responsible," and these are not matters of empirical science. Nevertheless, insofar as neuroscience reveals the causes of bad behavior, it provides an important new set of considerations to be weighed as we judge, decide how and why to punish, and try to prevent antisocial behavior.

Neuroscience has identified brain circuits involved in motivation and self-control. It has also revealed that many criminal offenders have abnormalities of these circuits, with either particularly strong motivations, for example violent urges or drug cravings, or particularly weak self-control, or both. On the face of things this would seem to excuse the criminal behavior in these cases as, given a certain kind of brain, certain behaviors are inevitable. In addition to throwing a monkey wrench into our traditional conceptions of agency and responsibility by "blaming" the brain rather than the person, this perspective also opens up a new range of solutions to the problem of crime, based on correcting the dysfunctional brain.

The readings in this section address the implications of neuroscience for our understanding of moral and legal responsibility, for the assessment of an individual's degree of criminal responsibility, and for the rehabilitation of offenders. We begin with a brief review of the highlights of the neuroscience of responsible behavior, specifically motivation, self-control, empathy, and moral reasoning.

The Neuroscience of Responsible Behavior

Self-control and motivation are the "stop" and "go" systems, respectively, that underlie all human behavior. Motivational systems are primarily subcortical, evolutionarily old parts of us concerned with wants and needs related to survival, including the four F's of survival: feeding, fighting, fleeing, and sexual reproduction. Although each of these drives is controlled by complex systems involving multiple parts of the brain and neurotransmitter systems, all have some relation to the neurotransmitter dopamine, mentioned in chapter 2. In addition to playing a crucial role in attention, dopamine is also involved in the anticipation and enjoyment of rewarding stimuli as diverse as chocolate, a romantic partner, or points in a video game. Indeed, it is this motivational role of dopamine that gives the classic stimulants, used for the treatment of attention deficit–hyperactivity disorder (ADHD) and for attention enhancement, their abuse potential. Recreational drugs all increase dopamine activity, either directly or indirectly, and heavy use leads to long-term changes in the dopamine system that alter motivational systems.

The prefrontal cortex has inhibitory and excitatory connections to many of the subcortical structures involved in motivation (Chow & Cummings, 2006) and is the main source of self-control in the human brain. Its role was first deduced by observing the behavior of prefrontal-damaged patients. Polite, conscientious people can become socially inappropriate and impulsive after prefrontal damage. The nineteenth century railway worker Phineas Gage, transformed from model employee to fighting, cursing ne'er-do-well after a steel rod was blown through his head and destroyed most of his ventromedial prefrontal cortex, is history's best known illustration of this (Damasio, Grabowski, Frank, Galaburda, & Damasio, 1994).

Functional brain imaging has confirmed the localization of self-control in prefrontal cortex, with ventral and medial areas being most directly involved (Aron, Robbins, & Poldrack, 2004; Fellows, 2007). Prefrontal cortex is implicated in normal individual differences in impulse control and the developmental changes normally seen in adolescence (Meyer-Lindenberg et al., 2006; Steinberg, 2008). Chronic use of alcohol and illegal drugs is associated with decreased prefrontal function (Jentsch & Taylor, 1999), and even normal allelic variation in genes related to dopamine impact prefrontal function as measured by behavioral tests and brain imaging (Caldu et al., 2007). Rates of neurologic abnormality in criminal offenders is high relative to the noncriminal population, and imaging studies have consistently shown increased likelihood of prefrontal dysfunction among such offenders (Popma & Raine, 2006). These findings are important for issues of responsibility and blame, as they extend the relevance of prefrontal function beyond exceptional cases of obvious injury and disease into the realm of normal variations in behavior.

Of course, self-control is not the only brain process that enforces morally responsible behavior. A good person is not simply one who manages to inhibit antisocial urges. Among the additional capacities keeping us on the side of the angels are the ability to feel empathy for others and the ability to discriminate right from wrong. Recent neuroscience research has revealed much about the brain bases of these abilities as well.

Empathy has been investigated in imaging studies where the subject witnesses the distress of another person while being scanned. It turns out that Bill Clinton's famous line, "I feel your pain," has a basis in neurophysiology: Some of the same brain systems activated by the first-hand experience of pain are also activated when viewing another person experiencing it (Decety & Jackson, 2004). Again, this neural phenomenon is relevant to variation in empathy among normal people: the more empathic the observer, the more the observed person's pain activates pain regions in the observer (Singer, 2006).

Finally, the ability to discriminate right from wrong courses of action is a form of decision-making with obvious relevance to moral behavior. Most research in this area has focused on understanding the qualitatively different ways we have of making these decisions, rather than distinguishing between quantitatively better or worse moral decision-making. The most fundamental qualitative distinction that has been mapped onto the brain is the distinction between gut-level reactions to the wrongness of a certain course of action and a more rational approach that weighs the benefits and dangers of each course (see Greene & Haidt, 2002, for a review). In general, people engage in both kinds of process, although individuals may be more inclined toward one or the other. Both approaches can result in morally good behavior. More relevant to the understanding of bad behavior is the question of whether some people's brains are less capable of activating either of these systems. Recent research has suggested that individuals higher in psychopathic tendencies (that is, more inclined to act antisocially and without remorse) show less activation of brain regions associated with gut responses to immorality (Glenn, Raine, & Schug, 2009).

If the brain is the origin of, or at least a required stop on the causal pathway to, responsible behavior, then it is also a place to intervene for the sake of encouraging responsible behavior. Of course, our understanding of real-world human behavior is not nearly advanced enough to dictate precise formulae for increasing people's empathy or conscientiousness or decreasing their aggressive tendencies while leaving other aspects of their personalities unchanged. Nevertheless, certain drugs can effect changes in the brain that result in improved behavior in some criminal offenders. Impulsivity can be reduced with stimulant medications, irritability has been reduced with a wide range of medication types, from antidepressants to anticonvulsants, and violent aggression can be curbed with SSRIs (Fava, 1997). Similarly, a range of psychopharmaceutical approaches have been explored for the treatment of sex offenders, with a degree of success (Briken & Kafka, 2007). Forensic psychopharmacology also encompasses the problems of drug-addicted criminal offenders and aims to medically reduce or prevent craving and relapse as a means of reducing recidivism (Bonnie, Chen, & O'Brien, 2008).

Neuroethics of Responsibility and Autonomy

The Readings

The staff working paper of the *President's Council on Bioethics* begins with an overview of the law's view of responsibility and the central role played by *mens rea* ("guilty mind") in determining criminal guilt and punishment. It then explains the criteria that brain images and other neuroscience evidence must meet to be admitted as evidence in a criminal trial, recounts some of the ways that neuroscience has been used, and anticipates some of the ways that it will likely be used in the future.

Greene and Cohen also start with the concept of *mens rea*, but rather than focus on the specifics of current legal practice, they address more general philosophical issues. Specifically, they point out that if we accept the materialist metaphysical foundations of cognitive neuroscience, in particular the assumption that there is nothing more to us than our brains, the result could revolutionize our conception of justice. The concept of free will is deeply ingrained in our intuitions about human nature and justice, yet it conflicts with materialist views of human nature. Although this conflict is fundamentally a metaphysical one, the three general stances toward this conflict that they review—determinism, libertarianism, and compatibilism—have differing implications for the more pragmatic matter of how and why to punish people. *Hyman*'s discussion of the neuroscience of addiction provides a concrete, real-world example that illustrates many of the issues and arguments reviewed by Greene and Cohen. He sketches out our emerging understanding of how drugs "hijack" the motivational systems of the brain and undermine self-control. He concludes with a somewhat ambivalent endorsement of the "disease model" and its deterministic view of behavior. Although this seems to him to be the correct view scientifically and philosophically, he suggests that it may nevertheless be in society's interest to hold addicts responsible for their actions.

Morse attempts to sort out the ways in which neuroscience information is, and is not, relevant to the criminal justice system in his "diagnostic note" on brain overclaim syndrome and humorously exposes some of the less than clear thinking that neuroscience has inspired in some legal commentators. He draws a distinction between the kind of sweeping reinterpretation of morality and law envisaged by Greene and Hyman and the more modest role of neuroscience as an ancillary source of evidence in criminal proceedings discussed in the staff working paper of the President's Council on Bioethics. In the latter vein, he takes us through the arguments and evidence considered by the Supreme Court in deciding whether adolescents have sufficient responsibility for their actions to be eligible for the death penalty, including the much publicized brain imaging evidence on prefrontal maturation. Finally, *Boire* points to an important consequence of our developing ability to modify antisocial and illegal behavior through brain interventions, namely the possibility that such interventions could be legally required and addresses some of the ethical issues that would be raised by such a requirement. Although some of these issues are specific to illegal drug use, most could be raised about any illegal behavior that can be diminished by psychopharmacology or other brain treatment methods.

Selected Cross-cutting Issues

Of the five sections of this book, the neuroethical issues of this section are probably the most challenging conceptually. There are at least three reasons for this, and pausing here to reflect on them may help to make the material that follows clearer. First, these issues concern metaphysical distinctions between free will, determinism, and compatibilism, concepts that have been addressed by some of the greatest philosophers of all time without a fully satisfactory resolution. Even if a contracausal notion of free will (that is, free will that operates outside the universe of physical causes) is pretty much off the table these days, the alternatives don't seem right to most people either. If we endorse determinism, then we are failing to account for our own compelling first-person sense of free will, as well as abandoning the socially vital and commonsense distinction between intentional actions and mere "happenings." If we embrace compatibilism, then we ought to be able to explain the difference between physical happenings that do and do not correspond with intentional acts, and as far as I can tell this essential groundwork has yet to be accomplished. In sum, if you read Greene and Cohen's wonderfully clear explication of free will, determinism, and compatibilism and end up feeling confused, it's not you and it's not Greene and Cohen; it's the nature of the problem.

A second contributing factor to the difficulty of the neuroethical issues discussed in these readings is the prominent role played by specific legal principles and precedents. For those of us not schooled in law, this constitutes a body of unfamiliar technical knowledge that must be learned at the same time we are trying to learn about the larger points being made with this material. Fortunately, the authors of the President's Council on Bioethics staff working paper as well as Morse and Boire all explain their legal terms, concepts, and cases very clearly, so one need only fight the temptation to skim the legalese sections and all will be explained!

Alas, there is a third factor that takes the confusions created by the first two and makes everything extra slippery. That third factor is the similarity in the language used to discuss two quite different issues, both of which can be expressed as "implications of neuroscience for responsibility." One issue has to do with the very coherence of the idea of responsibility in a material world. This is the metaphysical issue just discussed. However, this has minimal relevance to the real-world practice of law. No defense lawyer is going to face a judge and jury and say "given the system that was set in motion at the time of the Big Bang, my client could not have done otherwise, so you should find him—and every other defendant—not guilty." Greene and Cohen take the metaphysical issue as far as possible in the direction of real-world legal matters when they observe that our legal system must accord with people's intuitions about responsibility. As the public learns more about advances in the neuroscience of decision-making, antisocial behavior, and so forth, those intuitions may shift to favor a less retributive system of punishment.

The second way in which the words "implications of neuroscience for responsibility" can be interpreted is in terms of the specific ways in which the theories and methods of neuroscience can assist the legal system in determining who meets the *conventional, psychological* criteria for responsibility. As explained in the President's Council on Bioethics staff working paper and in Morse's paper, there are a number of ways that neuroscience could in principle shed light on the mental states and abilities that the law deems relevant to assessing responsibility. We can expect to see neuroscience increasingly used in this way in decisions on guilt and sentencing. Morse specifically addresses the confusion between the first and second interpretations of "implications of neuroscience for responsibility" when he talks of the confusion between external and internal critiques, although the bulk of his reading concerns the second.

Most of the readings in this section touch on one or more conditions that are associated with diminished self-control. Given the centrality of self-control to the assessment of criminal responsibility, such conditions are potentially exculpating. The neuroscience of these conditions is therefore potentially relevant to determining how much control someone had over their actions when they committed a crime. One such condition is immaturity. The role of neuroscience evidence in determining the capacity of adolescents for self-control is discussed at length by Morse in connection with *Roper v. Simmons*. Another is addiction. Two of the readings here highlight the diversity of ways in which the medicalized view of addiction as a brain disease may impact the law's treatment of addicts. Hyman's explanation of addiction as a learning process, in which the brain is rewired to treat drugs as more vital than food, shelter, or kin, indicates a rationale for viewing addiction as a state of diminished control. Although the consequences of the "neuroscience view" of addiction for determinations of guilt might appear favorable to an addict standing trial, the consequences for sentencing or legally mandated treatment of that addict may seem less appealing. With diminished responsibility goes diminished autonomy. If neuroscience succeeds in persuading society that an addict is not blameworthy but is merely operating with a malfunctioning brain, then it will have also persuaded society that its job is not to punish but to fix. We will feel justified in demanding that the addict submit to brain treatment. Boire raises a number of legal and ethical concerns regarding forcible pharmacotherapy for drug addiction.

Questions for Discussion

1. Imagine a legal system in which all behavior, criminal or otherwise, is exclusively viewed as the causal product of brain processes, devoid of any notion of personal agency or responsibility. What would the roles be of judge, jury, prosecuting and defense attorneys? Script some examples of how these figures would interact in the course of a trial. Note some statements or questions that would not be heard in such a trial.

2. There are already drugs that influence mental states that can play a role in antisocial behavior, for example self-control, aggression, interpersonal trust, and sex drive, as well as drug enjoyment and drug craving, and we can expect more in the future. In chapter 2 and associated readings we reviewed the potential risks and benefits of voluntary brain enhancement. What new risks and benefits come into play with forcible enhancement? Which ones are common to voluntary and forcible enhancement? Draft an "Offender's Bill of Brain Autonomy Rights." Would the offender's crime and likelihood of further crime influence how many of these rights she or he has?
3. Although Hyman points out that changes in the brains of addicts render them less able to chose abstention, he also states that "For many reasons, it may be wise for societies to err on the side of holding addicted individuals responsible for their behavior and to act as if they are capable of exerting more control than perhaps they can." What do you think these reasons are? Do you agree, and why or why not?

References

Aron, A. R, Robbins, T. W., & Poldrack, R. A. (2004). Inhibition and the right inferior frontal cortex. *Trends in Cogniive Sciences, 8*, 170–177.

Bonnie, R. J., Chen, D. T., & O'Brien, C. P. (2008). The Impact of Modern Neuroscience on Treatment of Parolees: Ethical Considerations in Using Pharmacology to Prevent Addiction Relapse. Retrieved Aug. 15, 2009 from http://www.dana.org/news/cerebrum/detail.aspx?id=13932.

Briken, P., & Kafka, M. P. (2007). Pharmacological treatments for paraphilic patients and sexual offenders. *Current Opinion in Psychiatry*, *20*, 609–613.

Caldú, X., Vendrell, P., Bartrés-Faz, D., Clemente, I., Bargallo, N., Jurado, M. A., et al. (2007). Impact of the COMT Val$^{108/158}$ Met and DAT genotypes on prefrontal function in healthy subjects. *NeuroImage*, *37*, 1437–1444.

Chow, T. W., & Cummings, J. L. (2006). Frontal-subcortical circuits. In B.L. Miller & J.L. Cummings (Eds.), *The human frontal lobes: Functions and disorders* (pp. 25–43). New York: Guilford Press.

Damasio, H., Grabowski, T., Frank, R., Galaburda, A. M., & Damasio, A. R. (1994). The return of Phineas Gage: Clues about the brain from the skull of a famous patient. *Science*, *264*, 1102–1105.

Decety, J., & Jackson, P. L. (2004). The functional architecture of human empathy. *Behavioral and Cognitive Neuroscience Reviews*, *3*, 71–100.

Fava, M. (1997). Drug treatment of pathologic aggression. *Psychiatric Clinics of North America*, *20*, 427–451.

Fellows, L. K. (2007). Advances in understanding ventromedial prefrontal function: The accountant joins the executive. *Neurology*, *68*, 991–995.

Glenn, A. L., Raine, A., & Schug, R. A. (2009). The neural correlates of moral decision-making in psychopathy. *Molecular Psychiatry, 14,* 5–6.

Greene, J., & Haidt, J. (2002). How (and where) does moral judgment work? *Trends in Cognitive Sciences,* 6(12), 517–523.

Jentsch, J. D., & Taylor, J. R. (1999). Impulsivity resulting from frontostriatal dysfunction in drug abuse: Implications for the control of behavior by reward-related stimuli. *Psychopharmacology, 146,* 373.

Meyer-Lindenberg, A., Buckholtz, J. W., Kolachana, B., Hariri, A.R., Pezawas, L., Blasi, G., et al. (2006). Neural mechanisms of genetic risk for impulsivity and violence in humans. *Proceedings of the National Academy of Sciences of the United States of America, 103,* 6269–6274.

Popma, A., & Raine, A. (2006). Will future forensic assessment be neurobiologic? *Child and Adolescent Psychiatric Clinics of North America, 15,* 429–444.

Singer, T. (2006). The neuronal basis and ontogeny of empathy and mind reading: Review of literature and implications for future research. *Neuroscience & Biobehavioral Reviews, 30,* 855–886.

Steinberg, L. (2008). A social neuroscience perspective on adolescent risk-taking. *Developmental Review, 28,* 78–106.

Reading 5.1

An Overview of the Impact of Neuroscience Evidence in Criminal Law[1]

The President's Council on Bioethics Staff

In approaching human behavior, science and ethics/law have different objectives and interests: science seeks to understand it, ethics seeks to judge it wisely. Law, the embodiment and teacher of many of the community's shared moral practices and norms, seeks to protect the community against dangerous or unacceptable behavior by judging misconduct and punishing offenders. Although understanding and judging are different activities, efforts to understand criminal behavior and its causes continue to exert an influence on how society deals with criminals, not only in considering guilt and innocence, but, for example, in sentencing, decisions about parole, and proposals for mandatory treatment, as well as in communal efforts to prevent people from becoming criminals in the first place. In previous generations, people looked to inheritance (genetics), anatomic features (phrenology), a history of emotional trauma or unresolved psychic conflicts (psychoanalysis), or socioeconomic deprivation (sociology and economics) to explain why some people commit crimes and others do not. Today and tomorrow, it seems, people will look increasingly to the brain (neuroscience). It is none too soon to begin to think about how neuroscience will and should affect our legal judgments and practices.

5.1.1 The Role of Moral Responsibility in Criminal Law

The two chief activities of the criminal law—determination of guilt and imposition of punishment—turn largely on questions of culpability and blameworthiness. Within the context of determining guilt, the doctrine of *mens rea* is primarily driven by questions of personal responsibility and intention; a failure to prove beyond a reasonable doubt that the defendant had the requisite culpable mental state can result in either acquittal or conviction on a lesser charge. Similarly, the affirmative defense of excuse directly implicates the question of moral culpability. Nowhere is this more dramatically demonstrated than in the context of the excuse defense of insanity, where the relevant inquiry often focuses on whether the defendant could distinguish right from wrong, or conform his behavior to an appropriate standard. The processes in place for determining criminal punishment (that is, sentencing) are likewise aimed largely at the question of moral responsibility: "The criminal law will generally only impose its retributive or deterrent sanctions upon those who are morally blameworthy—those who know they are doing wrong but nonetheless persist in wrongdoing."[2] There are procedures in place allowing litigants to introduce evidence of mitigation or aggravation, in their effort to influence the sentencing court's decision. Expert psychiatric testimony is frequently introduced intro criminal proceedings, both for determining guilt and for assigning punishment.

5.1.2 Determination of Guilt and the Doctrine of *Mens Rea*

Coke's maxim, "Actus Non Facit Reus Nisi Mens Sit Rea,"[3,4] is a seminal principle in American criminal law. With very few exceptions, criminal guilt cannot be established merely by demonstrating that a particular act was committed; it must also be shown that the defendant acted with the requisite intention for each element of the offense in question. The Model Penal Code (which has been adopted in one form or another by a majority of jurisdictions) sets forth four types of culpable mental states for purposes of *mens rea* analysis (listed from most culpable to least culpable): Purposefulness (acting with the conscious purpose to engage in specific conduct or to cause a specific result)[5]; Knowledge (awareness that one's conduct is of a particular nature, or the practical certainty that one's conduct will cause a specific result")[6]; Recklessness (conscious disregard for a substantial and unjustifiable risk")[7,8]; and Negligence (the creation of a substantial or known risk of which one ought to have been aware).[9,10] The failure to prove the requisite degree of culpability for any element of the charged crime will result in either a finding of not guilty or conviction on a less serious offense.

For example, the Model Penal Code establishes three types (degrees) of criminal homicide: murder, manslaughter, and negligent homicide.[11] The various types of criminal homicide share the same *actus reus*, namely, "caus[ing] the death of another human being," but are distinguishable based on whether the defendant did so purposely or knowingly (murder), recklessly (manslaughter), or negligently (negligent homicide). Thus, the moral gravity of the offense (and the severity of punishment) is deemed greater to the extent that it can be shown that the defendant's decision was based on cool calculation rather than a momentary rash impulse. Thus, if a person charged with murder can show that he was "acting under the influence of extreme mental or emotional disturbance for which there is a reasonable explanation or excuse,"[12,13] he will be found not guilty, or will be charged instead with manslaughter—a significantly less serious offense that carries a lesser punishment.

In this way, the doctrine of *mens rea* not only distinguishes among various degrees of moral culpability, but also makes room for the notion that under certain circumstances of provocation or mental abnormality, moral responsibility for one's actions is diminished, rendering one deserving of a more lenient punishment.

5.1.3 Determination of Guilt and the Defense of Excuse

The doctrine of excuse provides another clear illustration of the relationship between criminal law and moral responsibility. The law "excuses" criminal conduct if there are circumstances that negate the moral blameworthiness of the actor. In *Holloway v. U.S.* (1945), the United States Court of Appeals for the D.C. Circuit formulated the doctrine thusly: "Our collective conscience does not allow punishment where it cannot impose blame."[14] Renowned criminal law professor Sanford Kadish elaborated:

> To blame a person is to express moral criticism, and if the person's action does not deserve criticism, blaming him is a kind of falsehood, and is, to the extent the person is injured by being blamed, unjust to him. This lies behind the law's excuses.[15]

There are a variety of excuses that can serve as affirmative defenses. Two are particularly relevant to the current inquiry: insanity and infancy.

The species of excuse defense that perhaps most clearly illustrates the nexus between criminal law and moral responsibility is the insanity defense. The insanity defense rests on the premise that it is unjust to hold an irrational individual morally (and by extension, criminally) responsible. Again, the D.C. Circuit gave voice to this intuition in the *Holloway* case: "To punish a man who lacks the power to reason is as undignified and unworthy as punishing an inanimate object or animal. A man who cannot reason cannot be subject to blame."[16] There are a variety of competing approaches to the insanity defense. One test for insanity, the "M'Naghten rule," requires proof that "at the time of the committing of the act, the party accused was laboring under such a defect of reason, from disease of mind, as not to know the nature and quality of the act he was doing; or if he did know it, he did not know he was doing what was wrong."[17] An alternative approach is the "irresistible impulse" test, which asks whether mental disease impaired the defendant's ability to control his actions. The "Durham rule" holds that "an accused is not criminally responsible if his unlawful act was the product of mental disease or defect."[18] The American Law Institute (ALI) developed a test for insanity as part of its Model Penal Code, which holds that "a person is not responsible for criminal conduct if at the time of such conduct as a result of mental disease or defect he lacks substantial capacity to either appreciate the criminality of his conduct or to conform his conduct to the requirements of the law." Among these approaches, the M'Naghten and ALI rules are the overwhelming preferences for state legislatures. Almost all states have adopted one or the other on a nearly equal basis. Very few jurisdictions have adopted either the Durham rule or the irresistible impulse test. The federal test for insanity, established by the Insanity Defense Reform Act of 1984 (enacted in response to John Hinckley, Jr.'s acquittal by reason of insanity), requires the defendant to demonstrate by clear and convincing evidence that "at the time of commission of the acts constituting the offense, the defendant, as a result of severe mental disease or defect, was unable to appreciate the nature and quality or the wrongfulness of his acts."[19] The statute further specifies that "mental disease or defect does not otherwise constitute a defense."[20]

Despite the differences in the competing approaches to the excuse defense of insanity, they all originate from the intuition that an individual's moral blameworthiness (and thus criminal responsibility) is mitigated when his reason is impaired by mental defect or abnormality.

Another doctrine of excuse relevant to the current inquiry is the defense of infancy. This affirmative defense excuses criminal conduct of children below a certain age on the grounds that they are incapable of forming the requisite intent and are not susceptible to deterrence. The relevant age is prescribed by statute.

5.1.4 Imposition of Punishment

In addition to determining guilt, the process of imposing punishment is also animated by principles of moral responsibility. Criminal punishment, as stated above, aims at meting out justice, deterring future misconduct, and expressing society's condemnation for the transgression of its norms. The process of criminal sentencing finds both defendants and prosecutors making moral appeals to the sentencing authority, seeking lenience or arguing for severity. Defendants will regularly argue that they should be shown mercy on the grounds that they, while guilty as a matter of law, are the victims of harsh circumstances or misfortune—abuse, poverty, and the like. Defendants also often argue for leniency on the grounds that they do not pose a grave ongoing threat to society. Conversely, prosecutors commonly argue that severe punishments are in order based on the heinous nature of the original offense, or the defendant's propensity for future dangerousness. In cases involving the death penalty, this process is somewhat more formalized with a (statutorily defined) weighing of aggravating and mitigating factors. In such circumstances, as a matter of constitutional right (under the Eighth and Fourteenth Amendments), "the sentencer... [shall] not be precluded from considering, as a mitigating factor, any aspect of a defendant's character or record and any of the circumstances of the offense that the defendant proffers as a basis for a sentence less than death."[21]

5.1.5 Preliminary Procedural Issues

Before moving to the discussion of how neuroimaging evidence has been or may in the future be used within the aforementioned framework for the determination of guilt and imposition of punishment, it is useful to address very briefly the procedural threshold questions of admissibility of and constitutional entitlement to such evidence.

5.1.5.1 Scientific Evidence

There are essentially two standards for the admission of scientific evidence, both named for the cases in which they were enunciated: *Daubert* and *Frye*. Courts following the *Frye* approach admit only scientific testimony regarding theories and methodologies that have gained "general acceptance" among members of the relevant scientific field.[22] The *Daubert* rule rejects this singular reliance on "general acceptance" and instead vests the trial judge with the responsibility to assess whether the proffered expert testimony is both reliable and scientifically valid.[23] In evaluating "validity," courts are directed to consider a range of factors including general acceptance, peer review, error rates, and amenability to falsification.[24] The *Daubert* standard applies in federal court and a large array of state courts. The *Frye* standard controls in approximately 20 state jurisdictions.

If a court regards proffered neuroimaging evidence as too novel, it will sometimes deny its admission into evidence under the *Frye* standard, on the grounds that it has yet to achieve "general acceptability." Such was the case in *People v. Protsman*,[25] in which the defendant sought to admit positron emission tomography (PET) scan evidence and psychiatric testimony to the effect that he was suffering from decreased frontal lobe activity (due to traumatic brain injury) such that he could not formulate the requisite intent for first-degree murder. Because the evidence had not yet achieved general acceptance, it was not admissible. Similarly, in *People v. Chul Yum*,[26] the defendant (who had been convicted of second-degree murder) argued on appeal that the trial court erroneously refused to admit his proffered evidence of a single photon emission computed tomography (SPECT) brain scan that he claimed showed diminished activity in his left temporal lobe and damage caused by brain trauma, causing him to kill his mother and sister. Because of the novelty of the diagnostic approach (i.e., using a SPECT scan to diagnose brain trauma and posttraumatic stress disorder), the defendant failed to satisfy the court that the scientific evidence was "generally accepted."

There are, however, many noteworthy examples of cases in which neuroimaging evidence has met the requisite standards for scientific testimony and has been admitted (both from jurisdictions that follow *Daubert* rather than *Frye*, and in *Frye* jurisdictions where the judge was convinced that the proffered evidence met the "general acceptance" standard). These cases are set forth throughout the discussion below.

5.1.5.2 Right to Present Evidence

The United States Supreme Court has ruled that "the Constitution requires that a State provide access to a psychiatrist's assistance" when the question of a criminal defendant's sanity is being litigated.[27] At least one court has held that this right extends to the provision of neuroimaging tests; in *People v. Jones*, the appeals court reversed the defendant's murder conviction on the grounds that he was denied neurologic testing that supported his defense that he was suffering from brain damage that impaired his "ability to think quickly and flexibly" and "ability to perceive risk."[28] Additionally, it was reported that in 1998, a trial judge allocated money for a computed tomography (CT) scan and magnetic resonance imaging (MRI) for the defendant Jeremy Strohmeyer, on trial for the murder of a 7-year-old child in Las Vegas, Nevada.[29]

5.1.6 Neuroimaging and Criminal Law

There are a few noteworthy instances in which neuroimaging evidence has been introduced at the guilt phase of the criminal process to support claims of lack of requisite culpable mental state or excuse defenses based on insanity. These are detailed below, by category.

5.1.6.1 Determination of Guilt: Negation of Mens Rea

There are several cases in which defendants sought to admit neuroimaging evidence that they were incapable of formulating the requisite culpable mental state and were thus entitled to acquittal or conviction on a less serious charge. Some defendants have succeeded in getting this evidence before the jury (a noteworthy achievement in itself), but in the main, this approach has not been overwhelmingly successful. That said, it is difficult to develop a metric for success in this context, as many of these cases ended in plea bargains rather than convictions or acquittals. In any event, there does seem to be a significant (and growing) volume of cases in which neuroimaging evidence is adduced to negate *mens rea*.

In *United States v. Erskine*,[30] the defendant (who had been convicted of making false statements to an official of a federally insured bank) argued on appeal that the court erroneously prevented him from introducing testimony and a brain scan that he claimed showed that he lacked the mental capacity to formulate the specific intent to "influence a bank" (a statutory element of the crime with which he was charged). The U.S. Court of Appeals for the Ninth Circuit agreed that he was entitled to introduce such evidence on the issue of specific intent (though it expressed no opinion as to the probative nature of such evidence). Thus, the defendant's conviction was reversed.[31] Similarly, in 1995, former United Way executive William Aramony (charged with numerous counts of embezzlement from the charity fund), introduced neuroimaging evidence in support of his claim that he was suffering from "brain atrophy" and thus unable to satisfy the requisite intent requirement to commit embezzlement. Shortly after this evidence was introduced, he secured a favorable plea bargain.

Other defendants have not been as successful in demonstrating a lack of *mens rea* by appeal to neuroimaging evidence. In *State v. Anderson*,[32] the defendant presented to the jury expert testimony (supported by neuroscience evidence) that brain-damage–induced depression and paranoia precluded him from being able to premeditate and deliberate in a manner sufficient to justify the charge of first-degree murder. The jury was not persuaded and found him guilty on all counts. In *U.S. v. Mezvinsky*,[33] the court held that the defendant (who had been indicted on 66 counts of fraud and related offenses) was not entitled to introduce PET scan evidence in support of his claim that he was incapable of deception (the requisite *mens rea* for his charges).

5.1.6.2 Determination of Guilt: Excuse Defense of Insanity

There are a few very-high-profile examples of the introduction of neuroimaging evidence as an adjunct to a claim of not guilty by reason of insanity. For example, in *U.S. v. Hinckley*,[34] the defendant (who attempted to assassinate President Reagan) presented CT scan evidence showing "atrophy" of the brain. The neuroradiologist for the defense testified that the degree of atrophy was abnormal and possibly indicated the presence of organic brain disease. Another witness for the defense (a psychiatrist) testified that the evidence of atrophy increased the statistical likelihood that the defendant was suffering from schizophrenia. The court ultimately admitted this evidence in order to give the jury "all possibly relevant evidence bearing on cognition, volition, and capacity" in considering the defendant's insanity defense. The defendant was found not guilty by reason of insanity.

In another high-profile case, *People v. Weinstein*,[35] the defendant (accused of strangling and defenestrating his wife) successfully introduced PET scan images, which he asserted showed reduced brain function in and around an arachnoid cyst in his frontal lobe. The evidence was presented in support of the defense's theory that Weinstein was not responsible for his actions due to mental disease or defect. The prosecution vigorously opposed the admission of this evidence and moved to exclude it. Shortly after the judge ruled it to be admissible, the prosecution quickly agreed to negotiate a plea bargain for a reduced charge of manslaughter. It seems reasonable to infer that the prosecution was concerned that the images would be persuasive to jurors at trial.

One might speculate that as the ability to perform neurologic testing for the biological correlates of schizophrenia and related disorders grows, so too will the incidence of defendants moving for the admission of such evidence in support of their claims of insanity. That said, the insanity defense is rarely invoked, and is even more rarely successful.

5.1.6.3 Imposition of Punishment

Whereas there are few reported cases in which defendants have secured acquittals on the strength of neuroimaging evidence, defendants have enjoyed some measure of success in the context of sentencing. For example, such evidence has been introduced as an adjunct to support a plea for leniency or claim of mitigating circumstances.

In early 2004, MRI and PET scan evidence helped to defeat two separate death sentences for Simon Pirela. In April 1983, Mr. Pirela received a death sentence after being convicted of murder (*Commonwealth v. Pirela*); and in May 1983 Pirela received another death sentence, in a separate murder trial (*Commonwealth v. Morales*).[36] When the *Morales* sentence was vacated (due to reversible error for prosecutorial misconduct) and resentencing ordered, attorneys for Pirela introduced MRI and PET scans as evidence in support of mitigating factors of diminished capacity, brain damage, and mental impairment generally. The jury recommended unanimously that Pirela be resentenced to life in prison rather than executed.[37] On appeal for Pirela's second death sentence, attorneys for the defendant used the same PET and MRI scans to support the defense claim that Pirela was mentally retarded, thus requiring the court to vacate the death sentence and impose life imprisonment.[38,39] The judge in *Pirela* noted that the expert testimony on the brain scans, combined with neuropsychologists' testimony, "was quite convincing."

Similarly, in *McNamara v. Borg*,[40] PET scan evidence was introduced in support of the defendant's mitigation claim that he was suffering from schizophrenia. The defendant was sentenced to life imprisonment rather than execution. According to postsentencing interviews, jurors acknowledged that they were significantly influenced by the neuroimaging evidence in their decision to spare the defendant's life.

There is at least one case in which the failure to allow neuroimaging evidence at the sentencing phase of trial was held to be reversible error. In *Hoskins v. State*,[41] the Florida Supreme Court vacated the defendant's death sentence and remanded the case for a new penalty proceeding so that the defendant would have an opportunity to present a PET scan showing a brain abnormality.

In contrast with the foregoing, there are numerous examples of cases in which the defense's mitigation arguments supported by neuroimaging evidence were not persuasive to the jury. For example, in *People v. Kraft*,[42] the defendant (convicted of 16 counts of murder and assorted other crimes) introduced PET scan images during his mitigation case, which experts testified were consistent with obsessive-compulsive disorder. The jury was unmoved by this evidence and sentenced Kraft to death. Similarly, in *People v. Holt*,[43] the defendant (convicted of murder, robbery, rape, and other crimes) introduced PET scan images and an electroencephalogram (EEG) showing abnormalities in both temporal lobes and damage to the cingulate gyrus region of the brain, which experts testified was consistent with aberrant sexual behavior. The jury was not persuaded by this mitigation evidence and sentenced the defendant to death.

It bears noting that prosecutors might someday also seek to introduce neuroimaging evidence in the sentencing phase of the criminal process. Such evidence might prove useful in demonstrating aggravating factors such as future dangerousness, drug or alcohol addiction, and the like in an effort to secure a more severe punishment or even the denial or revocation of parole. To this point, there are no known cases in which neuroimaging data have been used in determining fitness for parole.

5.1.7 Conclusion

The determination of moral responsibility is an integral function of the criminal law, both at the guilt and punishment phases. Science-based testimony (usually psychiatric) has already established a legitimate place in both phases. Neuroimaging technology, still in its infancy, has already had a modest impact on this process. At the guilt phase, neuroimaging evidence has been marshaled in support of claims of insufficient *mens rea* and as an adjunct to the insanity defense. At the punishment phase, neuroimaging evidence has been introduced to support claims of mitigation, with a noteworthy measure of success.

Whereas the success rate of arguments supported by neuroimaging evidence in the criminal context is somewhat mixed, the large number of cases in which such evidence is presented is striking. Many courts have admitted this evidence, and more will follow as the science underlying neuroimaging becomes more "generally accepted" (for purposes of the *Frye* test). The incidence of neuroimaging evidence is also likely to increase as more courts follow the lead of *People v. Jones* and come to regard access to neuroimaging as part and parcel of the psychiatric testimony to which the defendant is constitutionally entitled. Also, as neuroimaging evidence becomes more widespread and word of related successes becomes well known, criminal defense attorneys may come to regard it as standard practice. Indeed, there is a handful of cases that have held that the failure of defense attorneys to investigate and present evidence of their client's organic brain damage constituted *ineffective assistance of counsel*—a notoriously high threshold to satisfy.[44]

Looking forward, it seems that criminal sentencing, where the defense counsel is afforded wide latitude to introduce any and all information that might persuade the jury to act leniently, presents the most fertile opportunity for argument supported by neuroimaging. Particularly if the following anecdote, as reported by the *Los Angeles Times*, proves to be true:

Jurors can be dazzled by the display. Christopher Plourd, a San Diego criminal defense attorney, remembers well the first time he use PET scans in the early 1990s during a murder trial. "Here was this nice color image we could enlarge, that the medical expert could point to," Plourd said. "It documented that this guy had a rotten spot in his brain. The jury glommed onto that."[45]

Notes

1. Editor's note: This reading is an abridged version of a working paper prepared by the staff of the President's Council on Bioethics as an aid to the Council's discussions and appears online at http://bioethicsprint.bioethics.gov/background/neuroscience_evidence.html with a note stating that it does not represent the official views of the Council or of the United States Government.

2. *In re Devon T*, 584 A.2d 1287 (Md. App. 1991).

3. Edward Coke, The Third Part of the Institutes of the Law of England 107 (1644).

4. Literally, "the act is not criminal unless the mind is criminal."

5. Model Penal Code 2.02(2)(a).

6. *Id.* at (b).

7. *Id.* at (c).

8. "The risk must be of such a nature and degree that, considering the nature and purpose of the actor's conduct and the circumstances known to him, its disregard involves a gross deviation from the standard of conduct that a law-abiding person would observe in the actor's situation." MPC 2.02(2)(c).

9. *Id.* at (d).

10. "The risk must be of such a nature and degree that the actor's failure to perceive it, considering the nature and purpose of his conduct and the circumstances known to him, involves a gross deviation from the standard of care that a reasonable person would observe in the actor's situation." MPC 2.02(2)(d).

11. *Id.* at 210.1–210.4.

12. *Id.* at 210.3(1)(b).

13. "The reasonableness of the such explanation or excuse shall be determined from the viewpoint of a person in the actor's situation under the circumstances as he believes them to be." MPC 210(1)(b).

14. 148 F.2d 665, 65–67 (D.C. Cir. 1945).

15. Sanford Kadish, "Excusing Crime," 75 California Law Review 257, 263–65 (1987).

16. 148 F.2d 665, 65–67 (D.C. Cir. 1945).
17. *M'Naghten's Case*, 8 Eng. Rep. 718, 722 (1843).
18. *Durham v. United States*, 214 F.2d 862 (D.C. Cir. 1954).
19. 18 U.S.C. 17 (1992).
20. *Id.*
21. *Lockett v. Ohio*, 438 U.S. 586, 604–05 (1978).
22. *Frye v. United States*, 293 F. 1013, 1014 (1923).
23. *Daubert v. Merrell Dow Pharms., Inc.*, 509 U.S. 579, 592–94 (1993).
24. *See id.* at 593; *see also* Jennifer Kulynych, "Psychiatric Neuroimaging Evidence: A High-Tech Crystal Ball?" 49 Stanford Law Review 1249, 1263 (1997).
25. 88 Cal. App. 4th 509 (2001).
26. 111 Cal. App. 4th 635 (2003).
27. *Ake v. Oklahoma*, 470 U.S. 68, 74 (1985).
28. 620 NYS2d 656 (1994).
29. See Las Vegas Review-Journal 8/30/98; National Law Journal 9/21/98.
30. 588 F.2d 721 (9th Cir. 1978).
31. There is no published record of the disposition of this case on remand.
32. 79 S.W.3d 420 (Mo. 2002).
33. 206 F. Supp 2d 661 (D. Pa. 2002).
34. 525 F. Supp. 1324 (D.D.C. 1981).
35. 591 N.Y.S.2d 715 (1992).
36. "Morales" was an alias used by Simon Pirela.
37. 549 Pa. 400, 701 A.2d 516 (1997).
38. Jan. Term, 1983, No. 2143 (Phila C.P. Apr. 30, 2004).
39. In *Atkins v. Virginia*, the U.S. Supreme Court held that execution of the mentally retarded violates the 8th Amendment injunction against cruel and unusual punishment. 536 U.S. 304 (2002).
40. 923 F.2d 862 (9th Cir. 1991).
41. 735 So.2d 1281 (Fla. 1999).
42. 23 Cal. 4th 978 (2000).
43. 15 Cal. 4th 619 (1997).
44. See, e.g., *People v. Morgan*, 187 Ill. 2d 500 (Ill. 1999).
45. Eric Bailey, "California and the West; Defense Probing Brain to Explain Yosemite Killings; Crime: Cary Stayner is among a Number of Defendants Whose Lawyers are Looking for Physical Explanations for Brutal Murders," Los Angeles Times, Part A; Part 1; Page 3 (June 15, 2000).

Reading 5.2

For the Law, Neuroscience Changes Nothing and Everything[1]

Joshua Greene and Jonathan Cohen

The law takes a long-standing interest in the mind. In most criminal cases, a successful conviction requires the prosecution to establish not only that the defendant engaged in proscribed behavior, but also that the misdeed in question was the product of *mens rea*, a "guilty mind." Narrowly interpreted, *mens rea* refers to the intention to commit a criminal act, but the term has a looser interpretation by which it refers to all mental states consistent with moral and/or legal blame (a killing motivated by insane delusional beliefs may meet the requirements for *mens rea* in the first sense, but not the second) (Goldstein, Morse, & Shapiro, 2003). Thus, for centuries, many legal issues have turned on the question: "What was he thinking?"

To answer this question, the law has often turned to science. Today, the newest kid on this particular scientific block is cognitive neuroscience, the study of the mind through the brain, which has gained prominence in part as a result of the advent of functional neuroimaging as a widely used tool for psychological research. Given the law's aforementioned concern for mental states, along with its preference for "hard" evidence, it is no surprise that interest in the potential legal implications of cognitive neuroscience abounds. But does our emerging understanding of the mind as brain really have any deep implications for the law? This theme issue is a testament to the thought that it might. Some have argued, however, that new neuroscience contributes nothing more than new details and that existing legal principles can handle anything that neuroscience will throw our way in the foreseeable future (Morse, 2004).

In our view, both of these positions are, in their respective ways, correct. Existing legal principles make virtually no assumptions about the neural bases of criminal behavior, and as a result they can comfortably assimilate new neuroscience without much in the way of conceptual upheaval: new details, new sources of evidence, but nothing for which the law is fundamentally unprepared. We maintain, however, that our operative legal principles exist because they more or less adequately capture an intuitive sense of justice. In our view, neuroscience will challenge and ultimately reshape our intuitive sense(s) of justice. New neuroscience will affect the way we view the law, not by furnishing us with new ideas or arguments about the nature of human action, but by breathing new life into old ones. Cognitive neuroscience, by identifying the specific mechanisms responsible for behavior, will vividly illustrate what until now could only be appreciated through esoteric theorizing: that there is something fishy about our ordinary conceptions of human action and responsibility, and that, as a result, the legal principles we have devised to reflect these conceptions may be flawed.

Our argument runs as follows: first, we draw a familiar distinction between the consequentialist justification for state punishment, according to which punishment is merely an instrument for promoting future social welfare, and the retributivist justification for punishment, according to which the principal aim of punishment is to give people what they deserve based on their past actions. We observe that the commonsense approach to moral and legal responsibility has consequentialist elements, but is largely retributivist. Unlike the consequentialist justification for punishment, the retributivist justification relies, either explicitly or implicitly, on a demanding—and some say overly demanding—conception of free will. We therefore consider the standard responses to the philosophical problem of free will (Watson, 1982). "Libertarians" (no relation to the political philosophy) and "hard determinists" agree on "incompatibilism," the thesis that free will and determinism are incompatible, but they disagree about whether determinism is true, or near enough true to preclude free will. Libertarians believe that we have free will because determinism is false, and hard determinists believe that we lack free will because determinism is (approximately) true. "Compatibilists," in contrast with libertarians and hard determinists, argue that free will and determinism are perfectly compatible.

We argue that current legal doctrine, although officially compatibilist, is ultimately grounded in intuitions that are incompatibilist and, more specifically, libertarian. In other words, the law *says* that it presupposes nothing more than a metaphysically modest notion of free will that is perfectly compatible with determinism. However, we argue that the law's intuitive support is ultimately grounded in a metaphysically overambitious, libertarian notion of free will that is threatened by determinism and, more pointedly, by forthcoming cognitive neuroscience. At present, the gap between what the law officially cares about and what people really care about is only revealed occasionally when vivid scientific information about the causes of criminal behavior leads people to doubt certain individuals' capacity for moral and legal responsibility, despite the fact that this information is irrelevant according to the law's stated principles. We argue that new neuroscience will continue to highlight and widen this gap. That is, new neuroscience will undermine people's commonsense, libertarian conception of free will and the retributivist thinking that depends on it, both of which have heretofore been shielded by the inaccessibility of sophisticated thinking about the mind and its neural basis.

The net effect of this influx of scientific information will be a rejection of free will as it is ordinarily conceived, with important ramifications for the law. As noted above, our criminal justice system is largely retributivist. We argue that retributivism, despite its unstable marriage to compatibilist philosophy in the letter of the law, ultimately depends on an intuitive, libertarian notion of free will that is undermined by science. Therefore, with the rejection of commonsense conceptions of free will comes the rejection of retributivism and an ensuing shift toward a consequentialist approach to punishment (i.e., one aimed at promoting future welfare rather than meting out just deserts). Because consequentialist approaches to punishment remain viable in the absence of commonsense free will, we need not give up on moral and legal responsibility. We argue further that the philosophical problem of free will arises out of a conflict between two cognitive subsystems that speak different "languages": the "folk psychology" system and the "folk physics" system. Because we are inherently of two minds when it comes to the problem of free will, this problem will never find an intuitively satisfying solution. We can, however, recognize that free will, as conceptualized by the folk psychology system, is an illusion and structure our society accordingly by rejecting retributivist legal principles that derive their intuitive force from this illusion.

5.2.1 Two Theories of Punishment: Consequentialism and Retributivism

There are two standard justifications for legal punishment (Lacey, 1988). According to the forward-looking, consequentialist theory, which emerges from the classic utilitarian tradition (Bentham, 1982), punishment is justified by its future beneficial effects. Chief among them are the prevention of future crime through the deterrent effect of the law and the containment of dangerous individuals. Few would deny that the deterrence of future crime and the protection of the public are legitimate justifications for punishment. The controversy surrounding consequentialist theories concerns their serviceability as *complete* normative theories of punishment. Most theorists find them inadequate in this regard (e.g., Hart, 1968), and many argue that consequentialism fundamentally mischaracterizes the primary justification for punishment, which, these critics argue, is retribution (Kant, 2002). As a result, they claim, consequentialist theories justify intuitively unfair forms of punishment, if not in practice then in principle. One problem is that of Draconian penalties. It is possible, for example, that imposing the death penalty for parking violations would maximize aggregate welfare by reducing parking violations to near zero. But, retributivists claim, whether or not this is a good idea does not depend on the balance of costs and benefits. It is simply wrong to kill someone for double parking. A related problem is that of punishing the innocent. It is possible that, under certain circumstances, falsely convicting an innocent person would have a salutary deterrent effect, enough to justify that person's suffering, etc. Critics also note that, so far as deterrence is concerned, it is the *threat* of punishment that is justified and not the punishment itself. Thus, consequentialism might justify letting murderers and rapists off the hook so long as their punishment could be convincingly faked.

The standard consequentialist response to these charges is that such concerns have no place in the real world. They say, for example, that the idea of imposing the death penalty for parking violations to make society an overall happier place is absurd. People everywhere would live in mortal fear of bureaucratic errors, and so on. Likewise, a legal system that deliberately convicted innocent people and/or secretly refrained from punishing guilty ones would require a kind of systematic deception that would lead inevitably to corruption and that could never survive in a free society. At this point critics retort that consequentialist theories, at best, get the right answers for the wrong reasons. It is wrong to punish innocent people, etc., because it is fundamentally unfair, not because it leads to bad consequences in practice. Such critics are certainly correct to point out that consequentialist theories fail to capture something central to commonsense intuitions about legitimate punishment.

The backward-looking, retributivist account does a better job of capturing these intuitions. Its fundamental principle is simple: In the absence of mitigating circumstances, people who engage in criminal behavior *deserve* to be punished, and that is why we punish them. Some would explicate this theory in terms of criminals' forfeiting rights, others in terms of the rights of the victimized, whereas others would appeal to the violation of a hypothetical social contract, and so on. Retributivist theories come in many flavors, but these distinctions need not concern us here. What is important for our purposes is that retributivism captures the intuitive idea that we legitimately punish to give people what they deserve based on their past actions—in proportion to their "internal wickedness," to use Kant's (2002) phrase—and not, primarily, to promote social welfare in the future.

The retributivist perspective is widespread, both in the explicit views of legal theorists and implicitly in common sense. There are two primary motivations for questioning retributivist theory. The first, which will not concern us here, comes from a prior commitment to a broader consequentialist moral theory. The second comes from scepticism regarding the notion of desert, grounded in a broader scepticism about the possibility of free will in a deterministic or mechanistic world.

5.2.2 Free Will and Retributivism

The problem of free will is old and has many formulations (Watson, 1982). Here is one, drawing on a more detailed and exacting formulation by Peter Van Inwagen (1982): determinism is true if the world is such that its current state is completely determined by (i) the laws of physics and (ii) past states of the world. Intuitively, the idea is that a deterministic universe starts however it starts and then ticks along like clockwork from there. Given a set of prior conditions in the universe and a set of physical laws that completely govern the way the universe evolves, there is only one way that things can actually proceed.

Free will, it is often said, requires the ability do otherwise (an assumption that has been questioned; Frankfurt, 1966). One cannot say, for example, that I have freely chosen soup over salad if forces beyond my control are sufficient to necessitate my choosing soup. But, the determinist argues, this is precisely what forces beyond your control do—always. You have no say whatsoever in the state of the universe before your birth; nor do you have any say about the laws of physics. However, if determinism is true, these two things together are sufficient to determine your choice of soup over salad. Thus, some say, if determinism is true, your sense of yourself and others as having free will is an illusion.

There are three standard responses to the problem of free will. The first, known as hard determinism, accepts the incompatibility of free will and determinism (incompatibilism), and asserts determinism, thus rejecting free will. The second response is libertarianism (again, no relation to the political philosophy), which accepts incompatibilism, but denies that determinism is true. This may seem like a promising approach. After all, has not modern physics shown us that the universe is *in*deterministic (Hughs, 1992)? The problem here is that the sort of indeterminism afforded by modern physics is not the sort the libertarian needs or desires. If it turns out that your ordering soup is completely determined by the laws of physics, the state of the universe 10,000 years ago, *and* the outcomes of myriad subatomic coin flips, your appetizer is no more freely chosen than before. Indeed, it is *randomly* chosen, which is no help to the libertarian. What about some other kind of indeterminism? What if, somewhere deep in the brain, there are mysterious events that operate independently of the ordinary laws of physics and that are somehow tied to the will of the brain's owner? In light of the available evidence, this is highly unlikely. Say what you will about the "hard problem" of consciousness (Shear, 1999), there is not a shred of scientific evidence to support the existence of *causally effective* processes in the mind or brain that violate the laws of physics. In our opinion, any scientifically respectable discussion of free will requires the rejection of what Strawson (1962) famously called the "panicky metaphysics" of libertarianism.[2]

Finally, we come to the dominant view among philosophers and legal theorists: compatibilism. Compatibilists concede that some notions of free will may require indefensible, panicky metaphysics, but maintain that the kinds of free will "worth wanting," to use Dennett's (1984) phrase, are perfectly compatible with determinism. Compatibilist theories vary, but all compatibilists agree that free will is a perfectly natural, scientifically respectable phenomenon and part of the ordinary human condition. They also agree that free will can be undermined by various kinds of psychological deficit (e.g., mental illness or "infancy"). Thus, according to this view, a freely willed action is one that is made using the right sort of psychology—rational, free of delusion, etc.

Compatibilists make some compelling arguments. After all, is it not obvious that we have free will? Could science plausibly deny the obvious fact that I am free to raise my hand *at will*? For many people, such simple observations make the reality of free will non-negotiable. But at the same time, many such people concede that determinism, or something like it, is a live possibility. And if free will is obviously real, but determinism is debatable, then the reality of free will must not hinge on the rejection of determinism. That is, free will and determinism must be compatible. Many compatibilists sceptically ask what would it mean to give up on free will. Were we to give it up, wouldn't we have to immediately reinvent it? Does not every decision involve an implicit commitment to the idea of free will? And how else would we distinguish between ordinary rational adults and other individuals, such as young children and the mentally ill, whose will—or whatever you want to call it—is clearly compromised? Free will, compatibilists argue, is here to stay, and the challenge for science is to figure out how exactly it works and not to peddle silly arguments that deny the undeniable (Dennett, 2003).

The forward-looking–consequentialist approach to punishment works with all three responses to the problem of free will, including hard determinism. This is because consequentialists are not concerned with whether anyone is really innocent or guilty in some ultimate sense that might depend on people's having free will, but only with the likely effects of punishment. The retributivist approach, by contrast, is plausibly regarded as requiring free will and the rejection of hard determinism. Retributivists want to know whether the defendant truly *deserves* to be punished. Assuming one can deserve to be punished only for actions that are freely willed, hard determinism implies that no one really deserves to be punished. Thus, hard determinism combined with retributivism requires the elimination of all punishment, which does not seem reasonable. This leaves retributivists with two options: compatibilism and libertarianism. Libertarianism, for reasons given above, and despite its intuitive appeal, is scientifically suspect. At the very least, the law should not depend on it. It seems, then, that retributivism requires compatibilism. Accordingly, the standard legal account of punishment is compatibilist.

5.2.3 Neuroscience Changes Nothing

The title of a recent paper by Stephen Morse (2004), "New Neuroscience, Old Problems," aptly summarizes many a seasoned legal thinker's response to the suggestion that brain research will revolutionize the law. The law has been dealing with issues of criminal responsibility for a long time; Morse argues that there is nothing on the neuroscientific horizon that it cannot handle.

The reason that the law is immune to such threats is that it makes no assumptions that neuroscience, or any science, is likely to challenge. The law assumes that people have a general capacity for rational choice. That is, people have beliefs and desires and are capable of producing behavior that serves their desires in light of their beliefs. The law acknowledges that our capacity for rational choice is far from perfect (Kahneman & Tversky, 2000), requiring only that the people it deems legally responsible have a *general* capacity for rational behavior.

Thus, questions about who is or is not responsible in the eyes of the law have and will continue to turn on questions about rationality. This approach was first codified in the *M'Naghten* standard according to which a defense on the ground of insanity requires proof that the defendant labored under "a defect of reason, from disease of the mind" (Goldstein, 1967). Not all standards developed and applied since *M'Naghten* explicitly mention the need to demonstrate the defendant's diminished rationality (e.g., the *Durham* standard; Goldstein, 1967), but it is generally agreed that a legal excuse requires a demonstration that the defendant "lacked a general capacity for rationality" (Goldstein et al., 2003). Thus, the argument goes, new science can help us figure out who was or was not rational at the scene of the crime, much as it has in the past, but new science will not justify any fundamental change in the law's approach to responsibility unless it shows that people in general fail to meet the law's very minimal requirements for rationality. Science shows no sign of doing this, and thus the basic precepts of legal responsibility stand firm. As for neuroscience more specifically, this discipline seems especially unlikely to undermine our faith in general minimal rationality. If any sciences have an outside chance of demonstrating that our behavior is thoroughly irrational or arational, it is the ones that study behavior directly rather than its proximate physical causes in the brain. The law, this argument continues, does not care if people have "free will" in any deep metaphysical sense that might be threatened by determinism. It only cares that people in general are minimally rational. So long as this appears to be the case, it can go on regarding people as free (compatibilism) and holding ordinary people responsible for their misdeeds while making exceptions for those who fail to meet the requirements of general rationality.

In light of this, one might wonder what all the fuss is about. If the law assumes nothing more than general minimal rationality, and neuroscience does nothing to undermine this assumption, then why would anyone even *think* that neuroscience poses some sort of threat to legal doctrines of criminal responsibility? It sounds like this is just a simple mistake, and that is precisely what Morse contends. He calls this mistake "the fundamental psycholegal error" which is "to believe that causation, especially abnormal causation, is *per se* an excusing condition" (Morse, 2004, p. 180). In other words, if you think that neuroscientific information about the causes of human action, or some particular human's action, can, by itself, make for a legitimate legal excuse, you just do not understand the law. Every action is caused by brain events, and describing those events and affirming their causal efficacy is of no legal interest in and of itself. Morse continues, "[The psycholegal error] leads people to try to create a new excuse every time an allegedly valid new 'syndrome' is discovered that is thought to play a role in behaviour. But syndromes and other causes do not have excusing force unless they sufficiently diminish rationality in the context in question" (Morse, 2004, p. 180).

In our opinion, Morse and like-minded theorists are absolutely correct about the relationship between current legal doctrine and any forthcoming neuroscientific results. For the law, as written, neuroscience changes nothing. The law provides a coherent framework for the assessment of criminal responsibility that is not threatened by anything neuroscience is likely to throw at it. But, we maintain, the law nevertheless stands on shakier ground than the foregoing would suggest. The legitimacy of the law itself depends on its adequately reflecting the moral intuitions and commitments of society. If neuroscience can change those intuitions, then neuroscience can change the law.

As it happens, this is a possibility that Morse explicitly acknowledges. However, he believes that such developments would require radical new ideas that we can scarcely imagine at this time (e.g., a new solution to the mind–body problem). We disagree. The seeds of discontent are already sown in commonsense legal thought. In our opinion, the "fundamental psycholegal error" is not so much an error as a reflection of the gap between what the law officially cares about and what people really care about. In modern criminal law, there has been a long tense marriage of convenience between compatibilist legal principles and libertarian moral intuitions. New neuroscience, we argue, will probably render this marriage unworkable.

5.2.4 What Really Matters for Responsibility? Materialist Theory, Dualist Intuitions, and the "Boys from Brazil" Problem

According to the law, the central question in a case of putative diminished responsibility is whether the accused was sufficiently rational at the time of the misdeed in question. We believe, however, that this is not what most people really care about, and that for them diminished rationality is just a presumed correlate of something deeper. It seems that what many people really want to know is: was it really *him*? This question usually comes in the form of a disjunction, depending on how the excuse is constructed: was it *him*, or was it his *upbringing*? Was it *him*, or was it his *genes*? Was it *him*, or was it his *circumstances*? Was it *him*, or was it his *brain*? But what most people do not understand, despite the fact that naturalistic philosophers and scientists have been saying it for centuries, is that there is no "him" independent of these other things. (Or, to be a bit more accommodating to the supernaturally inclined, there is no "him" independent of these things that shows any sign of affecting anything in the physical world, including his behavior.)

Most people's view of the mind is implicitly *dualist* and *libertarian* and not *materialist* and *compatibilist*. Dualism, for our purposes, is the view that mind and brain are separate, interacting, entities.[3] Dualism fits naturally with libertarianism because a mind distinct from the body is precisely the sort of nonphysical source of free will that libertarianism requires. Materialism, by contrast, is the view that all events, including the operations of the mind, are ultimately operations of matter that obeys the laws of physics. It is hard to imagine a belief in free will that is materialist but not compatibilist, given that ordinary matter does not seem capable of supplying the nonphysical processes that libertarianism requires.

Many people, particularly those who are religious, are explicitly dualist libertarians (again, not in the political sense). However, in our estimation, even people who do or would readily endorse a thoroughly material account of human action and its causes have dualist, libertarian intuitions. This goes not only for educated people in general, but also for experts in mental health and criminal behavior. Consider, for example, the following remarks from Jonathan Pincus, an expert on criminal behavior and the brain.

> When a composer conceives a symphony, the only way he or she can present it to the public is through an orchestra... If the performance is poor, the fault could lie with the composer's conception, or the orchestra, or both... Will is expressed by the brain. Violence can be the result of volition only, but if a brain is damaged, brain failure must be at least partly to blame. (Pincus, 2001, p. 128.)

To our untutored intuitions, this is a perfectly sensible analogy, but it is ultimately grounded in a kind of dualism that is scientifically untenable. It is not as if there is *you*, the composer, and then *your brain*, the orchestra. You *are* your brain, and your brain is the composer and the orchestra all rolled together. There is no little man, no "homunculus," in the brain that is the real you behind the mass of neuronal instrumentation. Scientifically minded philosophers have been saying this *ad nauseam* (Dennett, 1991), and we will not belabor the point. Moreover, we suspect that if you were to ask Dr. Pincus whether he thinks there is a little conductor directing his brain's activity from within or beyond he would adamantly deny that this is the case. At the same time, though, he is comfortable comparing a brain-damaged criminal to a healthy conductor saddled with an unhealthy orchestra. This sort of doublethink is not uncommon. As we will argue in section 5.2.6, when it comes to moral responsibility in a physical world, we are all of two minds.

A recent article by Laurence Steinberg and Elizabeth Scott (Steinberg & Scott, 2003), experts respectively on adolescent developmental psychology and juvenile law, illustrates the same point. They argue that adolescents do not meet the law's general requirements for rationality and that therefore they should be considered less than fully responsible for their actions and, more specifically, unsuitable candidates for the death penalty. Their main argument is sound, but they cannot resist embellishing it with a bit of superfluous neuroscience.

> Most of the developmental research on cognitive and psychosocial functioning in adolescence measures behaviors, self-perceptions, or attitudes, but mounting evidence suggests that at least some of the differences between adults and adolescents have neuropsychological and neurobiological underpinnings. (Steinberg & Scott, 2003, p. 5.)

Some of the differences? Unless some form of dualism is correct, *every* mental difference and *every* difference in behavioral tendency is a function of some kind of difference in the brain. But here it is implicitly suggested that things like "behaviors, self-perceptions, or attitudes" may be grounded in something other than the brain. In summing up their case, Steinberg and Scott look toward the future.

> Especially needed are studies that link developmental changes in decision making to changes in brain structure and function . . . In our view, however, there is sufficient indirect suggestive evidence of age differences in capacities that are relevant to criminal blameworthiness to support the position that youths who commit crimes should be punished more leniently then their adult counterparts. (Steinberg & Scott, 2003, p. 9.)

This gets the order of evidence backwards. If what the law ultimately cares about is whether adolescents can behave rationally, then it is evidence concerning adolescent behavior that is *directly* relevant. Studying the adolescent brain is a highly *indirect* way of figuring out whether adolescents in general are rational. Indeed, the only way we neuroscientists can tell if a brain structure is important for rational judgment is to see if its activity or damage is correlated with (ir)rational *behavior*.[4]

If everyone agrees that what the law ultimately cares about is the capacity for rational behavior, then why are Steinberg and Scott so optimistic about neuroscientific evidence that is only indirectly relevant? The reason, we suggest, is that they are appealing not to a legal argument, but to a moral intuition. So far as the law is concerned, information about the physical processes that give rise to bad behavior is irrelevant. But to people who implicitly believe that real decision-making takes place in the mind, not in the brain, demonstrating that there is a brain basis for adolescents' misdeeds allows us to blame adolescents' brains instead of the adolescents themselves.

The fact that people are tempted to attach great moral or legal significance to neuroscientific information that, according to the letter of the law, should not matter, suggests that what the law cares about and what people care about do not necessarily coincide. To make this point in a more general way, we offer the following thought experiment, which we call "The Boys from Brazil problem." It is an extension of an argument that has made the rounds in philosophical discussions of free will and responsibility (Rosen, 2002).

In the film *The Boys from Brazil*, members of the Nazi old guard have regrouped in South America after the war. Their plan is to bring their beloved *führer* back to life by raising children genetically identical to Hitler (courtesy of some salvaged DNA) in environments that mimic that of Hitler's upbringing. For example, Hitler's father died while young Adolph was still a boy, and so each Hitler clone's surrogate father is killed at just the right time, and so on, and so forth.

This is obviously a fantasy, but the idea that one could, in principle, produce a person with a particular personality and behavioral profile through tight genetic and environmental control is plausible. Let us suppose, then, that a group of scientists has managed to create an individual—call him "Mr. Puppet"—who, by design, engages in some kind of criminal behavior: say, a murder during a drug deal gone bad. The defense calls to the stand the project's lead scientist: "Please tell us about your relationship to Mr. Puppet. . . ."

It is very simple, really. I designed him. I carefully selected every gene in his body and carefully scripted every significant event in his life so that he would become precisely what he is today. I selected his mother knowing that she would let him cry for hours and hours before picking him up. I carefully selected each of his relatives, teachers, friends, enemies, etc., and told them exactly what to say to him and how to treat him. Things generally went as planned, but not always. For example, the angry letters written to his dead father were not supposed to appear until he was fourteen, but by the end of his thirteenth year he had already written four of them. In retrospect I think this was because of a handful of substitutions I made to his eighth chromosome. At any rate, my plans for him succeeded, as they have for 95% of the people I've designed. I assure you that the accused deserves none of the credit.

What to do with Mr. Puppet? Insofar as we believe this testimony, we are inclined to think that Mr. Puppet cannot be held fully responsible for his crimes, if he can be held responsible for them at all. He is, perhaps, a man to be feared, and we would not want to return him to the streets. But given the fact that forces beyond his control played a dominant role in causing him to commit these crimes, it is hard to think of him as anything more than a pawn.

But what does the law say about Mr. Puppet? The law asks whether or not he was rational at the time of his misdeeds, and as far as we know he was. For all we know, he is psychologically indistinguishable from the prototypical guilty criminal and therefore fully responsible in the eyes of the law. But, intuitively, this is not fair.

Thus, it seems that the law's exclusive interest in rationality misses something intuitively important. In our opinion, rationality is just a presumed correlate of what most people really care about. What people really want to know is if the accused, as opposed to something else, is responsible for the crime, where that "something else" could be the accused's brain, genes, or environment. The question of someone's ultimate responsibility seems to turn, intuitively, on a question of internal versus external determination. Mr. Puppet ought not be held responsible for his actions because forces beyond his control played a dominant role in the production of his behavior. Of course, the scientists did not have complete control—after all, they had a 5% failure rate—but that does not seem to be enough to restore Mr. Puppet's free will, at least not entirely. Yes, he is as rational as other criminals, and, yes, it was his desires and beliefs that produced his actions. But those beliefs and desires were rigged by external forces, and that is why, intuitively, he deserves our pity more than our moral condemnation.[5]

The story of Mr. Puppet raises an important question: What is the difference between Mr. Puppet and anyone else accused of a crime? After all, we have little reason to doubt that (i) the state of the universe 10,000 years ago, (ii) the laws of physics, and (iii) the outcomes of random quantum mechanical events are together sufficient to determine everything that happens nowadays, including our own actions. These things are all clearly beyond our control. So what is the real difference between us and Mr. Puppet? One obvious difference is that Mr. Puppet is the victim of a diabolical plot whereas most people, we presume, are not. But does this matter? The thought that Mr. Puppet is not fully responsible depends on the idea that his actions were externally determined. Forces beyond his control constrained his personality to the point that it was "no surprise" that he would behave badly. But the fact that these forces are connected to the desires and intentions of evil scientists is really irrelevant, is it not? What matters is only that these forces are beyond Mr. Puppet's control, that they're not really *his*. The fact that someone could deliberately harness these forces to reliably design criminals is an indication of the strength of these forces, but the fact that these forces are being guided by other minds rather than simply operating on their own seems irrelevant, so far as Mr. Puppet's freedom and responsibility are concerned.

Thus, it seems that, in a very real sense, we are all puppets. The combined effects of genes and environment determine all of our actions. Mr. Puppet is exceptional only in that the intentions of other humans lie behind his genes and environment. But, so long as his genes and environment are intrinsically comparable with those of ordinary people, this does not really matter. We are no more free than he is.

What all of this illustrates is that the "fundamental psycholegal error" is grounded in a powerful moral intuition that the law and allied compatibilist philosophies try to sweep under the rug. The foregoing suggests that people regard actions only as fully free when those actions are seen as robust against determination by external forces. But if determinism (or determinism plus quantum mechanics) is true, then no actions are truly free because forces beyond our control are always sufficient to determine behavior. Thus, intuitive free will is libertarian, not compatibilist. That is, it requires the rejection of determinism and an implicit commitment to some kind of magical mental causation.[6]

Naturalistic philosophers and scientists have known for a long time that magical mental causation is a nonstarter. But this realization is the result of philosophical reflection about the nature of the universe and its governance by physical law. Philosophical reflection, however, is not the only way to see the problems with libertarian accounts of free will. Indeed, we argue that neuroscience can help people appreciate the mechanical nature of human action in a way that bypasses complicated arguments.

5.2.5 Neuroscience and the Transparent Bottleneck

We have argued that, contrary to legal and philosophical orthodoxy, determinism really does threaten free will and responsibility as we intuitively understand them. It is just that most of us, including most philosophers and legal theorists, have yet to appreciate it. This controversial opinion amounts to an empirical prediction that may or may not hold: As more and more scientific facts come in, providing increasingly vivid illustrations of what the human mind is really like, more and more people will develop moral intuitions that are at odds with our current social practices (see Robert Wright, 1994, for similar thoughts).

Neuroscience has a special role to play in this process for the following reason. As long as the mind remains a black box, there will always be a donkey on which to pin dualist and libertarian intuitions. For a long time, philosophical arguments have persuaded some people that human action has purely mechanical causes, but not everyone cares for philosophical arguments. Arguments are nice, but physical demonstrations are far more compelling. What neuroscience does, and will continue to do at an accelerated pace, is elucidate the "when," "where," and "how" of the mechanical processes that cause behavior. It is one thing to deny that human decision-making is purely mechanical when your opponent offers only a general, philosophical argument. It is quite another to hold your ground when your opponent can make detailed predictions about how these mechanical processes work, complete with images of the brain structures involved and equations that describe their function.[7]

Thus, neuroscience holds the promise of turning the black box of the mind into a *transparent bottleneck*. There are many causes that impinge on behavior, but all of them—from the genes you inherited, to the pain in your lower back, to the advice your grandmother gave you when you were six—must exert their influence through the brain. Thus, your brain serves as a bottleneck for all the forces spread throughout the universe of your past that affect who you are and what you do. Moreover, this bottleneck contains the events that are, intuitively, most critical for moral and legal responsibility, and we may soon be able to observe them closely.

At some time in the future we may have extremely high-resolution scanners that can simultaneously track the neural activity and connectivity of every neuron in a human brain, along with computers and software that can analyze and organize these data. Imagine, for example, watching a film of your brain choosing between soup and salad. The analysis software highlights the neurons pushing for soup in red and the neurons pushing for salad in blue. You zoom in and slow down the film, allowing yourself to trace the cause-and-effect relationships between individual neurons—the mind's clockwork revealed in arbitrary detail. You find the tipping-point moment at which the blue neurons in your prefrontal cortex out-fire the red neurons, seizing control of your pre-motor cortex and causing you to say, "I will have the salad, please."

At some further point this son of brainware may be very widespread, with a high-resolution brain scanner in every classroom. People may grow up completely used to the idea that every decision is a thoroughly mechanical process, the outcome of which is completely determined by the results of prior mechanical processes. What will such people think as they sit in their jury boxes? Suppose a man has killed his wife in a jealous rage. Will jurors of the future wonder whether the defendant acted in that moment *of his own free will*? Will they wonder if it was *really him* who killed his wife rather than his *uncontrollable anger*? Will they ask whether he *could have done otherwise*? Whether he really *deserves* to be punished, or if he is just a victim of unfortunate circumstances? We submit that these questions, which seem so important today, will lose their grip in an age when the mechanical nature of human decision-making is fully appreciated. The law will continue to punish misdeeds, as it must for practical reasons, but the idea of distinguishing the truly, deeply guilty from those who are merely victims of neuronal circumstances will, we submit, seem pointless.

At least in our more reflective moments. Our intuitive sense of free will runs quite deep, and it is possible that we will never be able to fully talk ourselves out of it. Next we consider the psychological origins of the problem of free will.

5.2.6 Folk Psychology and Folk Physics Collide: A Cognitive Account of the Problem of Attributive Free Will

Could the problem of free will just melt away? This question begs another: Why do we have the problem of free will in the first place? Why does the idea of a deterministic universe seem to contradict something important in our conception of human action? A promising answer to this question is offered by Daniel Wegner in *The Illusion of Conscious Will* (Wegner, 2002). In short, Wegner argues, we feel as if we are uncaused causers, and therefore granted a degree of independence from the deterministic flow of the universe, because we are unaware of the deterministic processes that operate in our own heads. Our actions appear to be caused by our mental states, but not by physical states of our brains, and so we imagine that we are metaphysically special, that we are nonphysical causes of physical events. This belief in our specialness is likely to meet the same fate as other similarly narcissistic beliefs that we have cherished in our past: that the Earth lies at the center of the universe, that humans are unrelated to other species, that all of our behavior is consciously determined, etc. Each of these beliefs has been replaced by a scientific and humbling understanding of our place in the physical universe, and there is no reason to believe that the case will be any different for our sense of free will. (For similar thoughts, see Wright, 1994, on Darwin's clandestine views about free will and responsibility.)

We believe that Wegner's account of the problem of free will is essentially correct, although we disagree strongly with his conclusions concerning its (lack of) practical moral implications (see below). In this section we pick up on and extend one strand in Wegner's argument (Wegner, 2002, pp. 15–28). Wegner's primary aim is to explain, in psychological terms, why we attribute free will to ourselves, why we feel free from the inside. Our aim in this section is to explain, in psychological terms, why we insist on attributing free will to *others*—and why scientifically minded philosophers, despite persistent efforts, have managed to talk almost no one out of this practice. The findings we review serve as examples of how psychological and neuroscientific data are beginning to characterize the mechanisms that underlie our sense of free will, how these mechanisms can lead us to assume free will is operating when it is not, and how a scientific understanding of these mechanisms can serve to dismantle our commitment to the idea of free will.

Looking out at the world, it appears to contain two fundamentally different kinds of entity. On the one hand, there are ordinary objects that appear to obey the ordinary laws of physics: things like rocks and puddles of water and blocks of wood. These things do not get up and move around on their own. They are, in a word, inanimate. On the other hand, there are things that seem to operate by some kind of magic. Humans and other animals, so long as they are alive, can move about at will, in apparent defiance of the physical laws that govern ordinary matter. Because things like rocks and puddles, on the one hand, and mice and humans, on the other, behave in such radically different ways, it makes sense, from an evolutionary perspective, that creatures would evolve separate cognitive systems for processing information about each of these classes of objects (Pinker, 1997). There is a good deal of evidence to suggest that this is precisely how our minds work.

A line of research beginning with Fritz Heider illustrates this point. Heider and Simmel (Heider & Simmel, 1944) created a film involving three simple geometric shapes that move about in various ways. For example, a big triangle chases a little circle around the screen, bumping into it. The little circle repeatedly moves away, and a little triangle repeatedly moves in between the circle and the big triangle. When normal people watch this movie they cannot help but view it in social terms (Heberlein & Adolphs, 2004). They see the big triangle as *trying* to harm the little circle, and the little triangle as trying to *protect* the little circle; and they see the little circle as *afraid* and the big triangle as *frustrated.* Some people even spontaneously report that the big triangle is a *bully.* In other words, simple patterns of movement trigger in people's minds a cascade of complex social inferences. People not only see these shapes as "alive." They see beliefs, desires, intentions, emotions, personality traits, and even moral blameworthiness. It appears that this kind of inference is automatic (Scholl & Tremoulet, 2000). Of course, you, the observer, know that it is only a film, and a very simple one at that, but you nevertheless cannot help but see these events in social, even *moral*, terms.

That is, unless you have damage to your amygdala, a subcortical brain structure that is important for social cognition (Adolphs, 1999). Andrea Heberlein tested a patient with rare bilateral amygdala damage using Heider's film and found that this patient, unlike normal people, described what she saw in completely asocial terms, despite that fact that her visual and verbal abilities are not compromised by her brain damage. Somehow, this patient is blind to the "human" drama that normal people cannot help but see in these events (Heberlein & Adolphs, 2004).

The sort of thinking that is engaged when normal people view the Heider–Simmel film is sometimes known as "folk psychology" (Fodor, 1987), "the intentional stance" (Dennett, 1987), or "theory of mind" (Premack & Woodruff, 1978). There is a fair amount of evidence (including the work described above) suggesting that humans have a set of cognitive subsystems that are specialized for processing information about intentional agents (Saxe, Carey, & Kanwisher, 2004). At the same time, there is evidence to suggest that humans and other animals also have subsystems specialized for "folk physics," an intuitive sense of how ordinary matter behaves. One compelling piece of evidence for the claim that normal humans have subsystems specialized for folk physics comes from studies of people with autism spectrum disorder. These individuals are particularly bad at solving problems that require "folk psychology," but they do very well with problems related to how physical objects (e.g., the parts of machine) behave (i.e., "folk physics") (Baron Cohen, 2000). Another piece of evidence for a "folk physics" system comes from discrepancies between people's physical intuitions and the way the world actually works. People say, for example, that a ball shot out of a curved tube resting on a flat surface will continue to follow a curved path outside the tube when in fact it will follow a straight path (McCloskey, Caramazza, & Green, 1980). The fact that people's physical intuitions are slightly, but systematically, out of step with reality suggests that the mind brings a fair amount of implicit theory to the perception of physical objects.

Thus, it is at least plausible that we possess distinguishable cognitive systems for making sense of the behavior of objects in the world. These systems seem to have two fundamentally different "ontologies." The folk physics system deals with chunks of matter that move around without purposes of their own according to the laws of intuitive physics, whereas the folk psychology system deals with unseen features of minds: beliefs, desires, intentions, etc. But what, to our minds, is a mind? We suggest that a crucial feature, if not the defining feature, of a mind (intuitively understood) is that it is an uncaused causer (Scholl & Tremoulet, 2000). Minds animate material bodies, allowing them to move without any apparent physical cause and in pursuit of goals. Moreover, we reserve certain social attitudes for things that have minds. For example, we do not resent the rain for ruining our picnic, but we would resent a person who hosed our picnic (Strawson, 1962), and we resent picnic-hosers considerably more when we perceive that their actions are intentional. Thus, it seems that folk psychology is the gateway to moral evaluation. To see something as morally blameworthy or praiseworthy (even if it is just a moving square), one has to first see it as "someone," that is, as having a mind.

With all of this in the background, one can see how the problem of attributive free will arises. To see something as a responsible moral agent, one must first see it as having a mind. But, intuitively, a mind is, among other things, an uncaused causer. Consequently, when something is seen as a mere physical entity operating in accordance with deterministic physical laws, it ceases to be seen, intuitively, as a mind. Consequently, it is seen as an object unworthy of moral praise or blame. (Note that we are not claiming that people automatically attribute moral agency to anything that appears to be an uncaused causer. Rather, our claim is that seeing something as an uncaused causer is a *necessary but not sufficient* condition for seeing something as a moral agent.)

After thousands of years of our thinking of one another as uncaused causers, science comes along and tells us that there is no such thing—that all causes, with the possible exception of the Big Bang, are caused causes (determinism). This creates a problem. When we look at people as physical systems, we cannot see them as any more blameworthy or praiseworthy than bricks. But when we perceive people using our intuitive, folk psychology we cannot avoid attributing moral blame and praise.

So, philosophers who would honor both our scientific knowledge and our social instincts try to reconcile these two competing outlooks, but the result is never completely satisfying, and the debate wears on. Philosophers who cannot let go of the idea of uncaused causes defend libertarianism, and thus opt for scientifically dubious, "panicky metaphysics." Hard determinists, by contrast, embrace the conclusions of modern science, and concede what others will not: that many of our dearly held social practices are based on an illusion. The remaining majority, the compatibilists, try to talk themselves into a compromise. But the compromise is fragile. When the physical details of human action are made vivid, folk psychology loses its grip, just as folk physics loses its grip when the morally significant details are emphasized. The problem of free will and determinism will never find an intuitively satisfying solution because it arises out of a conflict between two distinct cognitive subsystems that speak different cognitive "languages" and that may ultimately be incapable of negotiation.

5.2.7 Free Will, Responsibility, and Consequentialism

Even if there is no intuitively satisfying solution to the problem of free will, it does not follow that there is no correct view of the matter. Ours is as follows: When it comes to the issue of free will itself, hard determinism is mostly correct. Free will, as we ordinarily understand it, is an illusion. However, it does not follow from the fact that free will is an illusion that there is no legitimate place for responsibility. There are two general justifications for holding people legally responsible for their actions. The retributive justification, by which the goal of punishment is to give people what they really deserve, does depend on this dubious notion of free will. However, the consequentialist approach does not require a belief in free will at all. As consequentialists, we can hold people responsible for crimes simply because doing so has, on balance, beneficial effects through deterrence, containment, etc. It is sometimes said that if we do not believe in free will, then we cannot legitimately punish anyone and that society must dissolve into anarchy. In a less hysterical vein, Daniel Wegner argues that free will, while illusory, is a necessary fiction for the maintenance of our social structure (Wegner, 2002, ch. 9). We disagree. There are perfectly good, forward-looking justifications for punishing criminals that do not depend on metaphysical fictions. (Wegner's observations may apply best to the personal sphere: see below.)

The vindication of responsibility in the absence of free will means that there is more than a grain of truth in compatibilism. The consequentialist approach to responsibility generates a derivative notion of free will that we can embrace (Smart, 1961). In the name of producing better consequences, we will want to make several distinctions among various actions and agents. To begin, we will want to distinguish the various classes of people who cannot be deterred by the law from those who can. That is, we will recognize many of the "diminished capacity" excuses that the law currently recognizes such as infancy and insanity. We will also recognize familiar justifications such as those associated with crimes committed under duress (e.g., threat of death). If we like, then, we can say that the actions of rational people operating free from duress, etc., are free actions, and that such people are exercising their free will.

At this point, compatibilists such as Daniel Dennett may claim victory: "what more could one want from free will?" In a word: retributivism. We have argued that commonsense retributivism really does depend on a notion of free will that is scientifically suspect. Intuitively, we want to punish those people who truly deserve it, but whenever the causes of someone's bad behavior are made sufficiently vivid, we no longer see that person as truly deserving of punishment. This insight is expressed by the old French proverb: "to know all is to forgive all." It is also expressed in the teachings of religious figures, such as Jesus and Buddha, who preach a message of universal compassion. Neuroscience can make this message more compelling by vividly illustrating the mechanical nature of human action.

Our penal system is highly counterproductive from a consequentialist perspective, especially in the United States, and yet it remains in place because retributivist principles have a powerful moral and political appeal (Lacey, 1988; Tonry, 2004). It is possible, however, that neuroscience will change these moral intuitions by undermining the intuitive, libertarian conceptions of free will on which retributivism depends.

As advocates of consequentialist legal reform, it behooves us to briefly respond to the three standard criticisms levied against consequentialist theories of punishment. First, it is claimed that consequentialism would justify extreme over-punishing. As noted above, it is possible in principle that the goal of deterrence would justify punishing parking violations with the death penalty or framing innocent people to make examples of them. Here, the standard response is adequate. The idea that such practices could, in the real world, make society happier on balance is absurd. Second, it is claimed that consequentialism justifies extreme under-punishment. In response to some versions of this objection, our response is the same as above. Deceptive practices such as a policy of faking punishment cannot survive in a free society, and a free society is required for the pursuit of most consequentialist ends. In other cases consequentialism may advocate more lenient punishments for people who, intuitively, deserve worse. Here, we maintain that a deeper understanding of human action and human nature will lead people—more of them, at any rate—to abandon these retributivist intuitions. Our response is much the same to the third and most general criticism of consequentialist punishment, which is that even when consequentialism gets the punishment policy right, it does so for the wrong reasons. These supposedly right reasons are reasons that we reject, however intuitive and natural they may feel. They are, we maintain, grounded in a metaphysical view of human action that is scientifically dubious and therefore an unfit basis for public policy in a pluralistic society.

Finally, as defenders of hard determinism and a consequentialist approach to responsibility, we should briefly address some standard concerns about the rejection of free will and conceptions of responsibility that depend on it. First, does not the fact that you can raise your hand "at will" prove that free will is real? Not in the sense that matters. As Daniel Wegner (2002) has argued, our first-person sense of ourselves as having free will may be a systematic illusion. And from a third-person perspective, we simply do not assume that anyone who exhibits voluntary control over his body is free in the relevant sense, as in the case of Mr. Puppet.

A more serious challenge is the claim that our commitments to free will and retributivism are simply inescapable for all practical purposes. Regarding free will, one might wonder whether one can so much as make a decision without implicitly assuming that one is free to choose among one's apparent options. Regarding responsibility and punishment, one might wonder if it is humanly possible to deny our retributive impulses (Pettit, 2002; Strawson, 1962). This challenge is bolstered by recent work in the behavioral sciences suggesting that an intuitive sense of fairness runs deep in our primate lineage (Brosnan & De Waal, 2003) and that an adaptive tendency toward retributive punishment may have been a crucial development in the biological and cultural evolution of human sociality (Bowles & Gintis, 2004; Boyd, Gintis, Bowles, & Richerson, 2003; Fehr & Gachter, 2002). Recent neuroscientific findings have added further support to this view, suggesting that the impulse to exact punishment may be driven by phylogenetically old mechanisms in the brain (Sanfey, Rilling, Aronson, Nystrom, & Cohen, 2003). These mechanisms may be an efficient and perhaps essential device for maintaining social stability. If retributivism runs that deep and is that useful, one might wonder whether we have any serious hope of, or reason for, getting rid of it. Have we any real choice but to see one another as free agents who deserve to be rewarded and punished for our past behaviors?

We offer the following analogy: Modern physics tells us that space is curved. Nevertheless, it may be impossible for us to see the world as anything other than flatly Euclidean in our day-to-day lives. And there are, no doubt, deep evolutionary explanations for our Euclidean tendencies. Does it then follow that we are forever bound by our innate Euclidean psychology? The answer depends on the domain of life in question. In navigating the aisles of the grocery store, an intuitive, Euclidean representation of space is not only adequate, but probably inevitable. However, when we are, for example, planning the launch of a spacecraft, we can and should make use of relativistic physical principles that are less intuitive but more accurate. In other words, a Euclidean perspective is not necessary for *all* practical purposes, and the same may be true for our implicit commitment to free will and retributivism. For most day-to-day purposes it may be pointless or impossible to view ourselves or others in this detached sort of way. But—and this is the crucial point—it may not be pointless or impossible to adopt this perspective when one is deciding what the criminal law should be or whether a given defendant should be put to death for his crimes. These may be special situations, analogous to those routinely encountered by "rocket scientists," in which the counterintuitive truth that we legitimately ignore most of the time can and should be acknowledged.

Finally, there is the worry that to reject free will is to render all of life pointless: Why would you bother with anything if it has all long since been determined? The answer is that you will bother because you are a human, and that is what humans do. Even if you decide, as part of a little intellectual exercise, that you are going to sit around and do nothing because you have concluded that you have no free will, you are eventually going to get up and make yourself a sandwich. And if you do not, you have got bigger problems than philosophy can fix.

5.2.8 Conclusion

Neuroscience is unlikely to tell us anything that will challenge the law's stated assumptions. However, we maintain that advances in neuroscience are likely to change the way people think about human action and criminal responsibility by vividly illustrating lessons that some people appreciated long ago. Free will as we ordinarily understand it is an illusion generated by our cognitive architecture. Retributivist notions of criminal responsibility ultimately depend on this illusion, and, if we are lucky, they will give way to consequentialist ones, thus radically transforming our approach to criminal justice. At this time, the law deals firmly but mercifully with individuals whose behavior is obviously the product of forces that are ultimately beyond their control. Some day, the law may treat all convicted criminals this way. That is, humanely.

Notes

1. Editor's note: This is an abridged version of a reading that originally appeared in 2004 in *Philosophical Transactions of the Royal Society B, Special Issue on Law and the Brain,* volume 359, pages 1775–1785, and is used with permission. The authors thank Stephen Morse, Andrea Heberlein, Aaron Schurger, Jennifer Kessler, and Simon Keller for their input.

2. Of course, scientific respectability is not everyone's first priority. However, the law in most Western states is a public institution designed to function in a society that respects a wide range of religious and otherwise metaphysical beliefs. The law cannot function in this way if it presupposes controversial and unverifiable metaphysical facts about the nature of human action, or anything else. Thus, the law must restrict itself to the class of intersubjectively verifiable facts, that is, the facts recognized by science, broadly construed. This practice need not derive from a conviction that the scientifically verifiable facts are necessarily the only facts, but merely from a recognition that verifiable or scientific facts are the only facts upon which public institutions in a pluralistic society can effectively rely.

3. There are some forms of dualism according to which the mind and body, although distinct, do not interact, making it impossible for the mind to have any observable effects on the brain or anything else in the physical world. These versions of dualism do not concern us here. For the purposes of this paper, we are happy to allow the metaphysical claim that souls or aspects of minds may exist independently of the physical body. Our concern is specifically with interactionist versions of dualism according to which nonphysical mental entities have observable physical effects. We believe that science has rendered such views untenable and that the law, insofar as it is a public institution designed to serve a pluralistic society, must not rely on beliefs that are scientifically suspect (see previous endnote).

4. It is conceivable that rationality could someday be redefined in neurocognitive rather than behavioral terms, much as water has been redefined in terms of its chemical composition. Were that to happen, neuroscientific evidence could then be construed as more direct than behavioral evidence. But Steinberg and Scott's argument appears to make use of a conventional, behavioral definition of rationality and not a neurocognitive redefinition.

5. This is not to say that we could not describe Mr. Puppet in such a way that our intuitions about him would change. Our point is only that, when the details are laid bare, it is very hard to see him as morally responsible.

6. Compatibilist philosophers such as Daniel Dennett (2003) might object that the story of Mr. Puppet is nothing but a misleading "intuition pump." Indeed, this is what Dennett says about a similar case of Alfred Mele's (1995). We believe that our case is importantly different from Mele's. Dennett and Mele imagine two women who are psychologically identical: Ann is a typical, good person, whereas Beth has been brainwashed to be just like Ann. Dennett argues, against Mele, that if you take seriously the claim that these two are psychologically identical and properly imagine that Beth is as rational, open-minded, etc., as Ann, you will come to see that the two are equally free. We agree with Dennett that Ann and Beth are comparable and that Mele's intuition falters when the details are fleshed out. But does the same hold for the intuition provoked by Mr. Puppet's story? It seems to us that the more one knows about Mr. Puppet and his life the less inclined one is to see him as truly responsible for his actions and our punishing him as a worthy end in itself. We can agree with Dennett that there is a sense in which Mr. Puppet is free. Our point is merely that there is a legitimate sense in which he, like all of us, is not free and that this sense matters for the law.

7. We do not wish to imply that neuroscience will inevitably put us in a position to predict any given action based on a neurologic examination. Rather, our suggestion is simply that neuroscience will eventually advance to the point at which the mechanistic nature of human decision-making is sufficiently apparent to undermine the force of dualist/libertarian intuitions.

References

Adolphs, R. (1999). Social cognition and the human brain. *Trends in Cognitive Sciences, 3*, 469–479.

Baron Cohen, S. (2000). Autism: deficits in folk psychology exist alongside superiority in folk physics. In S. Baron Cohen, H. Tager Flusberg, & D. Cohen (Eds.), *Understanding other minds: Perspectives from autism and developmental cognitive neuroscience* (pp. 78–82). New York: Oxford University Press.

Bentham, J. (1982). *An introduction to the principles of morals and legislation.* London: Methuen.

Bowles, S., & Gintis, H. (2004). The evolution of strong reciprocity: Cooperation in heterogeneous populations. *Theoretical Population Biology, 65*, 17–28.

Boyd, R., Gintis, H., Bowles, S., & Richerson, P. J. 2003 The evolution of altruistic punishment. *Proceedings of the National Academy of Sciences of the United States of America, 100*, 3531–3535.

Brosnan, S. F., & De Waal, F. B. (2003). Monkeys reject unequal pay. *Nature, 425*, 297–299.

Dennett, D. C. (1984). *Elbow room: The varieties of free will worth wanting.* Cambridge, MA: MIT Press.

Dennett, D. C. (1987). *The intentional stance.* Cambridge, MA: MIT Press.

Dennett, D. C. (1991). *Consciousness explained.* Boston, MA: Little Brown and Co.

Dennett, D. C. (2003). *Freedom evolves.* New York: Viking.

Fehr, E., & Gachter, S. (2002). Altruistic punishment in humans. *Nature, 415*, 137–140.

Fodor, J. A. (1987). *Psychosemantics: The problem of meaning in the philosophy of mind.* Cambridge, MA: MIT Press.

Frankfurt, H. (1966). Alternate possibilities and moral responsibility. *Journal of Philosophy, 66*, 829–839.

Goldstein, A. S. (1967). *The insanity defense.* New Haven, CT: Yale University Press.

Goldstein, A. M., Morse, S. J., & Shapiro, D. L. (2003). Evaluation of criminal responsibility. In A. M. Goldstein (Ed.), *Forensic psychology* (Vol. 11, pp. 381–406). New York: Wiley.

Hart, H. L. A. (1968). *Punishment and responsibility.* Oxford, UK: Oxford University Press.

Heberlein, A. S., & Adolphs, R. (2004). Impaired spontaneous anthropomorphizing despite intact perception and social knowledge. *Proceedings of the National Academy of Sciences of the United States of America, 101*, 7487–7491.

Heider, F., & Simmel, M. (1944). An experimental study of apparent behavior. *American Journal of Psychology, 57*, 243–259.

Hughs, R. I. G. (1992). *The structure and interpretation of quantum mechanics.* Cambridge, MA: Havard University Press.

Kahneman, D., & Tversky, A. (Eds.). (2000). *Choices, values, and frames.* Cambridge, UK: Cambridge University Press.

Kant, I. (2002). *The philosophy of law: An exposition of the fundamental principles of jurisprudence as the science of right.* Union, NJ: Lawbook Exchange.

Lacey, N. (1988). *State punishment: Political principles and community values.* London and New York: Routledge & Kegan Paul.

McCloskey, M., Caramazza, A., & Green, B. (1980). Curvilinear motion in the absence of external forces: Naive beliefs about the motion of objects. *Science, 210*, 1139–1141.

Mele, A. (1995). *Autonomous agents: From self-control to autonomy.* Oxford, UK: Oxford University Press.

Morse, S. J. (2004). New neuroscience, old problems. In B. Garland (Ed.), *Neuroscience and the law: Brain, mind, and the scales of justice* (pp. 157–198). New York: Dana Press.

Pettit, P. (2002). *The capacity to have done otherwise. Rules, reasons, and norms: Selected essays*. Oxford, UK: Oxford University Press.

Pincus, J. H. (2001). *Base instincts: What makes killers kill?* New York: Norton.

Pinker, S. (1997). *How the mind works*. New York: Norton.

Premack, D., & Woodruff, G. (1978). Does the chimpanzee have a theory of mind? *Behavioral and Brain Sciences, 4*, 515–526.

Rosen, G. (2002). The case for incompatibilism. *Philosophy and Phenomenological Research, 64*, 699–706.

Sanfey, A. G., Rilling, J. K., Aronson, J. A., Nystrom, L. E., & Cohen, J. D. (2003). The neural basis of economic decision-making in the ultimatum game. *Science, 300*, 1755–1758.

Saxe, R., Carey, S., & Kanwisher, N. (2004). Understanding other minds: Liking developmental psychology and functional neuroimaging. *Annual Review of Psychology, 55*, 87–124.

Scholl, B. J., & Tremoulet, P. D. (2000). Perceptual causality and animacy. *Trends in Cognitive Sciences, 4*, 299–309.

Shear, J. (Ed.). (1999). *Explaining consciousness: The hard problem*. Cambridge, MA: MIT Press.

Smart, J. J. C. (1961). Free will, praise, and blame. *Mind, 70*, 291–306.

Steinberg, L., & Scott, E. S. (2003). Less guilty by reason of adolescence: Developmental immaturity, diminished responsibility, and the juvenile death penalty. *American Psychologist, 58*, 1009–1018.

Strawson, P. F. (1962). Freedom and resentment. *Proceedings of the British Academy, xlviii*, 1–25.

Tonry, M. (2004). *Thinking about crime: Sense and sensibility in American penal culture*. New York: Oxford University Press.

Van Inwagen, P. (1982). The incompatibility of free will and determinism. In G. Watson (Ed.), *Free will* (pp. 46–58). New York: Oxford University Press.

Watson, G. (Ed.) (1982). *Free will*. New York: Oxford University Press.

Wegner, D. M. (2002). *The illusion of conscious will*. Cambridge, MA: MIT Press.

Wright, R. (1994). *The moral animal: Evolutionary psychology and everyday life*. New York: Pantheon.

Reading 5.3

The Neurobiology of Addiction: Implications for the Voluntary Control of Behavior[1]

Steven E. Hyman

Cognitive and social neuroscience and studies of the pathophysiologic processes underlying neuropsychiatric disorders have begun to probe the mechanisms by which human beings regulate their behavior in conformity with social conventions and in pursuit of chosen goals—and the circumstances under which such "cognitive control" may be eroded (Miller & Cohen, 2001; Miller & D'Esposito, 2005; Montague, Hyman, & Cohen, 2004). The resulting ideas call into question folk psychology views on the voluntary control of behavior, that is, for the most part, we regulate our actions based on conscious "reasons." Even in health, critical processes that intervene between sensory inputs to the brain and the execution of actions, including processes that permit "top-down" or "cognitive" control of behavior, do not appear to depend on conscious exertion of will (Wenger, 2002). Challenges to folk psychology views of the voluntary control of behavior may be highlighted most vividly, however, by conditions such as addiction, in which the core symptoms reflect a failure of the underlying processes (Hyman, 2005; Kalivas & Volkow, 2005; Montague et al., 2004), which I refer to as *cognitive control.*

The major justification for demarcating neuroethics from the broader field of bioethics derives from the special status of the brain (Roskies, 2002), which is the causal underpinning of our conscious mental lives and of our behavior. This is not a reductionist claim. The structure and function of the brain is influenced not only by "bottom-up" factors such as genes, but also by top-down factors such as "lived experience" and context. Moreover, neuroscience does not obviate the need for social and psychological level explanations intervening between the levels of cells, synapses, and circuits and that of ethical judgments. Indeed, modern cognitive and social neuroscience (Cacioppo, Berntson, Adolphs, & Carter, 2002; Gazzaniga, 2004) are, in no small measure, attempts to mediate between understandings of the functioning of neural networks in one regard and of sensation, thought, and action in another. What neuroscience contributes to ethical discourse is mechanistic insight that constrains our interpretations of psychological observations and that suggests new explanatory frameworks for thought and behavior. Neuroscience should make it possible to ask how the nature of our brains shapes and constrains what we call *rationality*, and therefore, ethical principles themselves, and it should permit us to probe deeply into the nature of reason, emotion, and the control of behavior (Churchland, 2006). Having recently reviewed the neurobiology of addiction for clinicians (Hyman, 2005) and for neuroscientists (Hyman, Malenka, & Nestler, 2006), I would like to examine the implications of emerging ideas about reward, cognitive control, and the pathophysiology of addiction for insights into the voluntary control of behavior.

5.3.1 Addiction and Responsibility

The question of whether and to what extent an addicted individual is responsible for his or her actions remains a matter of unsettled debate. One proxy (albeit imperfect) for this question is disagreement as to whether addiction is best conceptualized as a brain disease (Leshner, 1997; McLellan, Lewis, O'Brien, & Kleber, 2000), as a moral condition (Satel, 1999), or as some combination of the two (Morse, 2004b). Those who argue for the disease model not only believe it is justified by empirical data but also see virtue in the possibility that a disease model decreases the stigmatization of addicted people and increases their access to medical treatments. Those who argue that addiction is best conceptualized as a moral condition are struck by the observation that drug seeking and drug taking involve a series of voluntary acts that often require planning and flexible responses to changing conditions—not simply impulsive or robotic acts. They worry that medicalization will lead addicted people to fatalism about their condition and to excuses for their actions rather than full engagement with treatment and rehabilitation and an effort to conform to basic societal expectations.

Current definitions of addiction come from medical texts and thus, not surprisingly, favor a disease model. Indeed, addiction looks very much like a disease (admittedly definitions of "disease" remain somewhat fuzzy). Addiction has known risk factors (family history, male sex) and a typical course and outcome: often a chronic course punctuated by periods of abstinence followed by relapse (Hser, Hoffman, Grella, & Anglin, 2001; McLellan et al., 2000). True, the precise alterations in physiology that account for the symptoms and course are not yet known with certainty, but there is little doubt in the scientific community that such mechanisms will be found (Chao & Nestler, 2004). Similarly the search for the precise genetic variants that confer familial risk is in its early days, but existing data from family, twin, and adoption studies convincingly argue that genes play a central role in vulnerability (Goldman, Oroszi, & Ducci, 2005).

What is more interesting is that modern definitions of addiction focus squarely on the issue of voluntary control. The current medical consensus is that the cardinal feature of addiction is compulsive drug use despite significant negative consequences (American Psychiatric Association, 1994). The term *compulsion* is imprecise, but at a minimum implies diminished ability to control drug use, even in the face of factors (e.g., illness, failure in life roles, loss of job, arrest) that should motivate cessation of drug use in a rational agent willing and able to exert control over behavior. The focus on "loss of control" is not derived primarily from a theory, but from extensive observation of the behavior of addicted individuals (O'Brien, Childress, Ehrman, & Robbins, 1998; Tiffany, 1990) and indeed recognition of the failure of previous definitions to capture clinical realities. The current focus on compulsive use as the defining features of addiction superseded previous views that focused on dependence and withdrawal. These previous views implied that addicted individuals take drugs to seek pleasure and avoid aversive withdrawal symptoms. Although the avoidance of withdrawal might create strong motivation to take drugs, this view does not imply a loss of voluntary control. This previous view failed on several counts. First, some highly addictive drugs such as cocaine and amphetamine may produce mild withdrawal symptoms and lack a physical withdrawal syndrome entirely. Moreover, the previous view does not explain the stubborn persistence of relapse risk long after detoxification, long after the last withdrawal symptom, if any, has passed, and despite incentives to avoid a resumption of drug use (Hyman, 2005).

Before discussing my views of the neural basis of addiction, I should stipulate that the science is in its early stages and that there is not yet a fully convincing theory of how addiction results from the interaction of risk factors, drugs, and the brain. Moreover, there are still disagreements at the theoretical level of what the existing data signifies for the mechanisms of addiction. (Compare, for example, Hyman, 2005; Koob & Le Moal, 2005; and Robinson & Berridge, 2003). This state of affairs invites skepticism from those wary of a disease model (Satel, 1999).

Nonetheless, we cannot select models of human behavior based on desired social implications, but must rely on the scientific evidence we have. Despite somewhat different views of mechanism, all current mainstream formulations agree that addiction diminishes voluntary behavioral control. At the same time, none of the current views conceives of the addicted person to be devoid of all voluntary control and thus absolved of all responsibility for self-control.

5.3.2 Neural Bases of Addiction and Self-Control

Short of being harshly coerced, severely psychotic, or significantly demented, what can it mean to say that a person cannot control his or her actions? An alcoholic must obtain money, go to the liquor store or otherwise obtain alcohol (perhaps carefully hidden from a spouse), and consume drinks. A heroin user may have to go to great lengths to obtain the drug, perhaps committing one or more crimes, before beginning the ritual that ends in self-injection. How can these extended chains of apparently voluntary acts be the result of compulsion? In my view, addictive drugs tap into and, in vulnerable individuals, usurp powerful mechanisms by which survival-relevant goals shape behavior (Hyman, 2005; Hyman et al., 2006).

Diverse organisms, including humans, pursue goals with positive survival value such as food, safety, and opportunities for mating; such goals act as "rewards" (Kelley & Berridge, 2002). Rewards are experienced as pleasurable and as motivating (they are desired). Environmental cues that predict their availability (e.g., the smell of baking bread) are rapidly learned and are imbued with incentive properties: they activate "wanting" and initiate behaviors aimed at obtaining the desired goal. Such goal-directed behaviors tend to increase in frequency over time (reinforcement) and to become highly efficient. Of course, rewarding goals for humans can vary enormously in immediacy, complexity, and motivational power, ranging from a well-liked food to seeing a favorite painting in a museum.

The brain has evolved several specialized mechanisms to maximize the ability of an organism to obtain rewards. There are mechanisms to provide internal representations of rewards and to assign them relative values compared with pursuing other possible goals; these mechanisms depend primarily on the orbital prefrontal cortex (Schoenbaum, Roesch, & Stalnaker, 2006). There are mechanisms that permit an organism to learn and to make relatively efficient and automatic sequences of actions to obtain specific rewards; these depend primarily on the dorsal striatum (Everitt & Robbins, 2005). Mechanisms of cognitive control support successful completion of goal-directed behaviors by maintaining the goal representation over time, suppressing distractions, and inhibiting impulsive actions that redirect the organism. Cognitive control is dependent on the prefrontal cortex and its connections to the striatum and thalamus. In humans, the capacity for cognitive control appears to be a relatively stable trait that is an important predictor of life success (Eigsti et al., 2006). Deficits in cognitive control play an important role in attention deficit–hyperactivity disorder (Vaidya et al., 2005) and may increase vulnerability to later substance misuse.

These circuits respond in a coordinated fashion to new information about rewards through the action of the neurotransmitter dopamine (Montague et al., 2004). Dopamine is released from neurons with cell bodies in the ventral tegmental area (VTA) and substantia nigra within the midbrain. These neurons project widely through the forebrain and can influence all of the circuits involved in reward-related learning, as well as in other aspects of cognition and emotion. Dopamine projections from the VTA to the nucleus accumbens bind the pleasurable (hedonic) response to a reward to desire and to goal-directed behavior (Berridge & Robinson, 1998; Everitt & Robbins, 2005). Dopamine projections from the VTA to the prefrontal cortex play a critical role in the assignment of value and in updating goal representations in response to the state of the organism (Montague et al., 2004). Dopamine projections from the substantia nigra to the dorsal striatum are critical for consolidating new behavioral responses so that reward-related cues come to activate efficient strategies to reach the relevant goal (Everitt & Robbins, 2005).

Addictive drugs are Trojan horses. Unlike natural rewards, addictive drugs have no nutritional, reproductive, or other survival value. However, all addictive drugs exert pharmacologic effects that cause release of dopamine. Moreover, the effects of addictive drugs on dopamine release are quantitatively greater than that produced by natural rewards under almost all circumstances.

Normally dopamine serves as a "learning signal" in the brain. Dopamine is released when a reward is new, better than expected, or unpredicted in a particular circumstance (Schultz, 2006; Schultz, Dayam, & Montague, 1997). When the world is exactly as expected, there is nothing new to learn; no new circumstances to connect either to desire or to action—and no increase in dopamine release. Because addictive drugs increase synaptic dopamine by direct pharmacologic action, they short circuit the normal controls over dopamine release that compare the current circumstance with prior experience. Thus, unlike natural rewards, addictive drugs always signal "better than expected." Neural circuits "overlearn" on an excessive and grossly distorted dopamine signal (Hyman, 2005; Hyman et al., 2006; Montague et al., 2004). Cues that predict drug availability such as persons, places, or certain bodily sensations gain profound incentive salience and the ability to motivate drug seeking. Because of the excessive dopamine signal in the prefrontal cortex (Volkow & Fowler, 2000), drugs become overvalued compared with all other goals. Rational goals such as self-care, working, parenting, and obeying the law are devalued. In addition, normal aspects of cognitive control weaken; even if the addicted person wants to "cut down," prepotent cue-initiated drug-seeking responses are extremely difficult to suppress. If the person is successful in delaying drug seeking (or is for external reasons unable to seek drugs), intense craving may result (Tiffany, 1990). Because the changes in synaptic weight and synaptic structure that underlie memory are among the longest-lived alterations in biology, the ability of drug-related cues to cause relapses may persist for many years, even a lifetime.

5.3.3 Conclusion

There remains much to learn about the pathophysiology of addiction. Currently, much research is attempting to demonstrate that drug-induced changes in synaptic connectivity and drug-induced changes in the expression of neuronal genes and proteins are causally involved in addiction-related behaviors (Chao & Nestler, 2004; Hyman et al., 2006). This model of pathogenesis, and the research on reward-related learning on which it rests, suggest highly plausible mechanisms by which addicted individuals may "lose control" over drug seeking and drug taking (Hyman, 2005; Hyman et al., 2006; Kalivas & Volkow, 2005; Montague et al., 2004). Mechanisms that evolved to motivate survival behaviors, the pursuit of natural rewards, are usurped by the potent and abnormal dopamine signal produced by addictive drugs. The result is a brain in which drug cues powerfully activate drug seeking, and in which attempts to suppress drug-seeking result in intense craving. This model does not, however, reduce addicted individuals to zombies who are permanently controlled by external cues. As overvalued as drugs become, as potent as the effects of drug cues on behavior, other goals are not extirpated. Perhaps in a drug-free context, perhaps with a good measure of initial coercion, perhaps with family, friends, and caregivers acting as external "prostheses" to strengthen and partially replace damaged frontal mechanisms of cognitive control, and often despite multiple relapses, addicts can cease drug use and regain a good measure of control over their drug taking. Our current models help explain why recovery is difficult and why relapses occur even long after detoxification and rehabilitation. The long experience of humanity with addiction does not counsel fatalism, but implacable efforts to overcome the behavioral effects of neural circuits hijacked by drugs. Finally, views based on cognitive neuroscience and studies of addiction pathogenesis suggest that some apparently voluntary behaviors may not be as freely planned and executed as they first appear. Such cognitive views have not yet penetrated folk psychology, and it is premature for these views to have any place in the courtroom (Greene & Cohen, 2004; Morse, 2004a). Nonetheless, these cognitive views deserve a place in current ethical discussions of personal responsibility. For many reasons, it may be wise for societies to err on the side of holding addicted individuals responsible for their behavior and to act as if they are capable of exerting more control than perhaps they can; however, if the ideas expressed in this review are right, it should be with a view to rehabilitation of the addicted person and protection of society rather than moral opprobrium.

Notes

1. Editor's note: This reading originally appeared in 2007 in the *American Journal of Bioethics-Neuroscience,* volume 7, pages 8–11, and is used with permission. Section headings were added by the editor.

References

American Psychiatric Association. (1994). *Diagnostic and statistical manual of mental disorders*, 4th edition. Washington, DC: American Psychiatric Association.

Berridge, K. C., & Robinson, T. E. (1998). What is the role of dopamine in reward: Hedonic impact, reward learning, or incentive salience? *Brain Research Reviews, 28*, 309–369.

Cacioppo, J. T., Berntson, G. G., Adolphs, R., & Carter, C. S. (Eds.). (2002). *Foundations in social neuroscience*. Cambridge, MA: The MIT Press.

Chao, J., & Nestler, E. J. (2004). Molecular neurobiology of drug addiction. *Annual Review of Medicine, 55*, 113–132.

Churchland, P. S. (2006). Moral decision-making and the brain. In J. Illes (Ed.), *Neuroethics: Defining the issues in theory, practice, and policy* (pp. 3–16). Oxford, UK: Oxford University Press.

Eigsti, I. M., Zayas, V., Mischel, W., Schoda, Y., Ayduk, O., Dadlani, M. B., et al. (2006). Predicting cognitive control from preschool to late adolescence and young adulthood. *Psychological Science, 17*(6), 478–484.

Everitt, B. J., & Robbins, T. W. (2005). Neural systems of reinforcement for drug addiction: From actions to habits to compulsion. *Nature Neuroscience, 8*, 1481–1489.

Gazzaniga, M. S. (Ed.). (2004). *The cognitive neurosciences III*, 4th edition. Cambridge, MA: The MIT Press.

Goldman, D., Oroszi, G., & Ducci, F. (2005). The genetics of addictions: Uncovering the genes. *Nature Reviews Genetics*, 6, 521–532.

Greene, J., & Cohen, J. (2004). For the law, neuroscience changes nothing and everything. *Philosophical Transactions of the Royal Society of London. Series B, Biological Sciences, 359*, 1775–1785.

Hser, Y. I., Hoffman, V., Grella, C. E., & Anglin, M. D. (2001). A 33-year follow-up of narcotics addicts. *Archives of General Psychiatry, 58*(5), 503–508.

Hyman, S. E. (2005). Addiction: A disease of learning and memory. *American Journal of Psychiatry, 162*(8), 1414–1422.

Hyman, S. E., Malenka, R. C., & Nestler, E. J. (2006). Neural mechanisms of addiction: The role of reward-related learning and memory. *Annual Review of Neuroscience 21*(29), 565–598.

Kalivas, P. W., & Volkow, N. D. (2005). The neural basis of addiction: A pathology of motivation and choice. *American Journal of Psychiatry, 162*(8), 1403–1413.

Kelley, A. E., & Berridge, K. C. (2002). The neuroscience of natural rewards: Relevance to addictive drugs. *Journal of Neuroscience, 22*, 3306–3311.

Koob, G. F., & Le Moal, M. (2005). *Neurobiology of addiction*. New York: Academic Press.

Leshner, A. I. (1997). Addiction is a brain disease, and it matters. *Science, 278*(5335), 45–47.

McLellan, A. T., Lewis, D. C., O'Brien, C. P., & Kleber, H. D. (2000). Drug dependence, a chronic medical illness: Implications for treatment, insurance, and outcomes evaluation. *Journal of the American Medical Association, 284*(13), 1689–1689.

Miller, B. T., & D'Esposito, M. (2005). Searching for "the top" in top-down control. *Neuron, 48*, 535–538.

Miller, E. K., & Cohen, J. D. (2001). An integrative theory of prefrontal cortex function. *Annual Review of Neuroscience, 24*, 167–202.

Montague, P. R., Hyman, S. E., & Cohen, J. D. (2004). Computational roles for dopamine in behavioural control. *Nature, 431*, 760–767.

Morse, S. J. (2004a). New neuroscience, old problems: Legal implications of brain science. *Cerebrum, 6*(4), 81–90.

Morse, S. J. (2004b). Medicine and morals, craving and compulsion. *Substance Use & Misuse, 39*(3), 437–460.

O'Brien, C. P., Childress, A. R., Ehrman, R., & Robbins, S. J. (1998). Conditioning factors in drug abuse: Can they explain compulsion? *Journal of Psychopharmacology, 12*, 15–22.

Robinson, T. E., & Berridge, K. C. (2003). Addiction. *Annual Review of Psychology, 54*, 25–53.

Roskies, A. (2002). Neuroethics for the new millennium. *Neuron, 35*(1), 21–23.

Satel, S. L. (1999). What should we expect from drug abusers? *Psychiatric Services, 50*(7), 861.

Schoenbaum, G., Roesch, M. R., & Stalnaker, T. A. (2006). Orbitofrontal cortex, decision-making and drug addiction. *Trends in Neurosciences 29*(2), 116–124.

Schultz, W. (2006). Behavioral theories and the neurophysiology of reward. *Annual Review of Psychology, 57*, 87–115.

Schultz, W., Dayan, P., & Montague, P. R. (1997). A neural substrate of prediction and reward. *Science, 275*, 1593–1599.

Tiffany, S. T. (1990). A cognitive model of drug urges and drug-use behavior: Role of automatic and nonautomatic processes. *Psychological Review, 97*, 147–168.

Vaidya, C. J., Bunge, S. A., Dudukovic, N. M., Zalecki, C. A., Elliott, G. R., & Gabrieli, J. D. (2005). Altered neural substrates of cognitive control in childhood ADHD: Evidence from functional magnetic resonance imaging. *American Journal of Psychiatry, 162*(9), 1605–1613.

Volkow, N. D., & Fowler, J. S. (2000). Addiction, a disease of compulsion and drive: Involvement of the orbitofrontal cortex. *Cerebral Cortex, 10*, 318–325.

Wenger, D. M. (2002). *The illusion of conscious will.* Cambridge, MA: The MIT Press.

Reading 5.4

Brain Overclaim Syndrome and Criminal Responsibility: A Diagnostic Note[1]

Stephen J. Morse

Brains do not commit crimes; people commit crimes. This conclusion should be self-evident, but, infected and inflamed by stunning advances in our understanding of the brain, advocates all too often make moral and legal claims that the new neuroscience does not entail and cannot sustain. Particular brain findings are thought to lead inevitably to moral or legal conclusions. Brains are blamed for offenses; agency and responsibility disappear from the legal landscape. For example, in *Roper v. Simmons*,[2] advocates for abolition of the death penalty for adolescents who committed murder when they were 16 or 17 years old argued that the demonstrated lack of complete myelination of the cortical neurons of the adolescent brain was reason to believe that 16- and 17-year-old murderers were insufficiently responsible to deserve capital punishment. These types of responses, I claim, are the signs of a disorder that I have preliminarily entitled brain overclaim syndrome (BOS).

This brief diagnostic note first lays the contextual foundation for how one should think about the relation of neuroscience to criminal responsibility. Then it attempts to identify the nature of the pathology, to offer the criteria for the diagnosis, to evaluate the disorder in the *Roper* arguments, and to suggest the route to total cure. Footnotes will be scarce. Most of this note is an extended conceptual argument or based on this investigator's first-hand clinical observations. Where it depends on assertions about the state of the science, it takes positions, albeit sometimes controversial, that have strong support. Trust me: I'm a doctor (of psychology).

5.4.1 Brain Overclaim Syndrome: The Signs and Symptoms

New, powerful scientific findings about the correlates and causes of behavior often have a potent and, alas, rationality-unhinging effect on the thinking of potential commentators (PCs). Most flow from misunderstanding the relation between brains and responsibility. This section attempts to catalogue those effects that I have identified to date, many of which are related to each other, but the list has no pretensions to being complete. After all, this is a preliminary diagnostic investigation and future investigators may discover hitherto unidentified signs and symptoms. The final pathway in all cases, however, is that more legal implications are claimed for the brain science than can be justified.

5.4.2 Confusion about the Relation between Brain and Complex, Intentional Action

For a materialist, the brain always plays a causal role in behavior. Despite all the astonishing recent advances in neuroscience, however, we still know woefully little about how the brain enables the mind, and especially about how consciousness and intentionality can arise from the complicated hunk of matter that is the brain. At a recent conference on the abnormal brain, the eminent philosopher of mind and action, John Searle, opened his keynote speech by telling the following anecdote.[3] Some years ago, Searle said, he decided to learn what the new neuroscience had to teach about the relation of brain to mind and action. He devoured the most important texts only to be dismayed that these texts did not all begin with a disclaimer that we do not know much about this relation yet. Just so.

Brain imaging studies have been the most potent pathogen causing BOS, so it is useful to say a few words about such studies. Imaging is at present very expensive and requires carefully chosen and cooperative subjects. Consequently, the number of experimental subjects and controls in any study tends to be small, and precise replications are infrequent. The problem of small samples will probably be remedied by advances in the efficiency of the technology of imaging—indeed, this is already happening for readings of activity at the surface of the brain—but for now it is a dominant feature of imaging studies.

Statistically valid findings are based on mean differences and do not imply that there is an absolutely clear distinction between the experimental and control groups. Usually there is substantial overlap, meaning that some individual experimental brains look like individual control brains and vice versa. For example, suppose the experimental hypothesis is that task X will cause brain region Y to be activated. After controlling for other variables that might cause Y to be activated in both the experimental and control conditions (the "subtraction" method), the investigators discover that there is still a difference: Y is activated statistically significantly more in the experimental subjects. Nonetheless, some experimental subjects will not have Y activated by X and some control subjects will. Therefore, one could not predict perfectly from the brain image whether the subject was an experimental or a control. The question would always be how much overlap there was between the two groups. The greater the overlap, the more difficult it would be to predict that subject's experimental or control status from the image.[4]

Discovering the neural correlates of mental phenomena does not tell us how these phenomena are possible. For example, we may be able to identify the neural correlates of consciousness, but we do not have a clue about how those parts of the brain make subjective experience possible. Moreover, the causation of virtually any complex behavior is affected by psychological and sociological variables, even when brain causation has been identified. For example, the brains of late adolescents are almost certainly the same around the globe—holding nutrition and the like constant—but the rates of behaviors associated with immature adolescent brains, such as impulsive criminality, vary widely from place to place and from time to time. Monolithic brain explanation of complex behavior is almost always radically incomplete.

Certain lesions can of course disable various human capacities, but few criminal responsibility cases in which the result is not already obvious based on behavioral evidence will involve a precise, identifiable neurologic mechanism that will demonstrate that criminal responsibility was not present. Further, current neuroscience cannot begin to demonstrate that our view of ourselves as generally conscious, intentional, and potentially rational agents is false.

Until we know vastly more than we do now, in most cases we will not be in a position to add much to assessing responsibility behaviorally in individual cases, and even less do we have the resources to mount a potentially convincing external critique of responsibility *vel non*. In individual cases in which neuroscientific findings could be demonstrated to be genuinely relevant and probative, they should of course be admissible consistent with the usual evidentiary standards for scientific evidence.

5.4.3 The Confusion of Internal and External Critiques

There are two ways that neuroscience might change our view of criminal responsibility, which I call the "internal" and "external" critiques. The latter is external in the sense that it challenges our entire framework for thinking about responsibility, whereas the former works within that framework. Consider two classes of internal critique. In the first are those cases in which the behavioral evidence concerning *prima facie* liability or an affirmative defense seems clear, and neuroscientific evidence demonstrates that appearances are deceptive. For example, neuroscience might indicate that a defendant who appeared to have been acting consciously was in fact acting in an unconscious or automatic state, such as sleepwalking or in the wake of physical trauma.

A second class of specific cases involves those in which the behavioral evidence is in doubt. For example, suppose the defendant had received a blow to the head not long before committing an offense and whether the defendant acted consciously is unclear. Neuroscience will rarely be dispositive in such cases because the relation of the brain to complex behavior is itself immensely complex and beyond all but the most general current understanding. Nonetheless, in some cases valid neuroscience will help the finder of fact resolve the legal issue, although caution must always be exercised because the neuroscientific evidence often will not be sufficiently contemporaneous to permit valid inferences about the time of the crime.

The external critique, in contrast, does not engage the question of whether a defendant was conscious, intentional, or rational when committing the crime, because it denies that deserved blame and punishment are conceptually and morally justified and coherent. This critique is based on the claims that determinism, whether based on neuroscience or any other science, is true or that our mental states play no role in explaining our behavior. The truth of either claim is allegedly inconsistent with the very possibility of robust responsibility. PCs in the grips of BOS often confuse the internal and external critiques. Causation is not *per se* an excusing condition and partial causation does not exist. If this is a causal world, as neuroscientists and I believe, then all phenomena are fully caused by their necessary and sufficient causal conditions. Partial knowledge about causation does not mean that there is partial causation. Causation is also not the equivalent of being subjected to compulsion, which exists when the agent is non-culpably faced with a normative hard choice. All behavior is caused, but not all behavior results from a threat at gunpoint or the equivalent. And to think that causation *per se* excuses is an external critique. If causation excuses *per se* is the equivalent of compulsion, responsibility as we know it is impossible. The discovery that the brain, including a brain abnormality, played some causal role in the production of what is undeniably human action does not lead to any legal conclusions about responsibility. The proper internal question is whether the neuroscience evidence helps to establish the presence or absence of action, mental states, or a genuine affirmative defense, such as lack of rational capacity.

5.4.4 Misunderstanding the Criteria for Responsibility

The criteria for responsibility are behavioral and normative, not empirically demonstrable states of the brain. Even if there were a perfect correlation between brain states and the behavioral criteria for responsibility, the brain states would be nothing more than evidence of the behavioral states. Such a correlation is a fantasy based on current knowledge and probably always will be when we are considering complex human actions. If the person meets the behavioral criteria for responsibility, the person should be held responsible, whatever the brain evidence may indicate, such as the presence of an abnormality. If the person does not meet the behavioral criteria, the person should be held not responsible, however normal the brain may look. Brains are not held responsible. Acting people are. To believe that brain evidence has more than simple evidentiary value for assessing responsibility is to misconceive the criteria for responsibility.

One could claim, of course, that normatively the law should adopt brain-based criteria for responsibility, but this would be a category mistake. Even if it is not, it amounts to an external critique. In the alternative, one could argue that the positive account of responsibility that I have presented is fundamentally incorrect and that the brain science is more relevant to the properly understood criteria of responsibility. Perhaps so, but this requires an argument to demonstrate that the account is wrong. Finally, one could argue that the behavioral and normative criteria should be different as a result of what we have learned from brain science and other disciplines. This would be an internal critique that would depend on a normative argument about the relevance of brain science to responsibility. If this argument went through, however, it would not undermine this diagnostic note's claims about the relevance of brain science to current, positive responsibility criteria, and, even then, the brain science would still be relevant only as evidence concerning the new, improved behavioral responsibility criteria.

5.4.5 The Confusion of the Normative and the Positive

Factual behavioral differences between people do not entail the necessity of differential legal treatment unless one is operating under a normative theory that indicates why the factual difference should make a legal difference. I am not suggesting that it is impossible to *derive* an "ought" from an "is," a contentious issue most famously addressed by David Hume. I am agnostic about this. I am claiming, however, that one cannot *assume* an "ought" from an "is." This requires an argument.

Suppose that we can reliably identify valid group differences, say, between men and women on measures of upper body strength, a capacity useful in some occupations, such as fighting fires. Should the ranks of firefighters be limited to men? Of course not, because we might decide that values of equality trump those of efficiency or because we think that we can individualize decisions about an applicant's ability to do the job. Even if proportionately fewer women might qualify, some surely will, and we would not be able to predict whether an applicant could do the job based solely on sex. Virtually no finding, no practice, however hoary, necessarily entails any normative outcome without an argument about why it should. This is as true of neuroscience evidence as of any other kind of scientific evidence. Neuroscience evidence may provide premises in normative arguments, but it does not alone entail conclusions. To think otherwise is to confuse the positive with the normative.

In conclusion, based on the foregoing confusions and others that may be identified, the final pathway, the final expression of BOS is to make claims about the relation of the brain to responsibility that cannot be sustained logically or empirically.

5.4.6 Evaluating *Roper*

Few if any responsible commentators who accept the coherence and validity of criminal responsibility ascriptions—the internalists—claim that most adolescent offenders commit their crimes in automatic states or without *mens rea*. Crimes committed impulsively, for example, are still committed consciously and intentionally. Nor do most commentators claim that late adolescent offenders do not know the nature and quality of their acts, do not know their acts are wrong, or act in response to duress. No evidence from the behavioral or neurosciences even hints that the contrary might be true. Rather, the claim is that culpable adolescents, whose behavior meets the *prima facie* case for guilt and who do not have an affirmative defense, are nonetheless less criminally responsible because they have insufficiently developed rationality. Thus, to be relevant, any evidence must be addressed to the 16 and 17 year olds' capacity for rationality, broadly speaking.

In *Thompson v. Oklahoma*,[5] the Supreme Court barred capital punishment of murderers who killed when they were 15 years old or younger, and in *Atkins v. Virginia*,[6] the High Court categorically prohibited capital punishment of convicted killers with mental retardation. Although the Court provided many reasons for its *Thompson* and *Atkins* holdings, crucial to both was the conclusion that younger adolescents and persons with retardation are categorically less culpable, less responsible, and therefore do not deserve capital punishment. The operative language in *Atkins* concerning culpability and responsibility is instructive. The Court wrote:

> Mentally retarded persons frequently know the difference between right and wrong.... Because of their impairments, however, by definition they have diminished capacities to understand and process information, to communicate, to abstract from mistakes and learn from experience, to engage in logical reasoning, to control impulses, and to understand the reactions of others.... Their deficiencies do not warrant an exemption from criminal sanctions, but they do diminish their personal culpability.
>
> [...]
>
> With respect to retribution—the interest in seeing that the offender gets his "just deserts"—the severity of the appropriate punishment necessarily depends on the culpability of the offender.[7]

All the criteria the Court mentions are behavioral (broadly understood to refer to cognitive and "control" functioning[8]) and their relevance to criminal responsibility is based on the relation to desert.

Advocates of abolition in *Roper* seized on this language to make similar arguments concerning 16- and 17-year-old murderers. Although apparently normal adolescents do not suffer from abnormal impairments, lack of full developmental maturation allegedly distinguishes them from adults on behavioral dimensions, such as the capacity for judgment, that are relevant to rationality and therefore to responsibility and desert.

What was striking and new about the argument in *Roper*, however, was that advocates of abolition used newly discovered neuroscientific evidence concerning the adolescent brain to bolster their argument that 16- and 17-year-old killers do not deserve to die. Editorial pages encouraged the High Court to consider the neuroscientific evidence to help it reach its decision. Although neuroscience evidence had been adduced in earlier, high-profile cases, such as the 1982 prosecution of John Hinckley, Jr. for the attempted assassination of President Reagan and others, *Roper* has been the most important case to propose use of the new neuroscience to affect responsibility questions generally. Indeed, the American Medical

Association, the American Bar Association, the American Psychiatric Association, and the American Psychological Association, among others, all filed or subscribed to *amicus* briefs urging abolition based in part on the neuroscience findings. The real question was whether and how the new neuroscience was relevant to responsibility ascriptions and just punishment for adolescent offenders (or anyone else).

Here is the opening of the summary of the *amicus* brief filed by, *inter alia*, the American Medical Association, the American Psychiatric Association, the American Academy of Child and Adolescent Psychiatry, and the American Academy of Psychiatry and the Law: "The adolescent's mind works differently from ours. Parents know it. This Court [the United States Supreme Court] has said it. Legislatures have presumed it for decades or more."[9]

Precisely. The brief points to evidence concerning impulsivity, poor short-term risk and long-term benefit estimations, emotional volatility, and susceptibility to stress among adolescents compared with adults. These are commonsense, "fireside" conclusions that parents and others have drawn in one form or another since time immemorial. In recent years, common sense has been bolstered by methodologically rigorous behavioral investigations that have confirmed ordinary wisdom. Most important, all these behavioral characteristics are clearly relevant to responsibility because they all bear on the adolescent's capacity for rationality. Without any further scientific evidence, advocates of abolition would have an entirely ample factual basis to support the types of moral and constitutional claims they made.

The *Roper* briefs were filled with discussion of new neuroscientific evidence that confirms that adolescent brains are different from adult brains in ways consistent with the observed behavioral differences that alone bear on culpability and responsibility. Assuming the validity of the neuroscientific evidence, what does it add? The rigorous behavioral studies already confirm the behavioral differences. No one thinks that these data are invalid because adolescent subjects are faking or for some other reason. The moral and constitutional implications of the data may be controversial, but the data are not. At most, the neuroscientific evidence provides a partial causal explanation of why the observed behavioral differences exist and thus some further evidence of the validity of the behavioral differences. It is only of limited and indirect relevance to responsibility assessment, which is based on behavioral criteria.

Advocates claimed, however, that the neuroscience confirmed that adolescents are insufficiently responsible to be executed, thus confusing the positive and the normative. The neuroscience evidence in no way independently confirms that adolescents are less responsible. If the behavioral differences between adolescents and adults were slight, it would not matter if their brains are quite different. Similarly, if the behavioral differences were sufficient for moral and constitutional differential treatment, then it would not matter if the brains were essentially indistinguishable.

Decisions regarding whether the mean differences are large enough and whether the overlap between the two populations is small enough to warrant treating adolescents differently categorically as a class rather than trying to individuate responsibility are normative, moral, political, social, and ultimately legal constitutional questions about which behavioral and neuroscience must finally fall silent. Even if there were virtually no behavioral or brain overlaps between, say, 16 and 17 year olds on the one hand and 18 and 19 year olds on the other, it would still not entail that we must categorize rather than individuate. After all, because there is overlap—indeed, substantial overlap in the groups just mentioned—we know that some 16 and 17 years olds will be behaviorally and neurologically indistinguishable from many 18 and 19 year olds. Finally, even if there were no behavioral or brain overlap whatsoever, it would still not entail that abolition was constitutionally mandated. As a normative matter, the Court could decide that 16 and 17 year olds are responsible enough to be executed despite all of them being less responsible than older murderers. Assuming the validity of the findings of behavioral and biological difference, the size of that difference entails no necessary moral or constitutional conclusions.

In the event, the *Roper* majority cited many reasons for its decision, including the abundant commonsense and behavioral science evidence that adolescents differ from adults. This evidence demonstrates, said the Court, "that juvenile offenders cannot with reliability be classified among the worst offenders," for whom capital punishment is reserved. The Court cited three differences: adolescents have "[a] lack of maturity and an underdeveloped sense of responsibility"[10]; adolescents are more "vulnerable or susceptible to negative influences and outside pressures, including peer pressure," a difference in part explained by the adolescent's weaker control or experience of control over his or her own environment; adolescents do not have fully formed characters.[11] As a result of these factors—all of which, we may note, are behavioral and all of which can be confirmed with behavioral evidence alone—juvenile culpability is diminished and the penological justifications for capital punishment apply to adolescents with "lesser force."[12] The Court's opinion thus reflects two conclusions: the group difference between the rationality of late adolescents and of adults is constitutionally significant for Eighth Amendment purposes and it is large enough to justify abandoning individualized decision-making concerning responsibility for the former.

Characteristically, the Court did not cite much evidence for the empirical propositions that supported its diminished culpability argument. What is notable, however, is that the Court did not cite *any* of the neuroscience evidence concerning myelination and pruning that the *amici* and others had urged them to rely on. It did cite six behavioral sources, five of which were high-quality behavioral science. Perhaps the neuroscience evidence actually played a role in the decision, as many advocates for the use of neuroscience would like to believe, but there is no evidence in the opinion to support this speculation.

As this note has argued, the behavioral science was crucial to proper resolution of the case and furnished completely adequate resources to decide the issue. The neuroscience was largely irrelevant. The reasoning of the case is consistent with this argument and the opinion showed no signs of brain overclaim syndrome. In my view, *Roper* properly disregarded the neuroscience evidence and thus did not provide unwarranted legitimation for the use of such evidence to decide culpability questions generally.

5.4.7 The Royal Road to Complete Recovery: Cognitive-Jurotherapy (CJ) for BOS

The signs and symptoms of BOS are all cognitive. I therefore propose that CJ is likely to be the best treatment. The therapeutic techniques, all of which require motivation, effort, and practice, follow directly from the signs and symptoms of BOS.

First, the Potential Commentator must have a good understanding of the relevance of the new neuroscience to complex behavior generally, including an understanding of the relevant literature in philosophy of mind. Reasonable minds can differ about the basic neuroscience and the philosophy of mind, of course, and disagreement is not a sign of BOS. But naive neuroscience and philosophy of mind and question-begging about these subjects are signs, although they are completely curable.

Second, the PC must understand whether his or her contribution is an internal or external criticism, and what type it is within the two broad domains. Confusion between and within the critical domains must at all costs be avoided. This is also simple enough if one understands the distinctions.

Third, the PC must be very clear about precisely what criteria for criminal responsibility he or she is using and about whether it is a positive account of the current state of the law or a proposed account of what the law should be. There may of course be disagreement about the current state of the law, and, once again, disagreement is not a sign or symptom of BOS. But using naive criteria and question-begging about the criteria, without argument, are signs. If the PC is offering a proposed account, the PC should set forth the argument for why the legal system should accept this account. In either case, the criteria should be clear enough to permit reasonably apparent conclusions about the relevance of brain evidence to those criteria.

Fourth, the PC must understand the positive/normative distinction, and if he or she wishes to use brain findings as premises in an argument for legal change, the normative reasons for preferring the change should be crisply identified.

All of the above is really just a "high falutin," partially tongue-in-cheek way of suggesting that people need to think more clearly and make more transparent, logical arguments about the relationship of anything to criminal responsibility. The question is why more PCs do not do this. I do not know the answer, but I suspect that two primary culprits are at work: intellectual naiveté and ideological blinders. Sophisticated, non-hand-waving treatment of these issues requires a lot of capital investment by lawyers in disciplines outside the law and by non-lawyers in the law. Many PCs do not have the capital, but this is easily remedied by appropriate investment.

Ideological blinders are harder to fix, and sometimes it is not clear what role ideology is playing. Is the PC making an argument he or she knows is not the best argument because it supports his or her position and it does (barely) pass the "smell test"? I suppose that this is less objectionable for a practitioner than for a scholar, although it is less objectionable even for scholars if they are openly engaging in advocacy. Or, is the desire to achieve a certain result so important that the PC does not even recognize that the argument deployed is weak? This is a problem for anyone.

My impression is that most people who wish to inject neuroscience into criminal responsibility assessments believe that the neuroscience must necessarily be exculpatory. We have seen that this does not follow, and, indeed, even if neuroscience could be demonstrated to be routinely relevant, it is a knife that cuts both ways. Unless one makes the fundamental psycholegal error of believing that causation *per se* excuses, it is clear that neuroscience might also be a means to inculpate. One is reminded of the analogy to DNA evidence. For the moment it is being used extensively to exculpate alleged murderers on death row, but as many have pointed out, if inaccuracy is the primary criticism of application of the death penalty, DNA could erode that critique and give new impetus to capital punishment.

As the biological and behavioral sciences offer ever more sophisticated understandings of normal and abnormal behavior alike, there will be constant pressure to use their findings to affect assessment of criminal responsibility and other legal doctrines. A lot will be at stake morally, politically, and legally, and much will be debatable. I hope, however, that this modest contribution will help identify and ameliorate a pathologic entity that can deleteriously affect the debate.

Notes

1. Editor's note: This reading was excerpted from a substantially longer article published in 2006 in the *Ohio State Journal of Criminal Law*, volume 3, pages 397–412, and is used with permission of the publisher and author. The omitted text primarily concerns the law's concept of responsibility and distinguishes among different types of critiques of that concept. The author's acknowledgments follow: "This diagnostic investigation was first reported at a conference on the mind of a child held at the Moritz College of Law at The Ohio State University in March 2005. I would like to thank Kate Federle and Joshua Dressler for their efforts. Stephanos Bibas, Ed Greenlee, and Dave Rudovsky provided invaluable help. As always, my personal attorney, Jean Avnet Morse, furnished sound and sober counsel and moral support."

2. 543 U.S. 551 (2005).

3. John Searle, Keynote Address at the Arizona State University College of Law Conference: The Abnormal Brain and Criminal Responsibility (Apr. 29, 2005).

4. At the conference at which this diagnostic note was first presented, I asked with what accuracy, based only on the images of myelination, one could accurately distinguish the individual brains of 16 and 17 year olds on the one hand and 18 and 19 year olds on the other. One neuroscientist claimed that the scientist could do this with great accuracy if the scientist were furnished with the sex and handedness of the subject. This claim would have been quite believable if the comparison groups were 13 and 14 year olds versus 25 and 26 year olds, but it seemed doubtful to me because development is continuous and the groups were so close in age. After the conference, I therefore asked the question of other equally credentialed and experienced neuroscientists and neuroanatomists. My informants uniformly agreed that they could not very accurately distinguish

16- and 17-year-old brains from 18- and 19-year-old brains. I do not know who is right. To the best of my knowledge, a study to determine accuracy of this type has not been performed, but the outcome would be very interesting.

5. 487 U.S. 815 (1988).

6. 536 U.S. 304 (2002).

7. *Id.* at 318–19; see also *Thompson*, 487 U.S. at 834–35.

8. I put "control" in scare quotes because I am highly skeptical about claims concerning lack of control as an independent mitigating or excusing condition. I argue that lack of control can always be reduced to a cognitive deficiency. Stephen J. Morse, Uncontrollable Urges and Irrational People, 88 *Va. L. Rev.* 1025, 1054–63 (2002).

9. Brief of the American Medical Association et al. as Amici Curiae in Support of Respondent at 2, *Roper v. Simmons*, 543 U.S. 551 (2005) (no. 03-633).

10. *Roper*, 543 U.S. at 569 (quoting *Johnson v. Texas*, 509 U.S. 350, 367 (1993)).

11. *Id.*

12. *Id.* at 571.

Reading 5.5

State-Imposed Brain Intervention: The Case of Pharmacotherapy for Drug Abuse[1]

Richard G. Boire

The compelled use of pharmacotherapy would raise a number of constitutional and other legal issues. Inasmuch as the U.S. government has adopted an illness metaphor and expressly analogized the use/abuse of drugs to cancer, the potential for constitutional violations is underscored. It is well known and widely accepted that treating cancer often requires drastic measures, which knowingly compromise the health of other body systems. When treating cancer, notes Susan Sontag, "[i]t is impossible to avoid damaging or destroying healthy cells (indeed, some methods used to treat cancer can cause cancer), but it is thought that nearly any damage to the body is justified if it saves the patient's life."[2] Inasmuch as substantial damage has already been done to the U.S. Constitution in order to fight the war on drugs,[3] all signs foreshadow a continued narrowing of our rights, justified as an unavoidable side-effect of waging war on the "cancer" of drug use/abuse.

Among the rights implicated by compulsory use of pharmacotherapy drugs is the right to provide informed consent before receiving medical treatment, the constitutional protections against cruel and unusual punishment, bodily integrity, and privacy, and the right to freedom of thought. Because our assessment leads us to conclude that three segments of society (prisoners, probationers, and public benefit recipients) are most likely to come under pressure for compulsory pharmacotherapy, we address their unique concerns within each section.

5.5.1 The Right to Informed Consent

The 2002 *National Drug Control Strategy* report coined the term "compassionate coercion" and promoted it as a key element for success. The White House press release announcing the report explained that in addition to pressure from family, friends, employers, and the community, "[c]ompassionate coercion also uses the criminal justice system to get people into treatment."[4]

How exactly "compassionate coercion" will work in practice has yet to be seen, but the term itself, especially when accompanied by statements like those surrounding the 2002 and 2003 reports, foreshadow interventionist government actions that would be at stark odds with a number of well-established legal rights.

The Supreme Court has recognized a constitutional right to bodily integrity that includes the right of a person to make voluntary and informed decisions about medical treatment.[5] The government's concept of "compassionate coercion" appears to turn upside-down the individual's right to informed consent. All fifty states have laws that protect informed consent. These laws require that before performing medical procedures or treatments, medical personnel must make certain disclosures to patients and obtain the patient's consent.[6]

In general, informed consent requires the satisfaction of two conditions. First, trained medical personnel must tell the person to be treated what alternative treatments exist, the benefits and dangers associated with the proposed treatment, and the disadvantages of forgoing treatment. Second, once the person has received all the relevant medical information, he or she must freely and voluntarily decide whether or not to undergo the treatment.[7] Coercion is anathema to informed consent, as emphasized by a U.S. Department of Health and Human Services regulation defining informed consent:

> [I]nformed consent means the knowing consent of an individual or his legally authorized representative, so situated as to be able to exercise free power of choice without undue inducement or any element of force, fraud, deceit, duress, or other form of constraint or coercion.[8]

Coercion, whether "compassionate" or otherwise, is still coercion. Indeed, "compassionate coercion" can be more insidious. As one of America's most prominent Supreme Court justices warned decades ago: "Experience should teach us to be most on our guard to protect liberty when the government's purposes are beneficent.... The greatest dangers to liberty lurk in insidious encroachment by men of zeal, well meaning but without understanding."[9]

Although it is a criminal offense to use or possess drugs like marijuana, opium, and cocaine for nonmedical purposes,[10] a person who desires medical treatment for his or her drug use does not forfeit the right to decide whether to utilize drug-based medical treatment. There is no "drug war exception" to informed consent requirements. The only exception to requirements for informed consent arises when a person has been declared mentally incompetent or is too young to make his or her own medical decisions. The overwhelming majority of people who use illegal drugs do not fall into either category and thus the doctrine of informed consent should stand as a strong barrier to coercive pharmacotherapy.

5.5.1.1 Prisoners

People serving time in prison—especially for drug offenses—would appear to be prime candidates for coercive pharmacotherapy. Prisoners are politically weak and generally regarded unsympathetically by the general populace. Further, prisoners appear to be one of the express targets for "compassionate coercion," which "uses the criminal justice system to get people into treatment."[11]

Although prisoners do not enjoy the same rights as nonincarcerated Americans, prisoners retain their right to give informed consent before being the subject of a medical procedure or treatment. Thus, unless a court has determined that a prisoner is mentally incompetent, the informed consent requirements discussed in the previous section retain their validity within the prison context.

Whereas no court, let alone the United States Supreme Court, has ruled on the circumstances in which a prisoner can be forced to undergo pharmacotherapy for illegal drug use, the Supreme Court has placed strict limits on when prison officials can force a prisoner to take psychotropic medication.[12]

A prisoner can be compelled to take psychiatric medication in only two circumstances. First if he suffers from a serious mental illness that renders him mentally incompetent to make his own medical decisions, prison medical authorities are permitted to forcibly treat the prisoner, so long as the treatment is in the best interests of the prisoner and complies with due process.[13] Second, a prisoner whose mental illness leads him or her to engage in dangerous behavior that threatens to harm other prisoners or prison staff may be forcibly treated with psychotropic medication.[14] This ruling is based on the unique safety and security issues within prisons.

These rulings instruct that in all but extraordinary circumstances (those meeting the factors noted in the above cases), prison authorities would be acting unlawfully if they were to compel a prisoner to take a pharmacotherapy drug against his or her will. The mere fact that a person has used illegal drugs is not regarded by clinicians as a "mental disorder," and an insufficient reason to find the person mentally incompetent. Thus, unless a prisoner is dangerous to others or is truly mentally incompetent, he or she has a right to refuse pharmacotherapy drugs and a right to give informed consent before receiving them.

5.5.1.2 Probationers

The overwhelming majority of people charged with violating federal or state drug prohibition laws are placed on probation, rather than incarcerated.[15] The United States Supreme Court has held that the purpose of probation in criminal cases is to provide a period of grace in order to aid the rehabilitation of an offender.[16] Most states have laws that require sentencing courts to impose various requirements on probationers, which must be satisfied in order to successfully complete probation and thereby avoid spending time in custody. So long as they are reasonably related to rehabilitation and are not blatantly unconstitutional, relatively few limitations exist on a trial judge's discretion to impose particular probation conditions. As one law review author noted:

> Courts have quite accurately described the scope of the sentencing court's discretion as "breathtaking," and commentators have observed that any legislative limitations on that discretion are "conspicuously absent." One recent media account suggested that the content of special conditions "is limited only by the sentencing judge's imagination."[17] [Footnotes omitted.]

People granted probation in drug cases are routinely required to: waive their Fourth Amendment rights by agreeing to be searched at any time; submit to regular and sometimes random drug testing; and successfully complete a drug treatment program.[18] Considered in light of the federal government's acknowledgment that "compassionate coercion also uses the criminal justice system to get people into treatment," some future courts may attempt to impose pharmacotherapy as a condition of granting probation in drug cases. Several states have enacted laws that authorize courts to impose the use of Norplant or Depo-Provera on grants of probation. Further, though no reliable statistics exist on its prevalence, some courts have conditioned a grant of probation for alcohol-related offenses on the probationer using Antabuse.[19]

Informed consent requires, at a bare minimum, adequate information about the possible risks and benefits of a given medical treatment, as well as an environment free of coercion. A criminal courtroom is an unlikely venue for satisfying either requirement for informed consent. Few judges, prosecutors, probation officers, or defense attorneys have the medical training necessary to make the required advisements to the defendant. Thus, a defendant who is offered probation on the condition that he or she undergo pharmacotherapy will likely be placed in the position of having to make a medical treatment decision without the appropriate information, thereby vitiating *informed* consent. Further, being forced to choose between imprisonment or "medical treatment" with a pharmacotherapy drug is inherently coercive. There are very few things that people will avoid more than going to jail or prison. Informed consent is incompatible with inherently coercive situations that force a person to barter his or her natural neuro- and biochemistry in exchange for freedom.

5.5.1.3 Public Assistance Recipients

The history of legislation seeking to link certain public assistance benefits with the use of the implantable contraceptive Norplant suggests that future legislation might be premised on conditioning certain public benefits, in particular, "welfare," on the use of pharmacotherapy. Such legislation, should it be introduced, would raise substantial informed consent concerns. The Center for Cognitive Liberty and Ethics (CCLE) can anticipate several versions of legislation that would connect public aid to pharmacotherapy. In order of increasing concern, these are (1) offering to reimburse public benefit recipients for the cost of undergoing pharmacotherapy; (2) offering a financial incentive (e.g., a "bonus" payment) for agreeing to undergo pharmacotherapy; (3) requiring pharmacotherapy in order to receive public aid.

Reimbursing public assistance recipients for the costs of voluntary pharmacotherapy would be good public policy, just as it is good public policy to provide low-income persons with drug treatment on demand. Another foreseeable form of public benefit legislation might offer a financial incentive, or bonus, for agreeing to undergo pharmacotherapy. In this scenario, the pubic aid recipient would receive the standard aid payment regardless of whether he or she underwent pharmacotherapy, but would receive an additional bonus payment if he or she agreed to undergo pharmacotherapy. Such a scheme would raise difficult informed consent issues. In order to obtain the added financial benefit, many low-income people, even those who do not use or desire to use illegal drugs, might decide to undergo pharmacotherapy.[20]

The most inherently coercive type of foreseeable legislation linking pharmacotherapy with public aid would be the direct conditioning of such aid on the use of a pharmacotherapy drug. Under this potential legislative scheme, only those who agreed to undergo pharmacotherapy would be eligible for public aid. Although less coercive than being physically forced to undergo pharmacotherapy, a parent who is dependent upon receiving Aid to Dependent Children in order to pay rent or buy food for his or her kids would undoubtedly feel powerless to refuse pharmacotherapy if it meant forfeiting the financial aid. Such a scheme would be overtly and intentionally premised on economic coercion, and would thus vitiate the possibility of free and uncoerced consent. Combined with the very limited, and sometimes completely absent, access to professional medical advice—making it difficult for the person to obtain information about the risks and effects of pharmacotherapy—such a legislative scheme would encourage the antithesis of informed consent.

5.5.2 Cruel and Unusual Punishment

Compelling a prisoner, parolee, or probationer to take a pharmacotherapy drug, assuming the person is not mentally incompetent or dangerous, is akin to torture or barbarism. It treats the person as a means, rather than an end, and ought to be considered cruel and unusual punishment.

There is an unfortunate worldwide history of prisoner-abuse, including within the United States.[21] In the 1920s, U.S. prisoners were routinely labeled as genetically unfit and then forcibly sterilized. Believing that such sterilization improved society, approximately 60,000 incarcerated or mentally handicapped people were sterilized in the United States between 1907 and the mid-1970s.[22]

The American eugenics movement reached its zenith in 1927 with the Supreme Court's decision in *Buck v. Bell*, wherein the Court upheld the sterilization of mentally challenged women as both constitutional and good for society.[23] The highest court of Maryland recently deplored this unfortunate chapter of American jurisprudence:

> [O]ur own use of prisoners, the institutionalized retarded, and the mentally ill to test malaria treatments during World War II was generally hailed as positive, making the war "everyone's war." Likewise, in the late 1940's and early 1950's, the testing of new polio vaccines on institutionalized mentally retarded children was considered appropriate. Utilitarianism was the ethic of the day.[24]

Not until 1942 did the United States Supreme Court hold that it was unconstitutional to permanently sterilize people convicted of criminal offenses.[25]

The Eighth Amendment prohibits "cruel and unusual punishment," and many state constitutions provide independent protections.[26] Prisoners, parolees, and probationers all come within the Eighth Amendment's protection.[27] Forcing such a person to undergo pharmacotherapy against his or her will, when other less invasive, less intrusive, and less coercive means are available for treating the person, is a form of torture and retribution. Blocking a person's brain receptors with a pharmacotherapy drug because their crime was filling those receptors with an illegal drug harkens back to archaic notions of retributive punishment such as "an eye for an eye, or a hand for a hand."

Pharmacotherapy is not without side effects, and these may well render its compulsory use on prisoners, parolees, or probationers "cruel and unusual."[28] Most of the pharmacotherapy drugs are so new that it has yet to be determined whether they will produce long-term side effects, or even what health risks may arise after several weeks, months, or years of use. Inasmuch as many of the pharmacotherapy drugs work by targeting parts of the brain, and others work by systemically altering a person's metabolism, the health risks associated with their use are potentially significant. Compelling a prisoner to use pharmacotherapy drugs would force that person to risk suffering side effects or other serious adverse reactions from the drug. This would be both psychologically and physiologically cruel.

Future proponents of compulsory pharmacotherapy within the criminal justice system will likely characterize pharmacotherapy as "rehabilitative" or "treatment-oriented" in nature, in an effort to distinguish it from "punishment." Although the term "pharmacotherapy" implies that the drugs provide a sort of "therapy," it would be superficial to conclude on that semantic basis that they could not be used for nontherapeutic purposes. Future legislation seeking to authorize the compulsory use of "pharmacotherapy" for some or all prisoners, parolees, or probationers under the guise of "treatment" or "rehabilitation" would not be immune to judicial scrutiny under the Eighth Amendment. The legislative classification of a statute as authorizing "therapy" or "treatment" is not conclusive in determining whether there has been a violation of the Eighth Amendment.[29]

Until 1973, "homosexuality" was listed as a psychiatric disorder in the *Diagnostic and Statistical Manual of Mental Disorders* (*DSM*). Up until June 2003, when the United States Supreme Court declared them unconstitutional,[30] thirteen states had laws making it a criminal offense to engage in consensual homosexual sex. In some of these states, people who admitted that they were homosexual, or who were "accused" of being gay or lesbian, were subject to involuntary confinement under mental health laws and subjected to "reparative therapy" designed to forcibly convert them into heterosexuals.[31] "Treatment," in addition to counseling, included penile plethysmograph shocks (electronic shock triggered by penile erection), drugging, and hypnosis. Some state laws even permitted the forcible sterilization of homosexuals.[32]

Drug use, like homosexuality, has a ubiquitous presence throughout history and across cultures.[33] Like homosexuality, drug use and drug prohibition is the subject of contention and controversy. The discussion of both topics is often influenced by ignorance, fear and avoidance, conflicting moral and religious dogmas, and contrasting political aims. History has a way of showing that the forced "treatments" of today will tomorrow be seen as cruel, unusual, and barbaric punishment.

5.5.3 Freedom of Thought

Consciousness may turn out to be the ultimate mystery, resistant to self-interrogation. Whatever may be at the roots of human consciousness, there is no debate that what, and how, a person thinks is deeply intertwined with his or her functional neurochemistry.[34] Simply put, controlling what chemicals can or cannot reach a person's brain synapses directly affects how that person thinks. As a result, compelling a person to use a pharmacotherapy drug not only implicates the person's traditional rights to bodily integrity and informed consent, it also implicates the fundamental right to freedom of thought.

Americans have always cherished freedom of thought. Although the phrase "freedom of thought" is not explicitly used in the United States Constitution, it has long been recognized as a fundamental right of equal stature to the express constitutional guarantees. As Supreme Court Justice Benjamin Cardozo observed, "freedom of thought . . . is the matrix, the indispensable condition, of nearly every other form of freedom. With rare aberrations a pervasive recognition of that truth can be traced in our history, political and legal."[35]

The Supreme Court has repeatedly recognized that freedom of thought is one of the most elementary and important rights inherent in the First Amendment. Without freedom of thought, freedom of speech is moot. You cannot express what you cannot think. Likewise, you can *only* express what you *can* think. Chemical manipulation of the brain, therefore, could become the ultimate prior restraint on speech.[36]

In *West Virginia State Board of Education v. Barnette*, 319 U.S. 624 (1943), the Supreme Court, in an 8–1 decision, invalidated a school requirement that compelled a flag salute on the ground that it was an unconstitutional invasion of "the sphere of intellect and spirit which it is the purpose of the First Amendment to our Constitution to reserve from official control."[37] The First Amendment, declared the Court, gives a constitutional preference for "individual freedom of mind" over "officially disciplined uniformity for which history indicates a disappointing and disastrous end."[38] At the center of our American freedom, is the "freedom to be intellectually and spiritually diverse."[39] "We can have intellectual individualism and the rich cultural diversities that we owe to exceptional minds," the Court explained, "only at the price of occasional eccentricity and abnormal attitudes."[40]

In *Wooley v. Maynard*, 430 U.S. 705 (1977), the Supreme Court invalidated a New Hampshire statute that required all noncommercial vehicle license plates to bear the state motto "Live Free or Die," finding the requirement inconsistent with "the right of freedom of thought protected by the First Amendment."[41]

In *Stanley v. Georgia*, 394 U.S. 557 (1969), the Supreme Court struck down a Georgia law that banned the private possession of obscene material, finding the law "wholly inconsistent with the philosophy of the First Amendment."[42] "Our whole constitutional heritage," explained the Court, "rebels at the thought of giving government the power to control men's minds."[43] Justice Harlan, concurring in *United States v. Reidel*, 402 U.S. 351 (1971), characterized the constitutional right protected in *Stanley* as "the First Amendment right of the individual to be free from governmental programs of thought control, however such programs might be justified in terms of permissible state objectives," and as the "freedom from governmental manipulation of the content of a man's mind. . . ."[44] If "[o]ur whole constitutional heritage rebels at the thought of giving government the power to control men's minds," as made clear by the United States Supreme Court, then our whole constitutional heritage must likewise rebel at the thought of giving government the power to compel a person to use a pharmacotherapy drug—a drug designed and intended to lock down certain receptor sites in the brain.

Inasmuch as one's thoughts and thought processes are the very core of one's individuality and the root of both freedom and responsibility, permitting the state to forcibly pierce a person's body to insert a pharmacotherapy drug that is designed to patrol or police that person's body for the purpose of controlling possible brain states grants the state the ultimate power over the individual. Such a power is incompatible with a democracy built upon the premise of individual freedom and limited government. It is a clear violation of the fundamental right to freedom of thought.

Notes

1. Editor's note: This reading was excerpted from a substantially longer report issued by the Center for Cognitive Liberty and Ethics in 2004 under the title "Threats to Cognitive Liberty: Psychopharmacology and the Drug War" and is used with permission of the CCLE and author. An extensively revised version of that report appeared in 2005 in the *Journal of Law and Health*, volume 19, pages 215–257, under the title "Neurocops: The politics of prohibition and the future of enforcing social policy from inside the body." The sections of these papers not included here primarily concerned the technology of pharmacotherapy and an analysis of U.S. drug policy.

2. Sontag, Susan. (2001). *Illness as Metaphor and AIDS and its Metaphors*, New York: Picador.

3. "A Wiser Course: Ending Drug Prohibition: A Report of The Special Committee on Drugs and the Law of the Association of the Bar of the City of New York, June 14, 1994." ["One of the more insidious effects of the "war on drugs" has been the gradual erosion of the rule of law and the public's civil liberties."] Available at http://www.drugtext.org/library/reports/nylawyer/nylawyer.htm.

4. "The President's National Drug Control Strategy." Press Release. (2002, February 12). Office of the White House Press Secretary. Retrieved April 8, 2004, from http://www.whitehouse.gov/news/releases/2002/02/20020212-2.html.

5. *Cruzan v. Director, Missouri Dept. of Health*, 110 S. Ct. 2841 (1990); See also *Washington v. Harper*, 494 U.S. 210, 221–22 (1990) (recognizing a "significant liberty interest in avoiding the unwanted administration of antipsychotic drugs under the Due Process Clause of the Fourteenth Amendment"); *Vitek v. Jones*, 445 U.S. 480, 494 (1980) (forced admission to mental hospital and behavior modification treatment implicate liberty interests).

6. Tomes, Jonathan P. *Informed Consent: A Guide for the Healthcare Professional*. Chicago: Probus Publishing Company, 1993. pp. 69–102, "Chapter 7, State Informed Consent Laws." Essay with citations to codes and cases in footnotes.

7. The American Medical Association's (AMA) "General Statement on Informed Consent," Code of Medical Ethics:

"The patient's right of self-decision can be effectively exercised only if the patient possesses enough information to enable an intelligent choice. The patient should make his or her own determination on treatment. The physician's obligation is to present the medical facts accurately to the patient or to the individual responsible for the patient's care and to make recommendations for management in accordance with good medical practice. The physician has an ethical obligation to help the patient make choices from among the therapeutic alternatives consistent with good medical practice. Informed consent is a basic social policy for which exceptions are permitted: (1) where the patient is unconscious or otherwise incapable of consenting and harm from failure to treat is imminent; or (2) when risk disclosure poses such a serious psychological threat of detriment to the patient as to be medically contraindicated. Social policy does not accept the paternalistic view that the physician may remain silent because divulgence might prompt the patient to forego needed therapy. Rational, informed patients should not be expected to act uniformly, even under similar circumstances, in agreeing to or refusing treatment." AMA Code of Ethics sec. E-8.08 (1981). Available at http://www.ama-assn.org/ama/pub/category/8488.html (for relevant text).

8. 39 Fed. Reg. 18913.

9. *Olmstead v. United States*, 277 U.S. 438, 479 (1928) (Brandeis, J., dissenting).

10. Depending on the circumstances, it is legal to use these drugs for medical purposes. Eight states currently allow sick people to use marijuana for medical purposes. Opium and cocaine are both Schedule II substances, which (like Ritalin) can be used with a doctor's prescription.

11. "The President's National Drug Strategy." Press Release. (2002, February 12). Office of the White House Press Secretary. Retrieved April 9, 2004, from http://www.whitehouse.gov/news/releases/2002/02/20020212-2.html.

12. *Vitek v. Jones*, 445 U.S. 480, 493–94 (1980) ["A criminal conviction and sentence of imprisonment extinguish an individual's right to freedom from confinement for the term of his sentence, but they do not authorize the State to . . . subject him to involuntary psychiatric treatment without affording him additional due process protections."]; *Rogers v. Okin*, 634 F.2d 650, 653 (1st Cir. 1980) ["[A] person has a constitutionally protected interest in being left free by the state to decide for himself whether to submit to the serious and potentially harmful medical treatment . . . as part of the penumbral right to privacy, bodily integrity, or personal security."]; *Runnels v. Rosendale*, 499 F.2d 733, 735 (9th Cir. 1974) [performing a hemorrhoidectomy without the prisoner's consent implicated the prisoner's right to refuse medical treatment]; *Riggins v. Nevada*, 504 U.S. 127, 134 (1992) ("The forcible injection of medication into a nonconsenting person's body . . . represents a substantial interference with that person's liberty. In the case of antipsychotic drugs . . . that interference is particularly severe. . . ."] (quoting *Washington v. Harper* (1990), 494 U.S. 210 at 229).

13. See *Sell v. United States*, 539 U.S. 166, 213 (2003) ["[e]very State provides avenues through which, for example, a doctor or institution can seek appointment of a guardian with the power to make a decision authorizing medication—when in the best interests of a patient who lacks the mental competence to make such a decision".]

14. *Washington v. Harper* (1990) 494 U.S. 210, 227. Federal regulations set forth specific procedures that must be followed before prisoners can be forced against their will to take a psychotropic medicine. See 28 CFR 549, et seq.

15. Probation and Parole in the United States, 2002. 8/03, NCJ 201135. Retrieved April 19, 2004, from http://www.ojp.usdoj.gov/bjs/abstract/ppus02.htm.

16. *Burns v. United States*, 287 U.S. 216, 220 (1932).

17. Horwitz, Andrew. (2000). Coercion, Pop-Psychology, and Judicial Moralizing: Some Proposals for Curbing Judicial Abuse of Probation Conditions. 57 *Wash & Lee L. Rev.* 75.

18. For men, the most frequent source of referral to drug treatment is through the criminal justice system. In 1998 (the latest year for which SAMHSA has statistics), some 39% of men, compared with 25% of women, entered treatment as the result of a judicial process. Sixty-two percent of adult men entering treatment for marijuana abuse were sent by the criminal justice system. (Figure 3. Criminal Justice Referrals, by Sex and Primary Substance: 1998. Available at http://www.samhsa.gov/oas/2k1/enterTX/enterTX.htm.)

19. In the late 1970s, the federal government funded pilot programs testing the use of Antabuse as a probation condition in some drunk-driving and public intoxication cases. A total of thirteen county court systems received funding under the "Demonstration Programs in Antabuse." See Springer, T.J. (1976, September), Program Level Evaluation of ASAP Diagnosis, Referral and Rehabilitation Efforts, Vol. I–Description of ASAP Diagnosis, Referral and Rehabilitation Functions. In Marco, Corey H. & Marco, Joni Michel. (1980, Winter.) Antabuse Medication in Exchange for a Limited Freedom—Is it Legal?. *American Jnl. of Law & Medicine*, 5 (4): 295–330, 330. [Concluding, "Because Antabuse's effects are highly invasive, and because the judicial context of such programs is inherently coercive, courts should not employ the drug at all. Instead they should adopt rehabilitation schemes . . . that attempt to re-educate alcohol offenders without the use of drugs."]

20. Illustratively, all fifty states currently have programs reimbursing poor people for the cost of surgery to implant Norplant. See Roberts, Dorothy E. (1995). The Only Good Poor Woman: Unconstitutional Conditions and Welfare. 72 *Denv. U.L. Rev.* 931, 931–32.

21. For discussions concerning informed consent and medical experimentation on prison populations, see Mark, Vernon H. & Neville, Robert. (1977). Brain Surgery in Aggressive Epileptics: Social and Ethical Implications. *Ethical Issues in Modern Medicine*. (eds. Robert Hunt & John Arras); Veatch, Robert M. (1977). *Case Studies In Medical Ethics*. pp. 267–71; Bernier, Barbara L. (1994). Class, Race, and Poverty: Medical Technologies and Socio-Political Choices. 11 *Harv. BlackLetter J.* 115.

22. Lombardo, Paul. Eugenic Sterilization Laws. Retrieved April 9, 2004, from http://www.eugenicsarchive.org/.

23. *Buck v. Bell*, 274 U.S. 200 (1927). This case gave rise to Justice Holmes's infamous quotation that "three generations of imbeciles are enough." Id. at 207. Evidence later indicated that Carrie Buck, the woman whose sterilization was upheld, was *not* mentally handicapped. Her child, who died at the age of eight, was a member of her school's honor roll.

24. *Grimes v. KennedyKrieger Inst., Inc.*, 366 Md. 29, 77, 782 A.2d 807, 836 (2001) (quoting Dr. George J. Annas, *Mengel's Birthmark: The Nuremberg Code in United States Courts*, 7 *J. Contemp. Health L. & Pol'y* 17, 24 (Spring 1991)).

25. *Skinner v. Oklahoma* 316 U.S. 535, 538 (1942). Despite the Skinner decision, a handful of states continue to have laws allowing for the compulsory sterilization of criminals or the mentally incompetent. See, e.g., MISS. CODE ANN.

§ 41-45-1 (1991); N.C. GEN. STAT. § 35-36 (1990); W.VA. CODE § 27-16-1 (1986). It is unlikely that these laws, if challenged, would withstand constitutional scrutiny.

26. The Eighth Amendment provides: "Excessive bail shall not be required, nor excessive fines imposed, nor cruel and unusual punishments inflicted." U.S. Const. amend. VIII. Many state constitutions also provide independent protections against cruel and unusual punishment.

27. See *State v. Brown*, 326 S.E.2d 410, 411 (S.C. 1985).

28. *Nelson v. Heyne* 355 F. Supp. 451, 455 (N.D. Ind. 1972) [holding it is cruel and unusual punishment to inject juveniles in a correctional institute with tranquilizing drugs that can have significant side effects].

29. See, e.g., *Trop v. Dulles*, 356 U.S. 86, 95 (1958) ["even a clear legislative classification of a statute as 'nonpenal' would not alter the fundamental nature of a plainly penal statute".]; *Knecht v. Gillman*, 488 F.2d 1136, 1139–40 (8th Cir. 1973) [noting that "the mere characterization of an act as 'treatment' does not insulate it from eighth amendment scrutiny," and that "neither the label which a State places on its own conduct, nor even the legitimacy of its motivation, can avoid the applicability of the Federal Constitution"].

30. See *Lawrence v. Texas* (2003) 123 S.Ct. 2472, 156 L.Ed.2d, 2003 U.S. LEXIS 5013. Available at http://www.supremecourtus.gov/opinions/02pdf/02-102.pdf.

31. See, e.g., Garland, J. A. (2001). The Low Road to Violence: Governmental Discrimination as a Catalyst for Pandemic Hate Crime. 10 *Law & Sex*. 1, 75–76 & nn.355–65; Chauncey, G., The Postwar Sex Crime Panic, in True Stories from the American Past (1993); Freedman, E. B., "Uncontrolled Desires": The Response to the Sexual Psychopath, 1920–1960; In *Passion and Power: Sexuality in History* (1989). An Alabama law "reform" commission announced that gay people are "persons with abnormal tendencies" who "have forfeited certain of their standings," and warned that Alabama would make itself "known as a place where it is tough for [such] persons." (Commission to Study Sex Offenses: Interim Report to the Alabama Legislature, June 12, 1967, at 5.n17.)

32. Painter, George. (2003). "The Sensibilities of Our Forefathers: The History of Sodomy Laws in the United States." Retrieved April 9, 2004, from http://www.sodomylaws.org/sensibilities/utah.htm.

33. Escohotado, Antonio. (1999). *A Brief History of Drugs: From the Stone Age to the Stoned Age*. Rochester: Park Street Press; Ott, Jonathan. (1993). *Pharmacotheon: Entheogenic drugs, their plant sources and history*. Kennewick, WA: Natural Products Co.

34. Altering a person's brain chemistry for the purpose of altering how that person thinks is the basis of a pharmaceutical sector with approximately $20 billion in global sales. The sale of Prozac and similar "antidepressant" drugs is currently one of the most profitable segments of the pharmaceutical drug industry. According to IMS Health, a 50-year-old company specializing in pharmaceutical market intelligence and analyses, "antidepressants, the #3-ranked therapy class

worldwide, experienced 18 percent sales growth in 2000, to $13.4 billion or 4.2 percent of all audited global pharmaceutical sales." IMS Health, "Antidepressants". Available at http://www.imshealth.com/public/structure/navcontent/1,3272,1034-1034-0,00.html (for summary). Sales of "antipsychotic" drugs are currently the eighth largest therapy class of drugs with worldwide sales of $6 billion in the year 2000, a 22% increase in sales over the previous year. See IMS Health, "Antipsychotics." Available at http://www.imshealth.com/public/structure/navcontent/1,3272,1035-1035-0,00.html (for summary). A report published by the Lewin Group in January 2000 found that within the Medicaid program alone, "Antidepressant prescriptions totaled 19 million in 1998 . . . [and] [a]ntipsychotic prescriptions totaled 11 million in 1998." Lewin Group. (2000, January). *Access and Utilization of New Antidepressant and Antipsychotic Medications*. The CCLE underscores that the development of such drugs is to be applauded for their potential to aid millions of suffering people who voluntarily use them.

35. *Palko v. Connecticut*, 302 U.S. 319, 326–27 (1937).

36. See argument presented in *amicus curiae* brief of the Center for Cognitive Liberty & Ethics, filed in *Sell v. United States* (Untied States Supreme Court, Case No. 02-5664). Available at http://supreme.usatoday.findlaw.com/supreme_court/briefs/02-5664/02-5664.ami.pet.ccle.pdf; Pynchon, Thomas. (1974). *Gravity's Rainbow*. Bantam Books. p. 293. "If they can get you asking the wrong questions, they don't have to worry about the answers."

37. *Id.* at 642.

38. *Id.* at 637.

39. *Id.* at 641.

40. *Id.* at 641–42.

41. *Id.* at 714.

42. *Id.* at 565–66.

43. *Id.* at 565.

44. *Id.* at 359 (Harlan J., concurring).

6

Brains and Persons

Many of the most difficult bioethical issues hinge on the question of who or what is a person. If we regard a human fetus as a person, then we cannot permit abortion. If we regard a patient in a permanent vegetative state as a person, then the withholding of nutrition and hydration is a deeply troubling act. If some animals are persons, as has been argued by animal rights advocates (e.g., Francione, 1993), then many of the ways in which animals are routinely treated in agriculture, research laboratories, and zoos are morally impermissible.

All of the moral consequences of personhood just described result from the special moral status of persons, as distinct from all other objects in the universe. Persons deserve protection from harm just because they are persons. Whereas we value objects for what they can do—a car because it transports us, a book because it contains information, a painting because it looks beautiful—the value of persons transcends their abilities, knowledge, or attractiveness. Persons have what Kant called *dignity*, meaning a special kind of intrinsic value that trumps the value of any use to which they could be put (Kant, 1996).

The special moral status of persons is a commonsensical ethical principle, and most of the time we have no problem applying it because it is obvious which entities are persons and which are not. But, as we just observed, some cases are tough calls. There is disagreement about the status of prenatal humans, severely brain-damaged humans, and our closest primate cousins. In addition, those working in the new field of roboethics have begun to wonder whether and when artificial intelligences might attain personhood. If we could agree on a set of objective, observable criteria for personhood, we would be much closer to settling many of the most difficult bioethical issues. The search for such criteria inevitably leads us to questions about the physical basis of the human mind.

The readings in this section all concern personhood, explicitly or implicitly, and explore some of the relations between brain function and personhood. We therefore begin with some background on the role of neuroscience in defining personhood.

Anchoring Personhood in Neuroscience

There is no universally agreed upon set of psychological traits that constitute personhood. Most people's lists include intelligence, rationality, self-awareness, cognition about the future, linguistic communication, mental states of all kinds, including mental states about other people's mental states, and all forms of consciousness. Each of these abilities is a function of the brain. The promise of a definition of personhood, anchored in the functioning of a physical, objectively observable system, has led many bioethicists to neuroscience.

In the context of the abortion debate, some bioethicists have proposed that the fetus should be protected after the point in prenatal development at which brain function begins (Jones, 1989). Of course, prenatal brain development is a gradual process and lacks the kinds of punctate, qualitative transition points that would most naturally be associated with the momentous transformation from nonperson to person. Furthermore, many of the milestones that have been proposed as marking a transition depend as much on our technologies for studying fetal brain function as on the fetal brain itself; more sensitive measures would indicate function at earlier gestational ages.

At the other end of the life span, the concept of *brain death* has met with more acceptance than *brain life* and is the basis for contemporary medical and legal definitions of death. However, brain death, which is typically taken to mean loss of clinically detectable function of the whole brain or loss of function of brain-stem structures, is not equivalent to death of the person. Personal death is generally associated with loss of higher cortical brain functions, which normally instantiate rationality, conscious self-awareness, and the other psychological traits just mentioned. Patients with extensive cortical damage but functioning brain stems are sometimes referred to with the potentially confusing terminology *cortically brain dead* but are better described as being in a *vegetative state*, because they are not dead by accepted medical and legal criteria. One might call them living nonpersons. Patients who seem vegetative at times but occasionally respond to stimuli in the environment are termed *minimally conscious*. Can neuroscience help determine the boundaries between death, mere biological life, and personhood?

One of the most promising applications of neuroscience to this problem has involved functional brain imaging of patients whose mental life, and hence personhood, are in question. This research has produced some striking findings, which have been widely reported in the popular press and widely discussed by bioethicists (e.g., Fins et al., 2008). The general idea behind all of this research is that imaging can provide us with information about mental activity without the need for overt communication. The specific logic linking imaging results to conclusions about conscious mental life varies across studies (see Farah, 2008, for a description of three general types of inference and an analysis of their assumptions and weaknesses).

In one of the best-known studies, Schiff and colleagues (2005) scanned subjects who were in a minimally conscious state (MCS) while presenting them with recordings of a relative telling a personally relevant story and with the same recording played backwards. Like the normal control subjects, the subjects in MCS activated a network of language-related areas in response to the meaningful recordings relative to the backwards recordings. In contrast, most imaging studies of vegetative patients have yielded little evidence of the kinds of neural processing associated with mental life.

A striking exception was the study by Owen and colleagues (2006) of a vegetative patient who later recovered but while meeting diagnostic criteria for the vegetative state showed patterns of brain activation indicative of language comprehension and voluntary mental imagery. One indication of preserved cognition in this patient was her increased brain activity when presented with sentences containing ambiguous words, in the same region as for normal subjects, consistent with the additional cognitive processing required for resolving the ambiguity of such sentences. In addition, when instructed to perform mental imagery tasks, her brain activity indicated that she understood the instructions and was able to comply. When asked to imagine playing tennis, she activated parts of the motor system, and when asked to imagine visiting each of the rooms of her home, she activated parts of the brain's spatial navigation system. Furthermore, her patterns of brain activation in these tasks were indistinguishable from those of normal subjects.

These and other studies have shown unexpectedly preserved neurocognitive processing in severely brain-damaged patients, including what would appear to be command following in the case of Owen et al. (2006). Ultimately, this type of research may be able to answer our questions about the presence and nature of cognitive processing in such patients. We will still be left, however, with the problem of deciding which cognitive processes are required for personhood and how fully present or functioning they must be in order to matter. A human with normal brain function may be easy to classify as a person, and a decorticate human may be equally easy to classify as a nonperson, but which cortical systems in which combinations are critical and how much functionality is required of each of those systems? For defining personhood, the devil is just as much present in the neurological details as in the psychological ones.

The Neuroethics of Personhood

The Readings

All of the readings in this section address some aspect of the relation between brains, on the one hand, and persons, on the other. In his review of brain death, *Laureys* clarifies the difference between being a living human and a living person. That some corpses are warm to the touch and may even move reminds us that the line between life and death can be drawn in different places and that current medical and legal definitions of death may not accord well with our intuitive understanding of it. Furthermore, just as a human need not be cold and inanimate to be dead, neither does a human need to have a mental life to be medically and legally alive by the current definition. Laureys' discussion of the vegetative state reveals that massive cortical damage may result in a damaged but living human body that apparently lacks mental activity and hence, in the view of many, personhood. *Farah and Heberlein* present evidence from psychology and neuroscience suggesting that people are innately predisposed to divide the world into persons and nonpersons and that the intuition that someone or something is a person may be triggered by certain simple features such as a face or even just eyes. These instinctual projections onto the humans, animals, and objects of the world undoubtedly complicate our attempts to reason objectively about the boundaries of personhood. The *Farah* reading on animal neuroethics explores the ways in which neuroscience may be able to illuminate the mental states of animals, including the mental state of suffering. This is a supremely morally relevant fact about animals and arguably a necessary, though not sufficient, aspect of personhood. In his brief essay on robots and electronically enhanced humans, *Perkowitz* reviews the history and current trends in the design and uses of "digital people"—artificial beings and partly artificial beings—and raises several ethical issues that arise as digital people enter society in a variety of different roles. These include issues concerning the personhood of those with either no human brain or a drastically altered human brain. They also include the role of nonhuman agents in warfare, where laws and mores concerning acts of aggression will have to come to terms with actors who may not be persons, and in the care of the ill and elderly, where the perceptual biases discussed by Farah and Heberlein may encourage the acceptance of caretaker machines. Finally, *Murphy* presents some of the historical and theological background to the question of what constitutes a person: Are persons entirely physical entities or do they also possess a nonphysical soul? She identifies cognitive neuroscience as threatening a widely held religious view of the person as having physical and nonphysical components but goes on to argue that this view is not, in fact, entailed by a careful reading of biblical scripture.

Selected Cross-cutting Issues

Not surprisingly, the topic of personhood is the primary site for the intersection of neuroethics and religion. What becomes of a person after death is an age-old question, which has traditionally been answered by religion. The nature of the answer depends on the religion's "anthropology," that is, the religion's view of what constitutes a person. Whereas non-human animals are virtually universally regarded as physical entities, without souls, human persons are regarded by many religions as having a soul that leaves the body at death. Of course, Murphy shows us that even the religious tradition best known for talk of "saving souls" has deep roots in physicalism. Furthermore, as Laureys explains, the special status of brain function as a physical criterion for human life is accepted within Jewish, Christian, and Islamic traditions.

The concept of a person is so intuitive that it is hard to see what the problem in understanding it could be, at least until one considers the kinds of atypical cases that occupy philosophers and legal scholars. Several of the readings in this section concern just such atypical cases. These include humans who may not be persons, and non-humans who may be persons. In the first category are the patients written about by Laureys, both brain dead and permanently vegetative. In the second case are the entities written about by Farah and Perkowitz. Many of the elements that seem central to personhood are more evident in certain animals than in vegetative patients. It may eventually come to pass that we will create machines with some or all of these characteristics. One might understandably react to some of these discussions by asking "are we discussing ethics or science fiction here?" The answer is that, for purposes of addressing moral dilemmas in the here and now, including those mentioned at the outset of this chapter, we need to clarify the concept of personhood and hypotheticals can help. Is "person" a real category of the world, and therefore worthy of guiding our moral decision making? If so, what are the essential criteria for personhood? By testing out candidate criteria with a range of real and hypothetical examples and comparing the results of our definitional efforts with the results of our intuition about the nature of persons, we may both hone our definitions and educate our intuitions.

Questions for Discussion

1. It seems unlikely that neuroscience will clarify the concept of personhood *per se*, that is, provide a better principled account of what makes someone or something a person. However, given a concept of personhood with certain defining criteria, neuroscience might help us resolve who meets those criteria for personhood and who does not, and this could have social and legal implications. Imagine the ways in which neuroscience might be used in end-of-life, animal welfare, and other policy areas by drafting sample guidelines and laws that explicitly incorporate ideas, methods, and measures from neuroscience. Do you believe that the current state of neuroscience supports some or all of the uses to which you have put it in your drafts? What further progress would be needed?
2. "Transhumanists" look forward to the day when technological enhancements will allow us to transcend the bounds of what has heretofore been considered human. Consider the ways in which extreme brain enhancement intersects with issues of personhood. For example, can you imagine enhancement leading to a superperson; that is, someone whose intrinsic moral value is even greater than a person's? What enhancements might be candidates for such transformations? Are there imaginable brain enhancements that would make someone arguably no longer a person, and what might those be?
3. Most discussions of personhood in bioethics focus on persons as patients in an agent–patient system; the question of who or what is a person is asked in order to know how we should treat the beings in question. Historically, however, personhood has also been linked to agency, especially by Kant. From this perspective, persons are agents capable of choosing their actions and therefore they bear moral responsibility for their actions. How well aligned are the concepts of person as patient

and agent, in terms of who or what is classified as a person, and in terms of the psychological criteria for personhood? Thinking back to the review of the neuroscience of responsible behavior in chapter 5, what do the neural underpinnings of these two senses of personhood have in common, and what is distinct?

References

Farah, M. J. (2008). That little matter of consciousness. *American Journal of Bioethics—Neuroscience, 8*, 17–19.

Fins, J. J., Illes, J., Bernat, J. L., Hirsch, J., Laureys, S., & Murphy, E. R. (2008). Consciousness, imaging, ethics, and the injured brain. *American Journal of Bioethics—Neuroscience, 8*, 3–12.

Francione, G. L. (1993). Personhood, property and legal competence. In P. Cavalieri & P. Singer (Eds.), *The great ape project* (pp. 248–257). New York: St. Martin's Griffin.

Jones, G. (1989). Brain birth and personal identity. *Journal of Medical Ethics, 15*, 173–178.

Kant, E. (1996). *Groundwork of the metaphysics of morals* (M. J. Gregor, Ed.). Cambridge, UK: Cambridge University Press.

Owen, A. M., Coleman, M. R., Boly, M., Davis, M. H., & Laureys, S. (2006). Detecting awareness in the vegetative state. *Science, 313*, 1402.

Schiff, N. D., Rodriguez-Moreno, D., Kamal, A., Kim, K. H., Giacino, J. T., Plum, F., & Hirsch, J. (2005). fMRI reveals large-scale network activation in minimally conscious patients. *Neurology, 64*, 514–523.

Reading 6.1

Death, Unconsciousness, and the Brain[1]

Steven Laureys

Throughout history, society and medicine have struggled with the definition and determination of death. In ancient Egypt and Greece, the heart was thought to create the vital spirits, and the absence of a heartbeat was regarded as the principal sign of death (Pernick, 1988). The first person to consider irreversible absence of brain function to be equivalent to death was Moses Maimonides (1135–1204), the foremost intellectual figure of medieval Judaism, who argued that the spasmodic jerking observed in decapitated humans did not represent evidence of life as their muscle movements were not indicative of presence of central control (Bernat, 2002a). However, it was not until the invention of the positive pressure mechanical ventilator by Bjorn Ibsen in the 1950s and the widespread use of high-tech intensive care in the 1960s that cardiac, respiratory, and brain function could be truly dissociated. Patients with severe brain damage could now have their heartbeat and systemic circulation provisionally sustained by artificial respiratory support. Such profound unconscious states had never been encountered before, as, until that time, all such patients had died instantly due to cessation of respiration.

The earliest steps toward a neurocentric definition of death were European (Lofstedt & von Reis, 1956; Wertheimer, Jouvet, & Descotes, 1959). In 1959, French neurologists Mollaret and Goulon first discussed the clinical, electrophysiologic, and ethical issues of what is now known as brain death, using the term *coma dépassé* ("irretrievable coma") (Mollaret & Goulon, 1959). Unfortunately, their paper was written in French and remained largely unnoticed by the international community. In 1968, the *Ad Hoc* Committee of Harvard Medical School, which included ten physicians, a theologian, a lawyer, and a historian of science, published a milestone paper defining death as irreversible coma ("A definition of irreversible coma," 1968). The report "opened new areas of law, and posed new and different problems for theologist and ethicist...it has made physicians into lawyers, lawyers into physicians, and both into philosophers" (Joynt, 1984). Some years later, neuropathologic studies showed that damage to the brain stem was critical for brain death (Mohandas & Chou, 1971). These findings initiated the concept of *brain stem death* (Pallis & Harley, 1996) and led U.K. physicians to define brain death as complete, irreversible loss of brain-stem function ("Criteria for the diagnosis of brain stem death," 1995; "Diagnosis of brain death," 1976): "if the brainstem is dead, the brain is dead, and if the brain is dead, the person is dead" (Pallis & Harley, 1996).

The tragic death of Terri Schiavo, misused by both "right-to-life" and "right-to-die" activists, recently illustrated to the world the difficulties that surround death in the vegetative state (Annas, 2005; Gostin, 2005; Quill & Shiavo; 2005). Many uneducated commentators have inaccurately referred to Schiavo's condition as "brain dead" or "neocortical dead," and her gravestone reads, "Departed This Earth February 25, 1990"—that is, the date on which her brain was damaged (although this was not total, and she was, therefore, not dead), whereas it was on March 31, 2005, that her entire brain died and her heart irreversibly stopped beating.

6.1.1 The Concept of Death

At present, the most accepted definition of death is the "permanent cessation of the critical functions of the organism as a whole" (Bernat, 1998; Loeb, 1916). The organism as a whole is an old concept in theoretical biology (1916) that refers to its unity and functional integrity—not to the simple sum of its parts—and encompasses the concept of an organism's critical system (Korein & Machado, 2004). Critical functions are those without which the organism as a whole cannot function: control of respiration and circulation, neuroendocrine and homeostatic regulation, and consciousness. Death is defined by the irreversible loss of all these functions.

Brain death means human death determined by neurologic criteria. It is an unfortunate term, as it misleadingly suggests that there are two types of death: "brain" death and "regular" death (Bernat, 2002a). There is, however, only one type of death, which can be measured in two ways—by cardiorespiratory or neurologic criteria. This misapprehension might explain much of the public and professional confusion about brain death. Bernat and colleagues have distinguished three levels of discussion: the definition or concept of death (a philosophical matter);

the anatomic criteria of death (a philosophical/medical matter); and the practical testing, by way of clinical or complementary examinations, that death has occurred (a medical matter) (Bernat, Culver, & Gert, 1981).

The brain-centered definition of human death has three formulations, known as whole brain, brain-stem, and neocortical death. Whole brain and brain-stem death are both defined as the irreversible cessation of the organism as a whole, but differ in their anatomic interpretation. Because many areas of the brain above the brain stem (including the neocortex, thalami, and basal ganglia) cannot be accurately tested for clinical function in a comatose patient, most bedside tests for brain death (such as cranial nerve reflexes and apnea testing) directly measure function of the brain stem alone. The neocortical formulation of death, which was proposed in the early days of the brain death debate (Veatch, 1975), advocates a fundamentally different concept of death: the irreversible loss of the capacity for consciousness and social interaction. By application of this consciousness- or personhood-centered definition of death, its proponents classify patients in a permanent vegetative state and anencephalic infants as dead. This most progressive and controversial concept of death is dealt with separately.

6.1.2 Vegetative State Is Not Brain Death

It might seem that the difference between brain death and the vegetative state is so fundamental that it need not be reviewed. However, in reality, both terms are all too often mixed up in the lay—and even medical—press. Part of this misunderstanding might have its origin in the interchangeable lay use of the terms *brain dead* and *vegetable* (Diringer & Wijdicks, 2001). This had already started when the *New York Times* (August 5, 1968) announced the Harvard criteria for brain death. In the accompanying editorial it read: “As old as medicine is the question of what to do about the human vegetable... Sometimes these living corpses have survived for years... It is such cases, as well as the need for organs to be transplanted that the Harvard faculty committee had in mind in urging that death be redefined as irreversible coma” (Diringer & Wijdicks, 2001). More recently, one study reported that slightly less than half of surveyed U.S. neurologists and nursing home directors believed that patients in a vegetative state could be declared dead (Payne, Taylor, Stock, & Sachs, 1996).

Both patients with brain death and those in a vegetative state are unconscious after severe brain injury. The first difference between the two is the time of diagnosis. Brain death can be diagnosed with an extremely high rate of probability within hours to days of the original insult, whereas diagnosing an irreversible vegetative state takes many months at best (3 months after a nontraumatic brain injury and 12 months after traumatic injury, as stated above) (The Multi-Society Task Force on PVS, 1994a). Unlike patients with brain death who are, by definition, comatose (that is, never show eye opening, even on noxious stimulation), patients in a vegetative state (who, it should be stressed, are not in a coma), classically have their eyes spontaneously open, which can be very disturbing to families and caregivers. Patients with brain death require controlled artificial ventilation, whereas patients in a vegetative state can breath spontaneously without assistance (even if during the acute stage ventilation must sometimes be artificially assisted). Unlike patients with brain death, those in a vegetative state have preserved brain-stem reflexes and hypothalamic functioning (for example, regulation of body temperature and vascular tone). At best, patients with brain death show a very limited set of body movements, which may be present in up to a third of patients (Saposnik et al., 2000, 2005). Patients in a vegetative state show a much richer array of motor activity, albeit always nonpurposeful, inconsistent, and coordinated only when expressed as part of subcortical, instinctively patterned, reflexive response to external stimulation: moving trunk, limbs, head, or eyes in meaningless ways and showing startle myoclonus to loud noises (Multisociety Task Force on PVS, 1994a). Finally, patients with brain death never show any facial expression and remain mute, whereas patients in a vegetative state may occasionally smile or cry, utter grunts and sometimes moan or scream (Jennett, 2002; Multisociety Task Force on PVS, 1994a).

6.1.3 Neocortical Death Myth

In 1971, Scottish neurologist Brierley and his colleagues urged that death be defined by the permanent cessation of "those higher functions of the nervous system that demarcate man from the lower primates" (Brierley, Graham, Adams, & Simpsom, 1971). This neocortical or higher brain death definition has been further developed by others, mainly philosophers (Gervais, 1986; Veatch, 1975), and its conceptual basis rests on the premise that consciousness, cognition, and social interaction, not the bodily physiologic integrity, are the essential characteristics of human life. The higher brain concept produces the neocortical death criterion, in which only the functions of the neocortex, not of the whole brain or of the brain stem, must be permanently lost. Clinical and confirmatory tests for neocortical death have never been validated as such.

Based on the neocortical definition of death, patients in a vegetative state following an acute injury or chronic degenerative disease and anencephalic infants are considered dead. Depending on how "irreversible loss of capacity for social interaction" (Veatch, 1976) is interpreted, even patients in a permanent "minimally conscious state" (Giacino et al., 2002), who, by definition, are unable to functionally communicate, could be regarded as dead. I argue that, despite its theoretical attractiveness to some, this concept of death cannot be reliably implemented using anatomic criteria nor in reliable clinical testing.

First, our current scientific understanding of the necessary and sufficient neural correlates of consciousness is incomplete at best (Baars, Ramsoy, & Laureys, 2003; Laureys, 2005). In contrast with brain death, for which the neuroanatomy and neurophysiology are both well established, anatomopathology, neuroimaging, and electrophysiology cannot, at present, determine human consciousness. Therefore, no accurate anatomic criteria can be defined for a higher brain formulation of death.

Second, clinical tests would require the provision of bedside behavioral evidence showing that consciousness has been irreversibly lost. There is an irreducible philosophical limitation in knowing for certain whether any other being possesses a conscious life (Chalmers, 1998). Consciousness is a multifaceted subjective first-person experience, and clinical evaluation is limited to evaluating patients' responsiveness to the environment (Majerus, Gill-Thwaites, Andrews, & Laureys, 2005). As previously discussed, patients in a vegetative state, unlike patients with brain death, can move extensively, and clinical studies have shown how difficult it is to differentiate "automatic" from "willed" movements (Prochazka, Clarac, Loeb, Rothwell, & Wolpaw, 2000). This results in an underestimation of behavioral signs of consciousness and, therefore, a misdiagnosis, which is estimated to occur in about one third of patients in a chronic vegetative state (Andrews, Murphy, Munday, & Littlewood, 1996; Childs, Mercer, & Childs, 1993). In addition, physicians frequently erroneously diagnose the vegetative state in elderly residents with dementia in nursing homes (Volicer, Berman, Cipolloni, & Mandell, 1997). Clinical testing for absence of consciousness is much more problematic than testing for absence of wakefulness, brain-stem reflexes, and apnea in whole brain or brain-stem death. The vegetative state is one end of a spectrum of awareness, and the subtle differential diagnosis between this and the minimally conscious state necessitates repeated evaluations by experienced examiners. Practically, the neocortical death concept also implies the burial of breathing "corpses."

Third, complementary tests for neocortical death would require provision of confirmation that all cortical function has been irreversibly lost. Patients in a vegetative state are not apallic, as previously thought (Ingvar, Brun, Johansson, & Samuelsson, 1978; Ore, Gerstenbrand, & Lucking, 1977) and may show preserved islands of functional pallium or cortex. Recent functional neuroimaging studies have shown limited, but undeniable, neocortical activation in patients in a vegetative state, disproving the idea that there is complete neocortical death in the vegetative state. However, as previously stated, results from these studies should be interpreted cautiously for as long as we do not fully understand the neuronal basis of consciousness. Again, complementary tests for proving the absence of the neocortical integration that is necessary for consciousness are, at present, not feasible and unvalidated.

The absence of whole brain function in brain death can be confirmed by means of cerebral angiography (nonfilling of the intracranial arteries), transcranial Doppler ultrasonography (absent diastolic or reverberating flow), brain imaging (absence of cerebral blood flow: hollow-skull sign) or electroencephalography (EEG; absent electrical activity). In contrast with brain death, in which prolonged absent intracranial blood flow proves irreversibility (Bernat, 2004), the massively reduced—but not absent—cortical metabolism observed in the vegetative state (Boly et al., 2004; De Volder et al., 1990; Laureys et al., 1999; Levy et al., 1987; Schiff et al., 2002; Tommasino, Grana, Lucignani, Torri, & Vazio, 1995) cannot be regarded as evidence for irreversibility. Indeed, fully reversible causes of altered consciousness, such as deep sleep (Maquet et al., 1997) and general anaesthesia (Alkire et al., 1995; Alkire et al., 1999; Alkire, Haier, Shah, & Anderson, 1997), have shown similar decreases in brain function, and the rare patients who have recovered from a vegetative state have been shown to resume near-normal activity in previously dysfunctional associative neocortex (Laureys, Faymonville, Moonen, Luxen, & Maquet, 2000; Laureys, Lemaire, Maquet, Phillips, & Franck, 1999).

However, proponents of the neocortical death formulation might counter-argue that because all definitions of death and vegetative state are clinical, finding some metabolic activity in functional neuroimaging studies does not disprove the concept (as these studies are measuring nonclinical activities), although this does contrast with the validated nonclinical laboratory tests used to confirm whole brain death.

Finally, proving irreversibility is key to any concept of death. The clinical testing of irreversibility has stood the test of time only in the framework of whole brain or brain-stem formulations of death. Indeed, since Mollaret and Goulon first defined their neurologic criteria of death more than 45 years ago (Mollaret & Goulon, 1959), no patient in apneic coma who was properly declared brain (or brain stem) dead has ever regained consciousness (Bernat, 2005; Pallis & Harley, 1996; Wijdicks, 2001a). This cannot be said for the vegetative state, in which permanent is probabilistic—the chances of recovery depend on a patient's age, etiology, and time spent in the vegetative state (The Multi-Society Task Force on PVS, 1994b). Unlike brain death, for which the diagnosis can be made in the acute setting, the vegetative state can only be regarded as statistically permanent after long observation periods, and even then there is a chance that some patients might recover. However, it should be stressed that many anecdotes of late recovery are difficult to substantiate, and it is often difficult to know how certain the original diagnosis was.

6.1.4 Religion and Death

Both Judaism and Islam have a tradition of defining death on the basis of absence of respiration, but brain death has now become an accepted definition of death for these religions (Beresford, 2001). The Catholic Church has stated that the moment of death is not a matter for the church to resolve. More than 10 years before the Harvard criteria were established, anesthesiologists who were concerned that new resuscitation and intensive care technologies designed to save lives sometimes appeared to only extend the dying process sought advice from Pope Pius XII. The pope, up-to-date with (even, surprisingly, in advance of) modern day medicine, ruled that there was no obligation to use extraordinary means to prolong life in critically ill patients (Pius XII, 1957). Therefore, withholding or withdrawing life-sustaining treatment from patients with acute irreversible severe brain damage became morally accepted.

With regard to life-prolonging treatments in chronic conditions such as the vegetative state, many have found it difficult to view artificial hydration and nutrition as extraordinary means. However, recent ethical and legal discussions have abandoned the extraordinary versus ordinary dichotomy in favor of disproportionate versus proportionate treatments. Many prominent progressive Catholic theologists have accepted the idea of therapeutic futility in patients in an irreversible vegetative state and have defended the decision to withdraw nutrition and hydration in well-documented cases (Schotsmans, 1993). Nevertheless, Pope John Paul II, addressing an international congress on the vegetative state in March 2004, considered that the cessation of artificial life-sustenance to patients in a permanent vegetative state could never be morally accepted, whatever the situation (Pope John Paul II, 2004). However, many of the meeting's invited neuroscientists had more nuanced viewpoints, and some Roman Catholic theologians considered it to be at variance with Christian tradition. The moral legitimacy to inquire about the duty to treat at all cost (that is, therapeutic obstinacy), which was accepted by the Catholic Church for acute cases of severe neurologic damage (irreversible coma) in 1957 (Pius XII, 1957), stands in contrast with the church's recent refusal to allow withdrawal of life-sustaining treatment in chronic cases (irreversible vegetative state) (Pope John Paul II, 2004). The official Catholic position de-emphasizes the reality of irreversibility in long-standing vegetative state and does not consider artificial nutrition and hydration to be treatments. So far, it has not changed practices in the United States, where withdrawal of life-sustaining treatment from patients in an irreversible vegetative state remains a settled view; a view that was endorsed by the U.S. Supreme Court in the case of Nancy Cruzan, and that is held by many other medical, ethical, and legal authorities (Gostin, 1997).

6.1.5 Death and the Law

Under the U.S. Uniform Determination of Death Act ("Uniform Determination of Death Act," 1997), a person is dead when physicians determine, by applying prevailing clinical criteria, that cardiorespiratory or brain functions are absent and cannot be retrieved (Beresford, 2001). The neurocentric definition is purposefully redundant, requiring a determination that "all functions of the entire brain, including the brain stem" have irreversibly ceased ("Uniform determination of Death Act," 1997). In 1971, Finland was the first European country to accept brain death criteria. Since then, all European Union countries have accepted the concept of brain death. However, although the required clinical signs are uniform, less than half the European countries that have accepted brain death criteria require technical confirmatory tests, and approximately half require more than one physician to be involved (Haupt & Rudolf, 1999). Confirmatory tests are not mandatory in many third-world countries because they are simply not available. In Asia, death based on neurologic criteria has not been uniformly accepted, and there are major differences in regulation. India follows the U.K. criteria of brain-stem death ("Transplantation of Human Organs Bill," 1992). China has no legal criteria, and there seems to be some hesitation among physicians to disconnect the ventilator in patients with irreversible coma (Diringer & Wijdicks, 2001). Japan now officially recognizes brain death, although the public remains reluctant—possibly as a result of the heart surgeon Sura Wada, who was charged with murder in 1968 after removing a heart from a patient who was allegedly not brain dead (Lock, 1999). Australia and New Zealand have accepted whole brain death criteria (Pearson, 1995).

Some legal scholars have also endorsed the neocortical definition of death (Smith, 1986; Stacy, 1992), but they have never convinced legislatures or courts. A physician who believes that a patient who is permanently unconscious but breathing is dead risks criminal prosecution or a civil claim for wrongful death if he or she acted on this belief (Beresford, 2001). A finding that consciousness is irreversibly lost will not, by itself, under any applicable medical practice guidelines or law, justify a diagnosis of death; evidence that brain-stem functions are absent is always required. However, withdrawing any treatment that is not considered to be of benefit to the patient is medically and legally accepted, and no doctor has ever been charged with murder for doing this in well-documented cases of patients in an irreversible vegetative state (Jennett, 2002). It should be noted, however, that N. Barber and R. Nejdl were charged with murder in California for withdrawing all treatment, including artificial hydration and nutrition, from a patient, Mr. Herbert, who had been comatose for 7 days. However, their case was dismissed before trial, and the patient's condition later evolved into an irretrievable vegetative state (California Court of Appeal, Second District, Division 2, 1983).

6.1.6 Ethics of Death and Dying

The debate on the need to withhold or withdraw "futile" life-prolonging treatments and the idea of "death with dignity" was started by intensive care physicians (not ethicists or lawyers) in the mid-1970s (Cassem, 1974). At present, almost half of all deaths in intensive care follow a decision to withhold or withdraw treatment (Smedira et al., 1990). There is no moral or legal distinction between withholding or withdrawing (Gillon, 1998).

As discussed above, a person who is brain dead is dead—disconnecting the ventilator will not cause him or her to die. Patients in a vegetative state are not dead, but when their situation becomes hopeless it can be judged unethical to continue their life-sustaining treatment. Unlike patients with brain death, patients in a vegetative state do not usually require ventilatory or cardiac support, needing only artificial hydration and nutrition. The internationally reported case of Terri Schiavo (Annas, 2005; Gostin, 2005; Quill & Shiavo, 2005) centered first on opposing opinions between her husband and parents about whether she would wish to continue living in such a severely disabled state and also on the lack of family consensus regarding her diagnosis of vegetative state. This case illustrated how hard it is for lay persons (and inexperienced physicians and policymakers) to accept the medically established ethical framework that justifies letting patients in an irremediable vegetative state die. Misinformation stemming from high-profile cases such as Schiavo's may increase societal confusion and consternation about end-of-life decision-making (Bernat, 2002c; Cranford, 2005; Jennett, 2002).

Stopping artificial nutrition and hydration to patients in a vegetative state is a complex issue, and it would be beyond the scope of this paper to cover all ethical, legal, and practical dilemmas involved (see Jennett's recent monograph for an in-depth account; Jennett, 2002). It should be stressed that "unless it is clearly established that the patient is permanently unconscious, a physician should not be deferred from appropriately aggressive treatment" (Council on Scientific Affairs & Council on Ethical and Judicial Affairs, 1990), and physicians also "have an obligation to provide effective palliative treatment" (Council on Ethical and Judicial Affairs, American Medical Association, 1992). Several U.S. (President's Commission for the Study of Ethical Problems in Medicine and Biomedical and Behavioral Research, 1983) medical societies and interdisciplinary bodies, including the American Medical Association (Council on Scientific Affairs & Council on Ethical and Judicial Affairs, 1990), the British Medical Association (British Medical Association, 2001), and the World Medical Association (World Medical Association, 1989) have asserted that surrogate decision makers and physicians with advance directives provided by patients have the right to terminate all forms of life-sustaining medical treatment, including hydration and nutrition, in patients in a permanent vegetative state.

The moral values that underlie these guidelines are the principles of autonomy, beneficence, non-maleficence, and justice (Beauchamp & Childress, 1979). Informed, mentally competent patients should consent to any treatment they receive and have the right to make choices regarding their bodies and lives. The primary factor determining the level of treatment provided for an incompetent patient should reflect the patient's personally expressed wishes for treatment in this situation. It should be noted that the principle of autonomy was developed as a product of the Enlightenment in Western culture and is not yet strongly emphasized beyond the United States and Western Europe (for example, in Japan; Asai et al., 1999). In the Western world, the main challenge for autonomy in justifying a right to refuse life-prolonging treatment comes from the vitalist religious view (mainly from orthodox Jews, fundamentalist Protestants, and conservative Roman Catholics) that holds that only God should determine when life ends.

In the past, physicians have interpreted beneficence to mean promotion of continued life, at almost any cost. With the advancement of medical technology, medicine is now ethically obliged not to promote life at all costs in a paternalistic way but rather to enable patients to choose what type of life represents a "good" life to them and what type of life does not. Medical choices should now depend on patients' individual values and can therefore be in disagreement with physicians' personal perceptions (Layon, D'Amico, Caton, & Mollet, 1990). If patients can no longer speak for themselves, having someone who knew them make decisions for them seems the best reasonable compromise. However, critics have argued that surrogate decisions are flawed. Most people would not want to continue living if they were in a vegetative state (Frankl, Oye, & Bellamy, 1989). However, severely disabled patients with brain damage seem to want to go on living (Homer-Ward, Bell, Dodd, & Wood, 2000; McMillan, 1997; McMillan & Herbert, 2004; Shiel & Wilson, 1998). Some studies have shown the limitations of spouses' predictions of patients' desires regarding resuscitation (Uhlmann, Pearlman, & Cain, 1988), and healthy people tend to underestimate impaired patients' quality of life (Starr, Pearlman, & Uhlmann, 1986).

The principle of justice, which includes equity, demands that an individual's worth not be judged solely on social status, nor on physical or intellectual attributes. Vulnerable patients, such as those who are non-communicative and have severe brain damage, those with other handicaps, and those who are very old or young, should not be treated any differently from healthy individuals. No person's life has more or less intrinsic value than the next. Concepts of justice should trump the claims of autonomy, based on a model of medical futility (Payne & Taylor, 1997).

Medical futility is defined as the situation in which a therapy that is hoped to benefit a patient's medical condition will predictably not do so on the basis of the best available evidence (exactly what probability threshold satisfies the standard of "ethical acceptability" is still under discussion) (Bernat, 2002b). Since the Multi-Society Task Force on PVS, we know that the chances of recovery after 3 months for nontraumatic and 12 months for traumatic cases are close to zero. Letting patients in a permanent vegetative state die, despite being ethically and legally justified, remains a complicated and sensitive issue for all those involved (Andrews, 2004).

Finally, the question remains about the mode of death. Stopping hydration and nutrition leads to death in 10–14 days (Cranford, 1984). Recent neuroimaging studies have concluded that patients in a vegetative state lack the neural integration that is considered necessary for pain perception (Laureys et al., 2002). Some, however, are in favor of injecting a lethal drug to quicken the dying process. At present, this practice can only be envisaged in countries or states in which euthanasia has been legalized (for example, Belgium, The Netherlands, and Switzerland) and only if patients have explicitly expressed this wish previously in living wills.

Patients in a vegetative state are not dead, even if their loss of consciousness results in our belief that they may be "as good as dead." However, letting patients in an irreversible vegetative state die can be the most humane option, just as abortion can be justified in, for example, cases of anencephaly, without needing the fetus to be declared dead. This is not a purely medical matter, but an ethical issue that is dependent on personal moral values, and we should accept deviating culture- and religion-dependent viewpoints.

6.1.7 Conclusions and Future Perspectives

In conclusion, brain death is death and irreversible vegetative state is not. Of the two bio-philosophical concepts of brain death (the "whole brain" and the "brain-stem" formulation), defined as the irreversible cessation of critical functions of the organism as a whole (that is, neuroendocrine and homeostatic regulation, circulation, respiration, and consciousness), the whole brain concept is most widely accepted and practiced. Since their first use in 1959 (Mollaret & Goulon, 1959), the neurocentric criteria of death—compared with the old cardiocentric criteria—are considered to be "among the safest medicine can achieve" (Lang, 1999). In those instances in which confirmatory tests for brain death are desirable, irreversibility can, at present, be more reliably demonstrated for the whole brain concept (for example, by measuring lack of intracranial blood flow) (Bernat, 2004), However, with future technological advances and a better understanding and identification of the human cerebral "critical system," the criteria might move further in the direction of brain-stem death (Bernat, 2002a).

In my view, neocortical death, as a confirmatory index for defining death, is conceptually inadequate and practically unfeasible. Clinical, electrophysiologic, neuroimaging, and postmortem studies now provide clear and convincing neurophysiologic and behavioral distinctions between brain death and the vegetative state. Similar lines of evidence also provide compelling data that neocortical death cannot be reliably demonstrated and is an insufficient criterion for establishing death.

Finally, death is a biological phenomenon for which we have constructed pragmatic medical, moral, and legal policies on the basis of their social acceptance (Bernat, 2001). The decision of whether a patient should live or die is a value judgment over which physicians can exert no specialized professional claim. The democratic traditions of our pluralistic society should permit personal freedom in patients' decisions to choose to continue or terminate life-sustaining therapy in cases of severe brain damage. Like most ethical issues, there are plausible arguments supporting both sides of the debate. However, these issues can and should be tackled without changes being made to the current neurocentric definition of death. The benefits of using living humans in a vegetative state as organ donors do not justify the harm to society that could ensue from sacrificing the dead donor principle (Bernat, 2001).

Many of the controversial issues relating to the death and end of life in patients with brain damage who have no hope of recovery result from confusion or ignorance on the part of the public or policymakers about the medical reality of brain death and the vegetative state. Therefore, the medical community should improve educational and public awareness programs on the neurocentric criteria and testing of death; stimulate the creation of advance directives as a form of advance medical care planning; continue to develop clinical practice guidelines; and more actively encourage research on physiologic effects and therapeutic benefit of treatment options in patients with severe brain damage.

What is the future of death? Improving technologies for brain repair and prosthetic support for brain functions (for example, stem cells, neurogenesis, neural computer prostheses, cryonic suspension, and nano-neurological repair) might one day change our current ideas of irreversibility and force medicine and society to once again revise its definition of death.

Notes

1. Editor's note: This reading is an abridged version of an article published in *Nature Reviews Neuroscience* in 2004, volume 6, pages 899–909, and is used with permission of the publisher and the author. The author is Research Associate at the Belgian Fonds Nationale de la Recherche Scientifique.

References

A definition of irreversible coma. (1968). Report of the Ad Hoc Committee of the Harvard Medical School to Examine the Definition of Brain Death. *JAMA, 205,* 337–340.

Alkire, M. T., Haier, R. J., Barker, S. J., Shah, N. K., Wu, J. C., & Kao, Y. J. (1995). Cerebral metabolism during propofol anesthesia in humans studied with positron emission tomography. *Anesthesiology, 82,* 393–403.

Alkire, M. T., Pomfrett, C. J., Haier, R. J., Gianzero, M. V., Chan, C. M., Jacobsen, B. P., et al. (1999). Functional brain imaging during anesthesia in humans: Effects of halothane on global and regional cerebral glucose metabolism. *Anesthesiology, 90,* 701–709.

Alkire, M. T., Haier, R. J., Shah, N. K., & Anderson, C. T. (1997). Positron emission tomography study of regional cerebral metabolism in humans during isoflurane anesthesia. *Anesthesiology, 86,* 549–557.

Andrews, K. (2004). Medical decision making in the vegetative state: Withdrawal of nutrition and hydration. *NeuroRehabilitation, 19,* 299–304.

Andrews, K., Murphy, L., Munday, R., & Littlewood, C. (1996). Misdiagnosis of the vegetative state: Retrospective study in a rehabilitation unit. *BMJ, 313,* 13–16.

Annas, G. J. (2005). "Culture of life" politics at the bedside—the case of Terri Schiavo. *N. Engl. J. Med., 352,* 1710–1715.

Asai, A., Maekawa, M., Akiguchi, I., Fukui, T., Miura, Y., Tanabe, N., et al. (1999). Survey of Japanese physicians' attitudes towards the care of adult patients in persistent vegetative state. *J. Med. Ethics, 25,* 302–308.

Baars, B., Ramsoy, T., & Laureys, S. (2003). Brain, conscious experience and the observing self. *Trends Neurosci., 26,* 671–675.

Beauchamp, T. L., & Childress, J. F. (1979). *Principles of biomedical ethics.* New York: Oxford University Press.

Beresford, H. R. (2001). In E. F. M. Wijdicks (Ed.), *Brain death* (pp. 151–169). Philadelphia: Lippincott Williams & Wilkins.

Bernat, J. L. (1998). A defense of the whole-brain concept of death. *Hastings Cent. Rep., 28*, 14–23.

Bernat, J. L. (2001). In E. F. M. Wijdicks (Ed.), *Brain death* (pp. 171–187). Philadelphia: Lippincott Williams & Wilkins.

Bernat, J. L. (Ed.). (2002a). In *Ethical issues in neurology* (pp. 243–281). Boston: Butterworth Heinemann. Bernat, J. L. (2002b). In J. L. Bernat (Ed.)., *Ethical issues in neurology* (pp. 215–239). Boston: Butterworth Heinemann.

Bernat, J. L. (2002c). In J. L. Bernat (Ed.)., *Ethical issues in neurology* (pp. 283–305). Boston: Butterworth Heinemann.

Bernat, J. L. (2004). On irreversibility as a prerequisite for brain death determination. *Advances in Experimental Medicine and Biology, 550*, 161–167.

Bernat, J. L. (2005). In S. Laureys (Ed.), *The boundaries of consciousness: Neurobiology and neuropathology* (pp. 369–379). Amsterdam, The Netherlands: Elsevier.

Bernat, J. L., Culver, C. M., & Gert, B. (1981). On the definition and criterion of death. *Arch. Intern. Med., 94*, 389–394.

Boly, M., Faymonville, M. E., Peigneux, P., Lambermont, B., Damas, P., Del Fiore, G., et al. (2004). Auditory processing in severely brain injured patients: Differences between the minimally conscious state and the persistent vegetative state. *Arch. Neurol., 61*, 233–238.

Brierley, J. B., Graham, D. I., Adams, J. H., & Simpsom, J. A. (1971). Neocortical death after cardiac arrest. A clinical, neurophysiological, and neuropathological report of two cases. *Lancet, 2*, 560–565.

British Medical Association. (2001). *Witholding or withdrawing life-prolonging medical treatment: Guidance for decision making*, 2nd edition. London, UK: BMJ Books.

California Court of Appeals, Second District, Division 2. (1983). Barber v. Superior Court of State of California; Nejdl v. Superior Court of State of California. *Wests Calif. Report., 195*, 484–494.

Cassem, N. H. (1974). Confronting the decision to let death come. *Crit. Care Med., 2*, 113–117.

Chalmers, D. J. (1998). The problems of consciousness. *Adv. Neurol., 77*, 7–16; discussion 16–18.

Childs, N. L., Mercer, W. N., & Childs, H. W. (1993). Accuracy of diagnosis of persistent vegetative state. *Neurology, 43*, 1465–1467.

Council on Ethical and Judicial Affairs, American Medical Association. (1992). Decisions near the end of life. *JAMA, 267*, 2229–2233.

Council on Scientific Affairs and Council on Ethical and Judicial Affairs. (1990). Persistent vegetative state and the decision to withdraw or withhold life support. *JAMA, 263*, 426–430.

Cranford, R. E. (1984). Termination of treatment in the persistent vegetative state. *Semin. Neurol., 4*, 36–44.

Cranford, R. (2005). Facts, lies, and videotapes: The permanent vegetative state and the sad case of Terri Schiavo. *J. Law Med. Ethics, 33*, 363–371.

Criteria for the diagnosis of brain stem death. (1995). Review by a working group convened by the Royal College of Physicians and endorsed by the Conference of Medical Royal Colleges and their Faculties in the United Kingdom. *J. R. Coll. Physicians Lond., 29*, 381–382.

De Volder, A. G., Goffinett, A. M., Bol, A., Michel, C., de Barsy, T., & Laterre, C. (1990). Brain glucose metabolism in postanoxic syndrome. Positron emission tomographic study. *Arch. Neurol., 47*, 197–204.

Diagnosis of brain death. (1976). Statement issued by the honorary secretary of the Conference of Medical Royal Colleges and their Faculties in the United Kingdom on 11 October 1976. *Br. Med. J., 2*, 1187–1188.

Diringer, M. N., & Wijdicks, E. F. M. (2001). In E. F. M. Wijdicks (Ed.), *Brain death* (pp. 5–27). Philadelphia: Lippincott Williams & Wilkins.

Frankl, D., Oye, R. K., & Bellamy, P. E. (1989). Attitudes of hospitalized patients toward life support: A survey of 200 medical inpatients. *Am. J. Med., 86*, 645–648.

Gervais, K. G. (1986). *Redefining death*. New Haven, CT: Yale University Press.

Giacino, J. T., Ashwal, S., Childs, N., Cranford, R., Jennett, B., Katz, D. I., et al. (2002). The minimally conscious state: Definition and diagnostic criteria. *Neurology, 58*, 349–353.

Gillon, R. (1998). Persistent vegetative state, withdrawal of artificial nutrition and hydration, and the patient's 'best interests.' *J. Med. Ethics, 24*, 75–76.

Gostin, L. O. (2005). Ethics, the constitution, and the dying process: The case of Theresa Marie Schiavo. *JAMA, 293*, 2403–2407.

Gostin, L. O. (1997). Deciding life and death in the courtroom. From Quinlan to Cruzan, Glucksberg, and Vacco—a brief history and analysis of constitutional protection of the "right to die." *JAMA, 278*, 1523–1528.

Haupt, W. F., & Rudolf, J. (1999). European brain death codes: A comparison of national guidelines. *J. Neurol., 246*, 432–437.

Homer-Ward, M. D., Bell, G., Dodd, S., & Wood, S. (2000). The use of structured questionnaires in facilitating ethical decision-making in a patient with low communicative ability. *Clin. Rehabil., 14*, 220.

Ingvar, D. H., Brun, A., Johansson, L., & Samuelsson, S. M. (1978). Survival after severe cerebral anoxia with destruction of the cerebral cortex: The apallic syndrome. *Ann. N. Y. Acad. Sci., 315*, 184–214.

Jennett, B. (2002). *The vegetative state. Medical facts, ethical and legal dilemmas.* Cambridge, UK: Cambridge University Press.

Joynt, R. J. (1984). Landmark perspective: A new look at death. *JAMA, 252*, 680–682.

Korein, J., & Machado, C. (2004). In C. Machado & D. Shewmon (Eds.), *Brain death and disorders of consciousness* (pp. 1–21). New York: Kluwer Academic/Plenum.

Laureys, S., Goldman, S., Phillips, C., Van Bogaert, P., Aerts, J., Luxen, A., et al. (1999). Impaired effective cortical connectivity in vegetative state: Preliminary investigation using PET. *Neuroimage, 9*, 377–382.

Laureys, S., Faymonville, M. E., Peigneux, P., Damas, P., Lambermont, B., et al. (2002). Cortical processing of noxious somatosensory stimuli in the persistent vegetative state. *NeuroImage, 17*(2), 732–741.

Laureys, S. (2005). The functional neuroanatomy of (un)awareness: Lessons from the vegetative state. *Trends Cogn. Sci., 9, 556–559.*

Laureys, S., Faymonville, M. E., Moonen, G., Luxen, A., & Maquet, P. (2000). PET scanning and neuronal loss in acute vegetative state. *Lancet, 355*, 1825–1826.

Laureys, S., Lemaire, C., Maquet, P., Phillips, C., & Franck, G. (1999). Cerebral metabolism during vegetative state and after recovery to consciousness. *J. Neurol. Neurosurg. Psychiatry, 67*, 121–122.

Layon, A. J., D'Amico, R., Caton, D., & Mollet, C. J. (1990). And the patient chose: Medical ethics and the case of the Jehovah's Witness. *Anesthesiology, 73*, 1258–1262.

Levy, D. E., Sidtis, J. J., Rottenberg, D. A., Jarden, J. O., Strother, S. C., Dhawan, V., et al. (1987). Differences in cerebral blood flow and glucose utilization in vegetative versus locked-in patients. *Ann. Neurol., 22*, 673–682.

Lock, M. (1999). In S. J. Youngner, R. M. Arnold, & R. Schapiro (Eds.), *The definition of death: Contemporary controversies* (pp. 239–256). Baltimore, MD: John Hopkins University Press.

Loeb, J. (1916). *The organism as a whole.* New York: G. P. Putnam's Sons.

Lofstedt, S., & von Reis, G. (1956). Intracranial lesions with abolished passage of X-ray contrast throughout the internal carotid arteries. *Pacing Clin. Electrophysiol., 8*, 199–202.

Majerus, S., Gill-Thwaites, H., Andrews, K., & Laureys, S. (2005). In S. Laureys (Ed.), *The boundaries of consciousness: Neurobiology and neuropathology* (pp. 397–413). Amsterdam, The Netherlands: Elsevier.

Maquet, P., Degueldre C, Delfiore G, Aerts J, Peters JM, & Luxen A, et al. (1997). Functional neuroanatomy of human slow wave sleep. *J. Neurosci., 17*, 2807–2812.

McMillan, T. M. (1997). Neuropsychological assessment after extremely severe head injury in a case of life or death. [Erratum: *Brain Inj., 11*, 775 (1997)] *Brain Inj., 11*, 483–490.

McMillan, T. M., & Herbert, C. M. (2004). Further recovery in a potential treatment withdrawal case 10 years after brain injury. *Brain Inj., 18*, 935–940.

Mohandas, A., & Chou, S. N. (1971). Brain death. A clinical and pathological study. *J. Neurosurg., 35*, 211–218.

Mollaret, P., & Goulon, M. (1959). Le coma dépassé. *Rev. Neurol., 101*, 3–15.

Ore, G. D., Gerstenbrand, F., & Lucking, C. H. (1977). *The apallic syndrome.* Berlin, Germany: Springer.

Pallis, C., & Harley, D. H. (1996). *ABC of brainstem death.* London, UK: BMJ.

Payne, S. K., & Taylor, R. M. (1997). The persistent vegetative state and anencephaly: Problematic paradigms for discussing futility and rationing. *Semin. Neurol., 17*, 257–263.

Payne, K., Taylor, R. M., Stocking, C., & Sachs, G. A. (1996). Physicians' attitudes about the care of patients in the persistent vegetative state: A national survey. *Ann. Intern. Med., 125*, 104–110.

Pearson, I. Y. (1995). Australia and New Zealand Intensive Care Society Statement and Guidelines on Brain Death and Model Policy on Organ Donation. *Anaesth. Intensive Care, 23*, 104–108.

Pernick, M. S. (1988). In R. M. Zaner (Ed.), *Death: Beyond whole-brain criteria* (pp. 17–74). Dordrecht, The Netherlands: Kluwer Academic.

Pius XII. (1957). Pope speaks on prolongation of life. *Osservatore Romano, 4*, 393–398.

Pope John Paul II. (2004). Address of Pope John Paul II to the participants in the International Congress on "Life-Sustaining Treatments and Vegetative State: Scientific Advances and Ethical Dilemmas," Saturday, 20 March 2004. *NeuroRehabilitation, 19*, 273–275.

President's Commission for the Study of Ethical Problems in Medicine and Biomedical and Behavioral Research. (1983). pp. 171–192. Washington, DC: U.S. Government Printing Office.

Prochazka, A., Clarac, F., Loeb, G. E., Rothwell, J. C., & Wolpaw, J. R. (2000). What do reflex and voluntary mean? Modern views on an ancient debate. *Exp. Brain Res., 130*, 417–432.

Quill, T. E. (2005). Terri Schiavo—a tragedy compounded. *N. Engl. J. Med., 352*, 1630–1633.

Saposnik, G., Bueri, J. A., Maurino, J., Saizar, R., & Garretto, N. S. (2000). Spontaneous and reflex movements in brain death. *Neurology, 54*, 221–223.

Saposnik, G., Maurino, J., Saizar, R., & Bueri, J. A. (2005). Spontaneous and reflex movements in 107 patients with brain death. *Am. J. Med., 118*, 311–314.

Schiff, N. D., Ribary, U., Moreno, D. R., Beattie, B., Kronberg, E., Blasberg, R., et al. (2002). Residual cerebral activity and behavioural fragments can remain in the persistently vegetative brain. *Brain, 125*, 1210–1234.

Schotsmans, P. (1993). The patient in a persistent vegetative state: An ethical reappraisal. *Int. J. Phil. Theol., 54*, 2–18.

Shiel, A., & Wilson, B. A. (1998). Assessment after extremely severe head injury in a case of life or death: Further support for McMillan. *Brain Inj., 12*, 809–816.

Smedira, N. G., Evans, B. H., Grais, L. S., Cohen, N. H., Lo, B., Cooke, M., et al. (1990). Withholding and withdrawal of life support from the critically ill. *N. Engl. J. Med., 322*, 309–315.

Smith, D. R. (1986). Legal recognition of neocortical death. *Cornell Law Rev., 71*, 850–888.

Stacy, T. (1992). Death, privacy, and the free exercise of religion. *Cornell Law Rev., 77*, 490–595.

Starr, T. J., Pearlman, R. A., & Uhlmann, R. F. (1986). Quality of life and resuscitation decisions in elderly patients. *J. Gen. Intern. Med., 1*, 373–379.

The Multi-Society Task Force on PVS. (1994a). Medical aspects of the persistent vegetative state (1). *N. Engl. J. Med., 330*, 1499–1508.

The Multi-Society Task Force on PVS. (1994b). Medical aspects of the persistent vegetative state (2). *N. Engl. J. Med., 330*, 1572–1579.

The Quality Standards Subcommittee of the American Academy of Neurology. (1995). Practice parameters for determining brain death in adults (summary statement). *Neurology, 45*, 1012–1014.

The transplantation of human organs bill. (1992). Republic of India, Bill No. LIX-C.

Tommasino, C., Grana, C., Lucignani, G., Torri, G., & Fazio, F. (1995). Regional cerebral metabolism of glucose in comatose and vegetative state patients. *J. Neurosurg. Anesthesiol., 7*, 109–116.

Uhlmann, R. F., Pearlman, R. A., & Cain, K. C. (1988). Physicians' and spouses' predictions of elderly patients' resuscitation preferences. *J. Gerontol., 43*, M115–M121.

Uniform determination of death act. (1997). *598 (West 1993 and West Suppl. 1997)* (Uniform Laws Annotated (U. L. A.).

Veatch, R. M. (1975). The whole-brain-oriented concept of death: An outmoded philosophical formulation. *J. Thanatol., 3*, 13–30.

Veatch, R. M. (1976). *Death, dying, and the biological revolution. Our last quest for responsibility.* New Haven, CT: Yale University Press.

Volicer, L., Berman, S. A., Cipolloni, P. B., & Mandell, A. (1997). Persistent vegetative state in Alzheimer disease. Does it exist? *Arch. Neurol., 54*, 1382–1384.

Wertheimer, P., Jouvet, M., & Descotes, J. (1959). A propos du diagnostic de la mort du sysème nerveux dans les comas avec arrêt respiratoire traités par respiration artificielle. *Presse Med., 67*, 87–88.

Wijdicks, E. F. M. (Ed.). (2001a). *Brain death.* Philadelphia: Lippincott Williams & Wilkins.

Wijdicks, E. F. M. (2001b). In E. F. M. Wijdicks (Ed.), *Brain death* (pp. 61–90). Philadelphia: Lippincott Williams & Wilkins.

Wijdicks, E. F. (2001c). The diagnosis of brain death. *N. Engl. J. Med., 344*, 1215–1221.

World Medical Association. (1989). Statement on persistent vegetative state. Adopted by the 41st World Medical Assembly, Hong Kong, September 1989. Available at http://www.wma.net/e/policy/p11.htm.

Reading 6.2

Personhood: An Illusion Rooted in Brain Function?[1]

Martha J. Farah and Andrea S. Heberlein

Personhood is a foundational concept in ethics, yet defining criteria have been elusive. Here we explore the possibility that personhood has been so hard to pin down because it is an illusion. According to this view, "persons" does not correspond with a category of objects in the world. Rather, it is an illusory concept that our brains have evolved to develop innately and project onto the world whenever triggered by stimulus features such as a human-like face, body, or contingent patterns of behavior. We review the evidence for the existence of an autonomous person network in the brain and discuss its implications for the field of ethics and for the implicit morality of everyday behavior.

In everyday life we have no problem deciding which entities to refer to as persons: human beings generally qualify and other things generally do not. Yet the attempt to specify criteria for personhood has occupied philosophers for centuries, without producing a consensus on the essential or defining characteristics of persons. In this reading we suggest that personhood continues to defy definition because it is not a real category of objects in the world. Instead, we suggest that the concept of personhood is the product of an evolved brain system that develops innately and projects itself automatically and irrepressibly onto the world.

Given that our brains were shaped by natural selection, we might expect a fairly good fit between normal human perceptions of the world and the objective physics of the world that is relevant to our survival. That is, there are good reasons to believe that our perceptions of the size, motion, temperature, and so forth of objects map onto the human-scale reality in fairly simple, lawful ways. If human survival depends not just on negotiating the physical world but also the social world, then we might expect our brains to have evolved some additional representational "vocabulary" beyond the kinds of physical predicates just discussed. And indeed, one of the most exciting developments in cognitive neuroscience is the discovery of brain systems that appear to be specialized for representing information about people. This research will be summarized next, followed by an analysis of its implications for our thinking about persons.

6.2.1 Evidence That We Are Hardwired to Represent Persons

The earliest clue that the organization of our brain representations carves the world into persons and non-persons came from studies of visual perception in brain-damaged patients. A rare disorder known as prosopagnosia consists of impaired visual recognition of the human face (see Farah, 2004, chapter 7). Prosopagnosia can be a relatively isolated impairment, that is, a prosopagnosic patient may fail to recognize faces but succeed in recognizing other equally challenging types of object, consistent with the existence of a specialized face recognition system that can be damaged selectively (Farah, Wilson, Drain, & Tanaka, 1995). The recognition of even animal faces may be spared in prosopagnosia, implying that the face recognition system is specialized for representing humans (McNeil & Warrington, 1993). The opposite pattern of visual recognition impairment has also been observed, namely generally poor object recognition with preserved face recognition, further strengthening the case for a distinct face recognition system (Feinberg, Schindler, Ochoa, Kwan, & Farah, 1994).

Functional neuroimaging of normal healthy individuals has confirmed the existence of a brain region specialized for human face recognition and localized it with greater precision than is possible with naturally occurring brain lesions (Kanwisher, McDermott, & Chun, 1997). The fusiform gyrus, on the ventral surface in the brain, is activated disproportionately by the sight of a human face, relative to many other types of visual stimulus materials. Although some controversy exists over whether this area is best described as responding to faces *per se* or to a set of perceptual and cognitive demands that are normally associated with face recognition (see Tarr & Gauthier, 2000, for a review), no one would deny that this area is normally recruited for human face recognition. Figure 6.2.1 shows the location of the fusiform gyrus in the human brain. Facial expressions of emotion, as well as vocally expressed emotion, activate additional brain areas including the amygdala (Phillips, Drevets, Rauch, & Lane, 2003) also shown in figure 6.2.1. Patients with bilateral amygdala damage are impaired in the perception of people's emotional states (e.g., Adolphs et al., 2005).

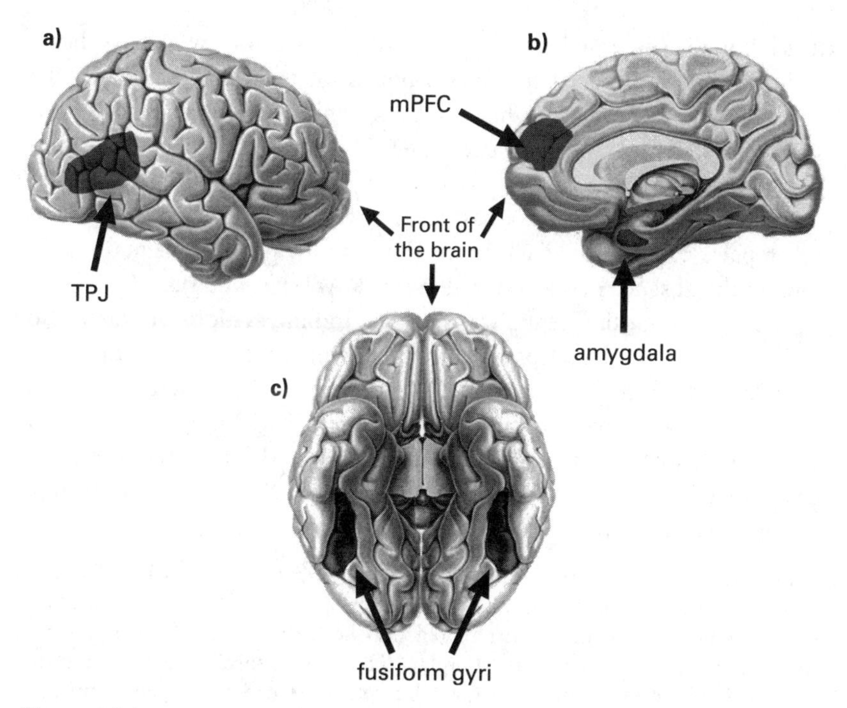

Figure 6.2.1
Three views of the human brain indicating the location of structures discussed in this article: (a) a lateral (side) view of the right hemisphere showing the temporoparietal junction (TPJ); (b) a medial (middle surface, between the two hemispheres) view of the right hemisphere showing the medial prefrontal cortex (mPFC) and the amygdala (which is buried inside the cortex but is here shown "glowing" through); (c) a ventral (bottom) view showing the fusiform gyrus.

Other perceptible aspects of people are also represented by distinct brain systems. Downing and his colleagues have shown that the sight of human bodies, with faces obscured, activates two distinct regions within the brain, one on the fusiform gyrus adjacent to, but distinct from, the face area (Peelen & Downing, 2005) and one on the lateral surface of the brain near the temporoparietal juncture (Downing, Jiang, Shuman, & Kanwisher, 2001), also shown in figure 6.2.1. Silhouettes and even stick figures of people activate these regions, but equally complex shapes that are not bodies do not.

Bodily movements activate another part of the temporoparietal junction, somewhat anterior to the body area (e.g., Grossman et al., 2000). Studies with "point light walker" stimuli have shown that this region is specialized for the representations of actions *per se* rather than the body;

these stimuli, generated by filming in darkness actors who have light-emitting diodes attached to various points on their bodies, convey the characteristic motion of a human body while excluding its other visual characteristics (Allison, Puce, & McCarthy, 2000). Parts of the temporoparietal junction are activated specifically by actions perceived to be goal-directed (Saxe, Xiao, Kovacs, Perrett, & Kanwisher, 2004), and other parts are activated when we think about people's mental states, even in the absence of visual input (Saxe & Wexler, 2005).

Thinking about the mental traits, states, and interactions of others also activates the medial prefrontal cortex, shown in Figure 6.2.1. In a pioneering study, Fletcher et al. (1995) compared the brain activity evoked by understanding two kinds of stories: those for which it was necessary to represent someone's mental state in order to understand the story, and those for which physical causation rather than psychology had to be represented. For example:

> A burglar who has just robbed a shop is making his getaway. As he is running home, a policeman on his beat sees him drop his glove. He doesn't know the man is a burglar, he just wants to tell him he dropped his glove. But when the policeman shouts out to the burglar "Hey, you! Stóp!", the burglar turns around, sees the policeman and gives himself up. He puts his hands up and admits that he did the break-in at the local shop.

> A burglar is about to break into a jeweller's shop. He skillfully picks the lock on the shop door. Carefully he crawls under the electronic detector beam. If he breaks this beam it will set off the alarm. Quietly he opens the door of the store-room and sees the gems glittering. As he reaches out, however, he steps on something soft. He hears a screech and something small and furry runs out past him, towards the shop door. Immediately the alarm sounds.

The brain activity associated with understanding the two types of story differed in the medial prefrontal cortex. A later study by Gallagher et al. (2000) replicated this localization with similar stories and also with nonverbal cartoons designed to vary in the degree to which they require the viewer to represent the psychology of others.

The medial prefrontal region has been found to represent other aspects of mental processes in a variety of very different task contexts. For example, Goel, Grafman, Sadato, and Hallett (1995) compared brain activity while participants judged whether Christopher Columbus would know how to use various objects such as a compact disc with brain activity during other kinds of judgments about the objects and found greater medial prefrontal activity when Columbus' knowledge was being considered. In a different type of task, Mitchell, Heatherton, and Macrae (2002) asked participants to decide whether a given adjective could ever be true of a given noun. In some cases the adjectives were psychological (e.g., assertive, energetic, fickle, nervous) and could only apply to people, and in other cases they were appropriate for fruits (e.g., sundried, seedless) or clothing (e.g., patched, threadbare). Accordingly, nouns were the first names of people, fruits, and articles of clothing. Patterns of brain activity associated with judgments about people and non-people were distinct, with a high degree of agreement between the areas associated with person processing in this study and in the previous ones, despite the very different type of task used.

More recently, Mitchell, Neil Macrae, and Banaji (2005) presented participants with photographs of people and objects and accompanied each photograph with a statement designed to create a positive or negative impression. For example, a picture of a person might be accompanied by the statement "promised not to smoke in his apartment since his roommate was trying to quit," and a picture of a car might be accompanied by "recently had new fog lights installed." In one condition, participants were told to form an impression of the people and objects based on the statements, and in another condition they were told to remember the sequence in which the statements were presented. The authors confirmed their prediction that impression formation instructions with face photographs would be associated with the most medial prefrontal activation, as these trials involved the most cognition about other persons.

A final example of evidence for dedicated brain systems for representing people comes from a game of "rock, paper, scissors" played in the scanner with a computer whose responses were randomly generated (Gallagher, Jack, Roepstorff, & Frith, 2002). Participants believed that the responses came from a human in one condition and from a computer in the other. When the conditions were compared, the medial prefrontal cortex was again found to be more active in the human condition.

The weight of the evidence, from a sizable literature only sampled here, clearly supports the conclusion that the human brain represents the appearance, actions, and thoughts of people in a distinct set of regions, different from those used to represent the appearance, movements, and properties of other entities. These regions together form a network that is sometimes referred to as "the social brain" (e.g., Adolphs, 2003; Brothers, 1990; Skuse, Morris, & Lawrence, 2003) but could equally well be termed a network for person representation.

6.2.2 The Autonomy of the Person Network

In addition to supporting the existence of a separate system for representing persons, recent neuroscience evidence suggests a surprising level of automaticity of person processing by this network, as well as a high degree of innateness. By automaticity we mean the tendency of the person network to be triggered by certain stimulus features even when we are aware that the stimulus is not a person. By innateness we mean the genetically preprogrammed nature of the system, without a need to learn that persons exist in the world. The autonomous development and functioning of the person network has important implications for how we think about persons. (Indeed, in some cases this irrepressible autonomy has implications for how we think about non-persons, too, as when we coax or curse at our computers.)

Early evidence for the automaticity of face recognition came from a prosopagnosic patient whose ability to process faces was actually improved when the faces were turned upside-down (Farah, Levinson, & Klein, 1995). As long as the faces were shown to him in a normal orientation, his damaged face recognition system interfered with his ability to perceive them; he was unable to "turn off" his face recognition and treat the faces like some other kind of object or pattern, even though treating them as faces was counterproductive.

Another manifestation of the automaticity of the person network is the ability of certain "trigger features" to engage it. Not just realistic depictions of people but also smiley faces and stick figures activate the system (Downing et al., 2001; Wright, Martis, Shin, Fischer, & Rauch, 2002). In other words, we need not believe that there is a person there to have our person network engaged. This is presumably the explanation of a recent finding in behavioral economics, that people adopt more generous strategies in a computer-run economic game when the computer screen happens to display a pair of cartoon eyes (Haley & Fessler, 2005). Indeed, the person network could be described as having a hair trigger for these visual features. Faces and bodies can activate the system even when we are not paying attention to them and even when we are unaware of them (Downing, Bray, Rogers, & Childs, 2004; Vuilleumier, 2000).

In addition to visual shape features such as static eyes, faces, and bodies, certain patterns of motion are also effective at engaging the system. In particular, contingent "behavior," by which a stimulus seems responsive to its environment, can evoke a sense of intentionality and personhood. In the famous animated film of Heider and Simmel (1944), two triangles and a circle move around the screen with motions that are inter-related, giving an impression of three entities interacting with motivations and intentions (see http://pantheon.yale.edu/~bs265/demos/causality.html). The automaticity of this attribution is apparent in the difficulty of describing this film without using psychological terms like *wants* and *tries* (Scholl & Tremoulet, 2000). This automaticity seems related to the triggering of the person network, in that a patient with complete bilateral amygdala degeneration described the film in purely physical terms (Heberlein & Adolphs, 2004).

Brain imaging studies of Heider and Simmel type animations show that all of the brain regions shown in Figure 6.2.1 are activated (Castelli, Happé, Frith, & Frith, 2000; Martin & Weisberg, 2003; Schultz et al., 2003). For example, in the study of Martin and Weisberg, two sets of animations were presented: both were composed of moving squares, triangles, and circles, which moved in a contingent interactive manner (e.g., as if dancing together or chasing each other) in the "social" set and in a manner consistent with mechanical motions (e.g., like billiard balls or objects on a conveyer belt) in the "mechanical" set. Despite the absence of anything resembling a human being in these animations, the former set and only that set activated the fusiform face area, amygdala, temporoparietal junction, and medial prefrontal cortex.

The evidence just reviewed indicates that the person network functions largely autonomously, independent of our conscious, rational beliefs about the nature of smiley faces or animated geometric shapes. Other evidence indicates that its development is also autonomous, in the sense that its specialization for persons comes about prior to experience with persons and other objects in the world.

Evidence for the innateness of the person–non-person distinction comes from the behavior of newborn infants. Mark Johnson and colleagues (1991) showed that newborns tested within 30 minutes of birth show a greater tendency to track moving facelike patterns with their eyes than other patterns of comparable complexity or symmetry. This implies that prior to virtually any opportunity to learn, the human brain is equipped with a general representation of the appearance of the human face. Another demonstration of innateness in person processing comes from the study of a boy who sustained visual cortical damage, including damage to the fusiform face area, in his first day of postnatal life (Farah, Rabinowitz, Quinn, & Liu, 2000). Despite his relatively preserved ability to recognize nonface objects, he never acquired the ability to recognize faces. In other words, a certain region of cortex is destined for face recognition as early as 1 day of age, and other regions, which are capable of recognizing inanimate objects, cannot take over this function. This striking absence of plasticity implies that the category of human face, as well as its representation by specific brain tissue, is determined essentially at birth.

Studies with older infants confirm that we distinguish between persons and non-persons, or more accurately between entities that do and do not possess the trigger features for the person network, as early as 3 months of age. Moving shapes whose motions are mutually contingent attract the attention of 3-month-olds more effectively than shapes moving in noncontingent ways (Rochat, Morgan, & Carpenter, 1997). A study with 12-month-olds found that they, like adults, tend to follow the "gaze" of football-shaped objects (i.e., look where the object seems to be looking) if the object has been seen moving in a contingent manner or if it has eye spots (Johnson, 2003).

Infants also implicitly attribute intentions to the behavior of persons at an early age. For example, Woodward (1998) used a habituation paradigm to probe 5-month-old infants' representations of two kinds of events, a person reaching around a barrier to retrieve an object and a mechanical "arm" doing the same thing. The barrier was then removed, and one of two things happened next: the reacher (person or machine) reached again for the object with either the same round-about trajectory or reached for it directly. Infants who saw the machine looked longer (evincing surprise) when it changed its trajectory from round-about to direct, but those who saw the person looked longer when the trajectory through space was the same as before. This implies that the infants' initial representation of the machine's action concerned its physical motion through space, so that a similar motion was less surprising, whereas their initial representation of the person's action concerned his or her intention to pick up the object, so that a direct retrieval was less surprising.

At the same age, infants grasp certain principles of the physical behavior of objects, including the need to traverse a continuous trajectory through space in going from one point to another, but seem to think of people as exempt from at least certain constraints of physical objects (Kulhmeier, Wynn, & Bloom, 2003). The authors interpret this as evidence for a distinction in the infant's mind between persons and things. Furthermore, because in reality humans do traverse continuous paths through space, this was clearly not a learned feature of persons. The authors suggest that it reflects the child's assumption that the important part of a person is the nonmaterial part and the resultant difficulty of thinking of people as physical objects (Bloom, 2004).

Another source of evidence for the innate nature of person representation in the brain comes from the study of individuals with autism and autistic spectrum disorders. Autism is a complex condition with a number of cognitive and affective components, but the core feature that distinguishes it from other developmental disorders is abnormal interpersonal behavior. There is a substantial genetic component to autism (Piven, 1997) and the social behaviors typical of it (Ronald, Happe, & Plomin, 2005). From infancy on, autistic individuals show an unusually low level of interest in other people, generally preferring to interact with inanimate objects. A retrospective study of home movies of first year birthday parties showed that this tendency was apparent well before the child was diagnosed (Osterling, Dawson, & Munson, 2002). Autistic children are sometimes described as treating people like objects, for example attempting to climb a conveniently located adult to get to a toy on a shelf. As adults, autistic individuals have difficulty anticipating the reactions of other people and understanding why others behave as they do. They have difficulty with tasks that require representing the mental states of others, for example understanding the first type of story quoted earlier (Happé, 1994).

Functional neuroimaging studies show that the brain regions normally activated for person representation are not activated in autistic participants (see Pelphrey, Adolphs, & Morris, 2004, for a review). For example, autistic participants do not show increased activation in the person network when viewing the kinds of animated shapes that evoke person-related cognition and neural activity in normal participants (Castelli, Frith, Happé, & Frith, 2002; Schultz et al., 2003). These participants also tend not to show activation in the fusiform face area when viewing human faces (Critchley et al., 2000; Schultz et al., 2000) and do not activate medial prefrontal cortex when reading stories involving other people's mental states (Happé et al., 1996). These studies suggest that the development of the person network is partly genetically determined and that autism represents an abnormality in this process.

In sum, we come into the world with a brain system genetically preprogrammed to represent persons as distinct from other kinds of object in the world. This system is surprisingly autonomous, in the sense that it is triggered by certain stimuli and can be difficult to suppress. It becomes active even when we know that the triggering stimulus is not a person, that is, when other parts of our brain represent the information that the stimulus is not a person, but an unrealistic drawing of a person or even a geometric shape. Indeed, it becomes active in the presence of triggering stimulus features even when irrelevant or downright counterproductive.

6.2.3 The Metaphysics of Personhood

Despite our intuitions that both plants and persons are "out there," in some similar sense of being natural kinds in the world, there are important differences between the two types of category. Science has found an objective basis for the distinction we make intuitively between plants and other multicellular organisms, but it has yet to identify useful criteria for personhood. We suggest that this is because the category "plant" has a kind of objective reality that the category "person" does not. In the previous section, we summarized evidence that the human brain is born equipped to treat certain types of stimuli—those with such trigger features as a human-like face or body or patterns of movement—in a special way. We perceive them and reason about them using a separate brain system, and do so innately, automatically, and irrepressibly. Our sense that the world contains two fundamentally different categories of things, persons and non-persons, may be a result of the periodic activation of this person network by certain stimuli rather than any fundamental distinction between the stimuli that do and don't tend to trigger it.

Of course, there must be some set of attributes in the world that determine whether or not the person network is triggered. Doesn't that imply that persons are "in the world" after all? To answer this question let us consider the relations between mental representation and reality for three categories: persons, plants, and phlogiston. Phlogiston is the name of a fluid that seventeenth and eighteenth century scientists believed was contained in combustible substances and that twentieth and twenty-first century philosophers have used to illustrate a point about theory change and word meaning. Combustion was thought to be a process by which phlogiston left the burning substance and was absorbed by the air. In conjunction with some other reasonable assumptions, the phlogiston theory was able to explain a number of different aspects of combustion. For example, the extinction of fires by limiting the air supply could be explained in terms of the air becoming saturated with phlogiston. We now understand that there is no such thing as phlogiston, and that combustion is part of a larger category of phenomena consisting of oxidation.

However, when phlogiston theorists perceived burning, their representations of phlogiston became active. These early scientists did not randomly or arbitrarily project a concept of phlogiston onto the world; there was of course some category of events in the world that corresponded in a systematic way with their representation of phlogiston. However, this category was based on relatively superficial perceptual features of the world (e.g., flames) combined in certain ways dictated by their theory and did not capture any of the deeper or more explanatory structure of nature. The point of this example is that mental representations can exist and be activated by stimuli in systematic ways without picking out fundamental categories of the natural world.

We do not believe that personhood is like phlogiston. Our evolved person representations are probably not as thoroughly wrong as the phlogiston theorists' representations of oxidation. Clearly, some things in the world have minds much like our own, and other things do not have minds. However there are also different degrees of mindedness, and perhaps even different kinds of minds (e.g., Brooks, 2002; Edelman, Baars, & Seth 2005). Furthermore, our intuitions about who or what has a mind are partly under the control of superficial and potentially misleading trigger features like eyes and faces. In this sense, our person representations do not reflect reality as accurately as our plant representations. We suggest that two features of person representation in the brain underlie this discrepancy.

The first relevant feature of the person network in the brain is its separateness from the systems representing other things. We suggest that this feature is responsible for the illusion that persons and non-persons are fundamentally different kinds of things in the world, despite our inability to draw a principled line between them. This illusion may come from the operation of two separate and incommensurate systems of representation in the brain for persons and for things in general, in contrast with a common distributed representation. Within a unitary distributed representation of color, for example, one could represent a shade that is red or orange, and if both are active, one would automatically be representing reddish orange (see O'Reilly and Munakata, 2000, for a discussion of distributed representations in the brain). Within a unitary distributed representation for shape with representations of bowl-like and cuplike forms, the simultaneous activation of both would represent an object with an in-between shape, a kind of large, wide cup. But what if red or bowlness are also represented by a separate system from other colors and other shapes? Then regardless of how much orange or cupness is being registered elsewhere in the brain, the sight of reddish orange or a large, wide cup will result in a representation of red or a bowl. These representations of red or bowl shape may be weaker than those engendered by a true red or a prototypical bowl shape, but they will nevertheless be weakly red as opposed to reddish orange, or weakly bowl-like as opposed to bowlish-cuplike.

Someone perceiving the world with such a system of representation would perceive both the continuities among colors and shapes, but also the existence of a divide between red things and nonred things, bowls and nonbowls. Such a person might say "On the one hand, I can't find a sensible place to draw a line between red and reddish-orange things. But on the other hand, it seems clear to me that some things have redness and some do not. Things may vary in how much redness they have, but by having redness they are fundamentally different from other things." Substitute the person system for the red system, and one gets the very intuition that has posed such a problem in philosophy and bioethics. This intuition could be expressed thus: "People, animals, and even computers may have varying amounts of intelligence, communication ability, self-awareness, etc. I can't find a sensible place to draw a line across the potential continuum of states linking, say, a healthy human and one in a vegetative state or linking a current-day computer and one endowed with human-like intelligence. Nevertheless, I have the sense that some beings have personhood and others do not."

The second relevant feature of the person network is its autonomy, its tendency to become activated by certain triggering stimuli (e.g., faces and contingent behavior) whether or not we believe there is actually a person there. Even if persons were like plants, and there were a clear objective basis for separating persons and non-persons, the relentless projection of personhood on the basis of fragmentary cues would lead to error and confusion on its own. For example, the human face is a powerful trigger cue that activates the whole person network, and this may be what makes it hard for many of us to dismiss the personhood of a vegetative patient or a fetus. If we had a plant network and it functioned similarly, we might feel the urge to sniff the flowers on a friend's Hawaiian print shirt or water carpets that are green.

Why would such a misleading system for person representation have evolved? The answer most likely concerns the intensely social nature of our species and also perhaps the rarity of ambiguous cases of personhood in our evolutionary history. Like other social species, our individual survival depends on relating successfully to our conspecifics. And more for us than for other species, this requires understanding the immensely complex behaviors that result from their beliefs, motivations, personalities, and so on. As the anthropologist Guthrie (1995) has observed, in discussing religious belief systems, the cost of attributing intentionality to some nonintentional systems may be less than the cost of failing to adopt the intentional stance toward some systems that are intentional. In other words, it may have been adaptive to err on the side of activating the personhood network too often.

Furthermore, the personhood network is an adaptation to an earlier world, which contained fewer ambiguous cases of personhood. Sonograms did not show us our fetuses, people did not live long enough to develop Alzheimer's disease, and vegetative states were fatal. It is interesting that infants and young children may be the one class of ambiguous cases that our ancestors did encounter on a regular basis, and for these cases it would be adaptive to attribute personhood even in the absence of intelligence, self-awareness, and so forth. Protohumans who accurately judged their offspring to be lacking in the various traits associated with personhood and accordingly treated them as non-persons would not have many surviving descendants!

If our analysis is correct, it suggests that personhood is a kind of illusion. Like visual illusions, it is the result of brain mechanisms that represent the world nonveridically under certain circumstances. Also like visual illusions, it is stubborn. Take the Hermann grid illusion, for example, in which a grid of white lines on a black background seems to have ghostly gray spots at the lines' intersections (see http://www.yorku.ca/eye/hermann.htm). We know that these spots are illusory, and that they result from interactions between the antagonistic center and surround compartments of the receptive fields of visual neurons. However, this knowledge does not make the spots go away! Similarly, knowing about the person network does not eliminate the sense that moving Heider and Simmel shapes have intentions.

6.2.4 Conclusion

The result of this analysis could be considered nihilistic. It does undercut ethical systems based on personhood and in particular suggests that difficult ethical issues should not be approached with the strategy of determining whether or not the parties involved are persons. If personhood is not really in the world, then there is no fact of the matter concerning the status of a given being as a person or not, and there is no point to the philosophical or bioethical program of seeking objective criteria for personhood more generally because there are none.

Where does this leave us? The answer is different for ethics, as a discipline, and for the everyday moral behavior of individuals. For ethics, the only alternative we can see is a shift to a more utilitarian approach. Rather than ask whether someone or something is a person, we should ask how much capacity there is for enjoying the kinds of psychological traits discussed earlier (intelligence, self-awareness, and so forth) and what the consequent interests of that being are. Of course, this requires deciding how these traits should be defined and ranked in importance and whether to take into account a being's potential, or only actual, status. In other words, many similar problems arise as in discussions of criteria for personhood. However, having understood the need to set aside intuitions about personhood and having avoided the distraction of seeking criteria for personhood, we can work more productively on assessing and protecting the interests of all.

In contrast, as individuals whose behavior includes countless implicit moral decisions each day, it matters little whether personhood is illusion or reality. We cannot reprogram ourselves to stop thinking in terms of persons, nor would we want to. It is thanks to this stubborn illusion that we persist in talking to our babies, who cannot understand what we are saying, but who clearly benefit from the social and linguistic stimulation. It is thanks to the personhood network's hair trigger that we slam on the brakes at the first glimpse of a human form in the road, rather than wait until our conscious mind has arrived at the belief that there is someone there. Although the concept of personhood may be bad metaphysics, and better suited to an earlier world, even today it gets us through. In this respect we are like the guy in the joke with the brother who thinks he's a chicken. When asked why he does not take his brother to a psychiatrist to be cured, he answers "because I need the eggs."

Notes

1. Editor's note: This reading was excerpted from a substantially longer article that appeared in 2006 in the *American Journal of Bioethics*, volume 7, pages 37–48, under the title "Personhood and neuroscience: Naturalizing or nihlating?" and is used with permission. The omitted sections included reviews of the concept of personhood in philosophy and bioethics and attempts to define personhood in terms of neuroscience. We thank the members of our lab group at the Center for Cognitive Neuroscience for helpful feedback. The writing of this article was supported by R21-DA01586, R01-HD043078, R01-DA18913, and

a postdoctoral fellowship through T32-NS07413 at the Children's Hospital of Philadelphia.

References

Adolphs, R. (2003). Cognitive neuroscience of human social behaviour. *Nature Reviews Neuroscience*, *4*, 165–178.

Adolphs, R., Gosselin, F., Buchanan, T. W., Tranel, D., Schyns, P., & Damasio, A. R. (2005). A mechanism for impaired fear recognition after amygdala damage. *Nature, 433*, 68–72.

Allison, T., Puce, A., & McCarthy, G. (2000). Social perception from visual cues: role of the STS region. *Trends in Cognitive Sciences*, 7, 267–278.

Bloom, P. (2004). *Descartes' baby: How the science of child development explains what makes us human.* New York: Basic Books.

Brooks, R. A. (2002). *Flesh and machines.* New York: Pantheon Books.

Brothers, L. (1990). The social brain: A project for integrating primate behavior and neurophysiology in a new domain. *Concepts in Neuroscience*, *1*, 27–51.

Castelli, F., Frith, C. D., Happé, F., & Frith, U. (2002). Autism, Asperger syndrome and brain mechanisms for the attribution of mental states to animated shapes. *Brain, 125*, 1839–1849.

Castelli, F., Happé, F., Frith, U., & Frith, C. (2000). Movement and mind: a functional imaging study of perception and interpretation of complex intentional movement patterns. *NeuroImage*, *12*, 314–325.

Critchley, H. D., Daly, E. M., Bullmore, E. T., Williams, S. C., Van Amelsvoort, T., Robertson, D. M., et al. (2000). The functional neuroanatomy of social behaviour: Changes in cerebral blood flow when people with autistic disorder process facial expressions. *Brain*, *123*, 2203–2212.

Downing, P. E., Bray, D., Rogers, J., & Childs, C. (2004). Bodies capture attention when nothing is expected. *Cognition*, *93*, B27–38.

Downing, P. E., Jiang, Y., Shuman, M., & Kanwisher, N. (2001). A cortical area selective for visual processing of the human body. *Science*, *293*, 2470–2473.

Edelman, D. B., Baars, B. J., & Seth, A. K. (2005). Identifying hallmarks of consciousness in non-mammalian species. *Consciousness and Cognition*, *14*, 169–187.

Farah, M. J. (2004). *Visual agnosia*, 2nd edition. Cambridge, MA: MIT Press.

Farah, M. J., Levinson, K. L., & Klein, K. L. (1995). Face perception and within-category discrimination in prosopagnosia. *Neuropsychologia*, *33*, 661–674.

Farah, M. J., Rabinowitz, C., Quinn, G., & Liu, G. (2000). Early commitment of the neural substrates of face recognition. *Cognitive Neuropsychology*, *17*, 117–123.

Farah, M. J., Wilson, K. D., Drain, H. M., & Tanaka, J.R. (1995). The inverted face inversion effect in prosopagnosia: Evidence for mandatory, face-specific perceptual mechanisms. *Vision Research*, *35*, 2089–2093.

Feinberg, T. E., Schindler, R. J., Ochoa, E., Kwan, P. C., & Farah, M. J. (1994). Associative visual agnosia and alexia without prosopagnosia. *Cortex*, *30*, 395–411.

Fletcher, P. C., Happe, F., Frith, U., Baker, S. C., Dolan, R. J., Frackowiak, R. S., & Frith, C. D. (1995). Other minds in the brain: A functional imaging study of "theory of mind" in story comprehension. *Cognition*, *57*, 109–128.

Gallagher, H., Happe, F., Brunswick, N., Fletcher, P., Frith, U., & Frith, C. (2000). Reading the mind in cartoons and stories: an fMRI study of 'theory of mind.' *Neuropsychologia*, *38*, 11–21.

Gallagher, H. L., Jack, A. I., Roepstorff, A., & Frith, C. D. (2002). Imaging the attentional stance in a competitive game. *NeuroImage*, *16*, 814–821.

Goel, V., Grafman, J., Sadato, N., & Hallett, M. (1995). Modeling other minds. *Neuroreport*, *6*, 1741–1746.

Grossman, E., Donnelly, M., Price, R., Pickens, D., Morgan, V., Neighbor, G., & Blake, R. (2000). Brain areas involved in perception of biological motion. *Journal of Cognitive Neuroscience*, *12*, 711–720.

Guthrie, S. E. (1995). *Faces in the clouds: A new theory of religion*. Oxford: Oxford University Press.

Haley, K. J., & Fessler, D. M. T. (2005). Nobody's watching? Subtle cues affect generosity in an anonymous economic game. *Evolution and Human Behavior*, *26*, 245–256.

Happé, F. G. (1994). An advanced test of theory of mind: understanding of story characters' thoughts and feelings by able autistic, mentally handicapped, and normal children and adults. *Journal of Autism and Developmental Disorders*, *24*, 129–154.

Happé, F., Ehlers, S., Fletcher, P., Frith, U., Johansson, M., Gillberg, C., et al. (1996). 'Theory of mind' in the brain. Evidence from a PET scan study of Asperger syndrome. *Neuroreport*, *8*, 197–201.

Heberlein, A. S., & Adolphs, R. (2004). Impaired spontaneous anthropomorphizing despite intact perception and social knowledge. *Proceedings of the National Academy of Sciences of the United States of America*, *101*, 7487–7491.

Heider, F., & Simmel, M. (1944). An experimental study of apparent behavior. *American Journal of Psychology*, *57*, 243–259.

Johnson, M. H., Dziurawiec, S., Ellis, H., & Morton, J. (1991). Newborns' preferential tracking of face-like stimuli and its subsequent decline. *Cognition*, *40*, 1–19.

Johnson, S. C. (2003). Detecting agents. *Philosophical Transactions of the Royal Society of London Series B Biological Sciences*, *358*, 549–559.

Kanwisher, N., McDermott, J., & Chun, M. M. (1997). The fusiform face area: A module in human extrastriate cortex specialized for face perception. *Journal of Neuroscience*, *17*, 4302–4311.

Kuhlmeier, V., Wynn, K., & Bloom, P. (2003). Attribution of dispositional states by 12-month-olds. *Psychological Science*, *14*, 402–408.

Martin, A., & Weisberg, J. (2003). Neural foundations for understanding social and mechanical concepts. *Cognitive Neuropsychology*, *20*, 575–587.

McNeil, J. E., & Warrington, E. K. (1993). Prosopagnosia: A face-specific disorder. *Quarterly Journal of Experimental Psychology: Human Experimental Psychology*, *46A*, 1–10.

Mitchell, J. P., Heatherton, T. F., & Macrae, C. N. (2002). Distinct neural systems subserve person and object knowledge. *Proceedings of the National Academy of Sciences of the United States of America*, *99*, 15238–15243.

Mitchell, J. P., Neil Macrae, C., & Banaji, M. R. (2005). Forming impressions of people versus inanimate objects: Social-cognitive processing in the medial prefrontal cortex. *Neuroimage*, *26*, 251–257.

O'Reilly, R. C., & Munakata, Y. (2000). *Computational explorations in cognitive neuroscience*. Cambridge, MA: MIT Press.

Osterling, J. A., Dawson, G., & Munson, J. A. (2002). Early recognition of 1-year-old infants with autism spectrum disorder versus mental retardation. *Development and Psychopathology*, *14*, 239–251.

Peelen, M. V., & Downing, P. E. (2005). Selectivity for the human body in the fusiform gyrus. *Journal of Neurophysiology*, *93*, 603–608.

Pelphrey, K., Adolphs, R., & Morris, J. P. (2004). Neuroanatomical substrates of social cognition dysfunction in autism. *Mental Retardation and Developmenal Disabilities Research Review*, *10*, 259–271.

Phillips, M. L., Drevets, W. C., Rauch, S. L., & Lane, R. (2003). Neurobiology of emotion perception I: The neural basis of normal emotion perception. *Biological Psychiatry*, *54*, 504–514.

Piven, J. (1997). The biological basis of autism. *Current Opinion in Neurobiology*, *7*, 708–712.

Rochat, P., Morgan, R., & Carpenter, M. (1997). Young infants' sensitivity to movement information specifying social causality. *Cognitive Development*, *12*, 537–561.

Ronald, A., Happe, F., & Plomin, R. (2005). The genetic relationship between individual differences in social and nonsocial behaviours characteristic of autism. *Developmental Science*, *8*, 444–458.

Saxe, R., & Wexler, A. (2005). Making sense of another mind: the role of the right temporo-parietal junction. *Neuropsychologia*, *43*, 1391–1399.

Saxe, R., Xiao, D. K., Kovacs, G., Perrett, D. I., & Kanwisher, N. (2004). A region of right posterior superior temporal sulcus responds to observed intentional actions. *Neuropsychologia*, *42*, 1435–1446.

Scholl, B. J., & Tremoulet, P. D. (2000). Perceptual causality and animacy. *Trends in Cognitive Sciences*, *4*, 299–309.

Schultz, R. T., Gauthier, I., Klin, A., Fulbright, R. K., Anderson, A. W., Volkmar, F., et al. (2000). Abnormal ventral temporal cortical activity during face discrimination among individuals with autism and Asperger syndrome. *Archives of General Psychiatry*, *57*, 331–340.

Schultz, R. T., Grelotti, D. J., Klin, A., Kleinman, J., Van der Gaag, C., Marois, R., & Skudlarski, P. (2003). The role of the fusiform face area in social cognition: Implications for the pathobiology of autism. *Philosophical Transactions of the Royal Society of London B Biological Sciences*, *358*, 415–427.

Skuse, D., Morris, J., & Lawrence, K. (2003). The amygdala and development of the social brain. *Annals of the New York Academy of Science*, *1008*, 91–101.

Tarr, M. J., & Gauthier, I. (2000). FFA: A flexible fusiform area for subordinate-level visual processing automatized by expertise. *Nature Neuroscience*, *3*, 764–769.

Vuilleumier, P. (2000). Faces call for attention: Evidence from patients with visual extinction. *Neuropsychologia*, *38*, 693–700.

Woodward, A. L. (1998). Infants selectively encode the goal object of an actor's reach. *Cognition*, *69*, 1–34.

Wright, C. I., Martis, B., Shin, L. M., Fischer, H., & Rauch, S. L. (2002). Enhanced amygdala responses to emotional versus neutral schematic facial expressions. *Neuroreport*, *13*, 785–790.

Reading 6.3

Animal Neuroethics and the Problem of Other Minds[1]

Martha J. Farah

The "problem of other minds" is a central problem in the philosophy of mind. It refers to the difficulty of knowing whether someone or something, other than oneself, has a mind. What is the relevance of the problem of other minds to ethics and neuroscience? Its relevance to ethics rests on the relation between moral standing and capacity for mental life, particularly the capacity to suffer. If a being is capable of suffering, then it deserves protection from suffering. How and whether we can know about the mental lives of others is therefore an epistemological question with direct relevance to ethics. The relevance of this question to neuroscience rests on the potential value of neuroscience evidence for informing us about a being's mental life. In this article I will argue that, within the context of a certain class of metaphysical assumptions concerning mind–brain relations, neuroscience evidence is different from the kinds of evidence traditionally used to infer mental life and that it is in principle more informative. I will then discuss the potential implications of neuroscience evidence for animal ethics.

6.3.1 From Behavior to Mental States: The Argument from Analogy

The problem of other minds is a consequence of mind–body dualism, specifically the idea that there is no necessary relation between physical bodies and their behavior, on the one hand, and mental processes, on the other. Descartes' famous "I think therefore I am" expresses a basis for certainty concerning the existence of our own mental life. But on what basis can we infer that other people have minds? Attempts to justify our belief that other people have minds have generally rested on a kind of analogy, also discussed by Descartes (1968) and emphasized by Locke (1959) and other British empiricists such as J.S. Mill (1979). The analogy uses the known relation between physical and mental events in oneself to infer the mental events that accompany the observable physical events for someone else. For example, as shown in Analogy 1, when I stub my toe, this causes me to feel pain, which in turn causes me to say "ouch!" When I see Joe stub his toe and say "ouch," I infer by analogy that he feels pain.

Analogy 1

You stub your toe	→ You feel pain	→ You say "ouch!"
Joe stubs his toe	→ *Joe feels pain*	→ Joe says "ouch!"

The problem with this analogy is that it begs the question. Why should I assume that the same behavioral–mental relations that hold in my case also hold in Joe's? Joe could be acting and not really feel pain. He could even be a robot without thoughts or sensations at all. The assumption that analogous behavior–mental state relations hold for other people is essentially what the analogy is supposed to help us infer.

The question of whether someone is actually in pain or is acting or a robot might seem academic. After all, common sense tells us that there are no robots human-like enough to fool us, and barring very special circumstances, there is little reason to suspect anyone of acting. However, the question is not academic when applied to non-human animals whose behavioral repertoires are different from ours and who lack language.

6.3.2 Behavior and Brain Activity as Evidence of Mental Life

Brain imaging may be able to provide evidence of mental life when behavior cannot. Here I will argue that brain activity is not simply a more sensitive measure of cognitive processing than behavior but is qualitatively different from behavior in the inferences that it allows. Consider the possibility that brain activity and behavior play analogous roles. The diagram of Analogy 2 illustrates this possibility, by replacing the "ouch" behavior of Analogy 1 with activation of the anterior cingulate cortex (ACC), part of the brain's pain network.

Analogy 2

You stub your toe	→ You feel pain	→ Your ACC activates
Joe stubs his toe	→ *Joe feels pain*	→ Joe's ACC activates

The problem with Analogy 2 is that it implies that feeling pain causes ACC activation, just as it causes saying "ouch." However, the relations between mental states and behaviors are different in kind from the relations between mental states and brain states. Mental states and behaviors are contingently related. What one means by a term like *feeling pain* is not a behavior, or even a behavioral disposition. Although this possibility was explored in earnest by some behaviorist philosophers several decades ago, for example by Ryle (1949), it is not now regarded as a viable approach to the meaning of mental state terms.

For purposes of knowing mental states, behavior is like an indicator light. Indicator lights can be disabled or disconnected, or they can be turned on by other means. Their relation to the thing indicated is contingent on being hooked up a certain way. Inferences based on indicator lights and, analogously, behavior are therefore fallible. In contrast, virtually all contemporary approaches to the mind–body problem regard the relation between mental states and brain states as noncontingent.

The predominant view of the relation between mental states and brain states in cognitive neuroscience and contemporary philosophy of mind is one of identity: mental states *are* brain states. According to one version of this view, "type identity," each type of mental event is a type of physical event (Churchland, 1988; Smart, 1959). According to a weaker version, "token identity," every instance of a mental event is an instance of a physical event. The most widely accepted version of token identity is based on "functionalism," which identifies the functional role of a physical state, in mediating between the inputs and outputs of the organism, as the determinant of its corresponding mental state (Block & Fodor, 1972). Functionalism has many versions of its own, some of which blur the line between type and token mind–brain identity theory (e.g., Armstrong, 1968; Lewis, 1980).

There is an alternative to mind–brain identity based on the idea that mental states "supervene" on brain states, which avoids substance dualism yet stops short of equating mental states with brain states (Davidson, 1970; Kim, 2005). Analogy 2 is incompatible with supervenience theories as well as identity theories. This is because, despite the nonidentity of mental and brain states according to supervenience theories, the relation between the two is stronger than mere causality. According to supervenience, mind–brain relations are noncontingent. In the words of Davidson, "there cannot be two events alike in all physical respects but differing in some mental respects [and] an object cannot alter in some mental respects without altering in some physical respects."

In sum, across all these different contemporary metaphysical positions on the mind–body problem, the relationship between mental states and brain states is not contingent, as with the causal relations in Analogy 1. For type identity theories as well as functionalist theories, the ACC activation of the example is identical to a pain. For supervenience theories, the ACC activation cannot exist without there being pain. Thus it makes more sense to diagram the inferences from brain activity to mental state as in Analogy 3. The gist of this analogy is that, however sure you are of the ACC activation in Joe's brain, you can be that sure that Joe is in pain. In sum, the argument from analogy with brain activity is immune to the alternative interpretations that plague the behavioral analogy.

Analogy 3

You stub your toe	→ You feel pain Your ACC activates
Joe stubs his toe	→ *Joe feels pain* Joe's ACC activates

6.3.3 The Problem of Other, Nonhuman, Minds

Non-human animals have limited communicative abilities, and this limitation deprives us of the usual methods for learning about their mental states. Although few people today would agree with Descartes' conclusion that animals lack mental states altogether (1968), most of us feel uncertain about the extent and nature of animals' mental lives. On the one hand, many of us anthropomorphize certain animals, especially our pets, attributing complex thoughts and expectations to them on the basis of what a human in the same situation might think. On the other hand, the mental life of animals is often treated by us as hypothetical, incomparably different from our own, or even nonexistent. How else to explain our acceptance of glue traps for rodents and boiled lobster dinners?

As shown in the diagram of Analogy 4, nonhuman animals present us with a version of the problem of other minds for which the usual problematic analogy is even more problematic because of differences between human and animal behavioral repertoires. Animals cannot talk, and may not even express distress in nonverbal ways that are analogous to ours. For example, they may not vocalize at all and may freeze rather than struggle when afraid.

Analogy 4

You stub your toe	→ You feel pain	→ You say "ouch!"
Bat stubs his toe	→ *Bat feels pain*	→ Bat squeaks

Can the neuroscience approach provide traction for exploring the mental life of other species? To a degree it already has, yet according to the current analysis it could provide even more. Ethicists have previously brought physiologic data to bear on the question of animal suffering, specifically the similarities between human and animal pain systems. For example, Singer (1990, pp. 12–13) quotes at length from the writings of a pain researcher to the effect that pain processing is a lower level brain function that differs little between humans and other animals. This use of physiologic data differs in two ways from the current one.

First, according to the current analysis, physiologic data are not simply one more source of evidence about a being's mental life, to be weighed together with behavioral evidence, valuable as they might be in that role. Rather, physiologic data can play a qualitatively different and more definitive role because of their noncontingent relation to mental states, as argued in the last section. In terms of the inferences diagrammed earlier, this is the difference between Analogies 2 and 3.

The second difference results from the relatively new ability of cognitive neuroscience to parse brain processes into psychologically and ethically meaningful categories. In the current case, it has revealed the neural basis of the distinction between what could be called "mere pain" and suffering. Pains vary along many dimensions, and one dimension of particular ethical relevance is the psychological quality of the pain (Dawkins, 1985; Dennett, 1996; Hardcastle, 1999). Some pain experiences are primarily physical whereas others are psychologically distressing. The latter, characterized by Dawkins as both unpleasant and intense, warrant the term *suffering*. The neural states corresponding with pain states appear to respect this important distinction, demarcating the physical and psychologic components of pain experience by the involvement of different brain areas.

Research with animals and humans has revealed a widespread network of brain areas that become active in response to pain-inducing stimuli, including thalamic and somatosensory cortical regions as well as regions further removed from the sensory input such as the insula and anterior cingulate cortex (Porro, 2003). When the physical intensity of pain is varied, for example in an imaging experiment by having human subjects touch a painfully hot surface that varies in temperature, the level of activity throughout this network varies (Becerra et al., 2001). Taking advantage of the human ability to report their mental states (and in principle the possibility of first person research in which one introspects on one's own mental states), it is possible to vary independently the physical and psychological dimensions of pain and map the brain states that correspond with each. Morphine, for example, is known to diminish the psychological component of pain. Patients commonly report that they still feel the "physical" pain but that they are less bothered by it. The same is reported by patients whose pain is treated with hypnotic suggestion. Both interventions have their neural effects primarily in the ACC (Kupers, Faymonville, & Laureys, 2005; Lidstone & Stoessl, 2007). When people who are not being subjected to pain are empathizing intensely with someone who is, their ACCs become activated in the absence of physical pain (Singer et al., 2004). These findings indicate that ACC activation reflects suffering rather than "mere" pain.

Shriver (2006) points out that mammals have ACCs and are thus neurally equipped for psychological as well as physical pain. Following Shriver, we can substitute an animal for Joe in Analogy 4. However, because brain states can only be as similar as the brains that have them, we must amend Analogy 4 to specify a human ACC in one's own case and an animal ACC in the animal's.

Analogy 5

You stub your toe	→ You feel pain
	Your human ACC activates
Bat stubs his toe	→ *Bat feels pain*
	Bat's bat ACC activates

This raises the question of how could one determine whether mind–brain relations established with one species' brain generalize to other species. Behind this question is a more fundamental one about how degrees and types of variation in brain states correspond with degrees and types of variation in mental states, a question that will arise even within a given species because no two brains are identical. In principle, one could manipulate human brains (including one's own) to systematically vary all the different biophysical characteristics by which brains differ, in order to discover what the relevant aspects of the brain state are for determining the mental state. Of course in practice this is not even remotely possible.

At best we can suppose that similarity of psychological state will fall off as similarity of brain state falls off, without knowing which aspects of brain state similarity are relevant or how sharply the one falls off relative to the other. Edelman, Baars, and Seth (2005) provide an example of the attempt to identify functional similarities in brain architecture across species, including nonmammalian species. Shriver (2006) attempts to address the problem of generalization from human to non-human in the case of pain by citing evidence that the ACC plays a similar role in rat and human pain experience (although this evidence is admittedly based on behavior, which the current appeal to brain evidence was intended to replace): LaGraize and colleagues (2004) compared the behavior of rats with and without lesions of the ACC when forced to choose between staying in the dark, which rats generally prefer, and avoiding electric shocks to their feet. All of the rats reacted similarly when shocked, by withdrawing the shocked foot, thus indicating preserved perception of pain. However, the lesioned rats were more willing to experience the shocks for the sake of staying in a dark region of the experimental apparatus. Like patients on morphine, they appeared to be less distressed by the pain. This implies that rat ACCs play a role similar to human ACCs.

Am I suggesting that neuroscience can tell us "what it is like to be a bat?" Yes and no. When Nagel (1974) framed this question, he chose the bat as his nonhuman animal because bats use echolocation to perceive the world, a sense that humans lack. Knowing what is it like to perceive the world with a sense we lack remains a problem, even with the help of neuroscience, because the neural systems that perform echolocation in bats have no obvious homolog in the human brain. However, given that we do share the same general pain physiology with bats, including an ACC, we can know certain things about what it's like to be one of them. Specifically, we can infer that to be a bat with an injured toe is more like being a human with an injured toe and no pain relief than it is like being a human with an injured toe who has been given morphine.

The problem of animal minds has not thus far figured prominently in the field of neuroethics. One reason may be that neuroethics is young and has yet to engage all of the subject matter that will eventually comprise the field. Another reason may be that the personal and political rancor associated with animal ethics has discouraged scholars from approaching this topic. Given the real-world importance of animal ethics, and the special role that neuroscience evidence can play in this endeavor, the study of animal neuroethics would seem to have great promise.

6.3.4 Assumptions and Conclusions

The idea that neuroscience can reveal ethically relevant information about non-human animals rests on a number of assumptions. One assumption that has not been examined in this article is that our ethical obligations toward a being depend at least in part on the mental life of that being. Although this assumption hardly needs defending, there is much more that could be said about which specific aspects of mental life have which specific ethical implications. Perhaps the most important further clarification concerns those aspects of mental life that obligate us to prevent suffering and those that obligate us to protect life.

The current article has focused on the question of whether another being has the capacity for relatively simple mental states, those with some consciously experienced content and affective valence. This mental capacity has more limited ethical implications than the mental capacity to conceive of oneself and one's life and have an explicit preference to continue living (Levy, 2008; Singer, 1999). The neuroscience evidence discussed so far pertains only to the capacity of animals to experience the former kind of mental state, and the relevant ethical implications are therefore confined to preventing suffering rather than protecting life. However, this is not an in-principle limitation of neuroscience data. Given the appropriate research program, there is no reason why we could not identify the neural systems, and states thereof, corresponding with the self-concept and the desire to continue living. This knowledge would have implications for many aspects of end-of-life decision making and might obligate us to refrain from killing certain animals.

Another assumption that deserves explicit discussion concerns the relation between cognitive processing, of the kind that cognitive neuroscientists correlate with brain activation in imaging experiments, and consciousness. This is an important assumption in the current context because our ethical concern is with conscious mental life, and conscious suffering in particular, rather than with unconscious information processing. Based on most of the views of mind–brain relations reviewed earlier, certain types or instances of neural processing are identical to, or are necessarily associated with, certain mental states, including conscious mental states. Therefore the problem is one of determining empirically which brain states correspond with which conscious mental states. This is not a trivial problem, but it is in principle solvable. Indeed, if one is willing to accept other normal humans' reports of conscious experience as evidence, we are on our way to solving it in practice. (Skeptics unwilling to accept others' reports of conscious experience would have to be scanned themselves, which could be done to verify specific findings but would not be feasible as a means of verifying all cognitive neuroscience knowledge.)

A final assumption concerns the accuracy and completeness of cognitive neuroscience. For purposes of exploring the in-principle prospects and limitations of neuroscience evidence as a solution to the problem of other minds, I have written as if we know the brain states associated with specific mental states. Unfortunately, this is not true. Although cognitive neuroscience has made tremendous progress in the past few decades, the current state of our knowledge is far from complete. For many mental states, including suffering, we have good working hypotheses about the brain regions that are relevant, but future research will undoubtedly call for the revision of some of these hypotheses. In addition, we know little about the specific mechanisms by which these regions implement the relevant mental states. "Activation" observed in brain imaging studies is closely related to neural activity measured at the single cell level, but does not map perfectly onto a specific aspect of neuronal behavior such as action potentials (Logothetis & Wandell, 2004). Furthermore, any single measure of brain activity, be it single cell or aggregate, electrical or chemical, will omit potentially important features of neuronal function. It is possible that activation as measured by our current methods is not diagnostic of the relevant neuronal activity and that under some circumstances it will be misleading. Knowing more about the specific computations performed by neurons in the brain regions implicated by brain imaging, including their interactions with neurons in other regions, will be particularly important as we attempt to evaluate cross-species homologies.

Notes

1. Editor's note: This reading was excerpted from an article that appeared in 2008 in the journal *Neuroethics,* volume 1, pages 9–18, under the title "Neuroethics and the problem of other minds: Implications of neuroscience for the moral status of vegetative patients and nonhuman animals," and is used with permission. The omitted section concerned the problem of other minds in relation to vegetative and minimally conscious patients. I thank Liz Camp, Neil Levy, Adrienne Martin, and Susan Schneider for their helpful comments on an earlier draft of the paper. The writing of this paper was supported by NIH grants R21-DA01586, R01-HD043078, R01-DA14129, and R01-DA18913 and by the MacArthur Project on Neuroscience and the Law.

References

Armstrong, D. M. (1968). *A materialistic theory of the mind.* London, UK: RKP.

Becerra, L. R., Breiter, H. C., Wise, R., Gonzalez, R. G., & Borsook, D. (2001). Reward circuitry activation by noxious thermal stimuli. *Neuron, 32,* 927–946.

Block, N., & Fodor, J. (1972). What psychological states are not. *Philosophical Review, 81*, 159–81.

Churchland, P. M. (1988). *Matter and consciousness*, 2nd edition. Cambridge, MA: MIT Press.

Davidson, D. (1970). Mental events. Reprinted in *Essays on actions and events*. Oxford: Clarendon Press.

Dawkins, M. S. (1985). The scientific basis for assessing suffering in animals. In P. Singer (Ed.), *In defense of animals* (pp. 27–40). New York: Basil Blackwell.

Dennett, D. C. (1996). *Kinds of minds: Toward an understanding of consciousness*. New York: Basic Books.

Descartes, R. (1968). *Discourse on method and the meditations* (trans. F. E. Sutcliffe). New York: Penguin, 1968.

Edelman, D. B, Baars, B. J., & Seth, A. K. (2005). Identifying hallmarks of consciousness in non-mammalian species, *Consciousness and Cognition, 14*, 169–187.

Hardcastle, V. G. (1999). *The myth of pain*. Cambridge, MA: MIT Press.

Kim, J. (2005). *Physicalism, or something near enough*. Princeton, NJ: Princeton University Press.

Kupers, R., Faymonville, M.E., & Laureys, S. (2005). The cognitive modulation of pain: Hypnosis- and placebo-induced analgesia. *Progress in Brain Research, 150*, 251–269.

LaGraize, S. C., LaBuda, C. J., Rutledge, M. A., Jackson, R. L., & Fuchs, P. N. (2004). Differential effect of anterior cingulate cortex lesion on mechanical hyperalgesia and escape/avoidance behavior in an animal model of neuropathic pain. *Experimental Neurology, 188*, 139–148.

Levy, N. (2008). *Cognition and Consciousness* (in press).

Lewis, D. (1980). Mad pain and Martian pain. In N. Block (Ed.), *Readings in the philosophy of psychology* (Vol. 1, pp. 216–222). Cambridge, MA: Harvard University Press.

Lidstone, S.C., & Stoessl, A.J. (2007). Understanding the placebo effect: Contributions from neuroimaging. *Molecular Imaging & Biology, 9*, 176–185.

Locke, J. (1959). *Essay concerning human understanding* (A.C. Fraser, Ed.). New York: Dover.

Logothetis, N.K., & Wandell, B.A. (2004). Interpreting the BOLD signal. *Annual Review of Physiology, 66*, 735–769.

Mill, J. S. (1979). Chapter XII: The psychological theory of the belief in matter, how far applicable to mind. In J. M. Robson (Ed.), *The collected works of John Stuart Mill, Volume IX—An examination of William Hamilton's philosophy and of the principal philosophical questions discussed in his writings*. (Introduction by A. Ryan). Toronto: University of Toronto Press. London: Routledge and Kegan Paul. Retrieved December 13, 2007, from http://oll.libertyfund.org/title/240/40871.

Nagel, T. (1974). What is it like to be a bat? *Philosophical Review, 83*, 435–450.

Porro, C. A. (2003). Functional imaging and pain: behavior, perception, and modulation. *Neuroscientist, 9*, 354–969.

Ryle, G. (1949). *The concept of mind.* Chicago: University of Chicago Press.

Shriver, A. (2006). Minding mammals. *Philosophical Psychology, 19*, 443–442.

Singer, P. (1990). *Animal liberation*, revised edition. New York: Random House.

Singer, P. (1999). *Practical ethics*, 2nd edition. Cambridge, UK: Cambridge University Press.

Singer, T., Seymour, B., O'Doherty, J., Kaube, H., Dolan, R. J., & Frith, C. D. (2004). Empathy for pain involves the affective but not sensory components of pain. *Science, 303*, 1157–1162.

Smart, J. J. C. (1959). Sensations and brain processes. *Philosophical Review, 68*, 141–156.

Reading 6.4

Digital People: Making Them and Using Them[1]

Sidney Perkowitz

Robots have played an increasingly prominent role in manufacturing for the past 50 years, and about a million industrial units are in use today worldwide. In a parallel development, the number of humans with bionic units, such as cochlear implants and artificial limbs, is also growing. The presence of these "digital people," a category that includes artificial and partly artificial beings, from mechatronic (mechanical plus electronic) robots to humans with bionic (biological plus electronic) implants, is rapidly increasing in industry and in society as a whole. Digital people represent a new technology that deserves serious attention.

The 1920s play *R. U. R.* (Rossum's Universal Robots) by the Czech author Karel Capek introduced the word *robot*, which comes from the Czech word *robota*, meaning forced labor. Capek foresaw the widespread use of robots, as he painted a picture of humanoid units made purely to serve humanity. At least two rudimentary humanoid mechatronic robots were built in the 1920s and 1930s (Elektro, a unit designed by the Westinghouse Corporation, was a hit at the 1939 New York World's Fair); however, true commercial use of robots began with the invention of a non-humanoid type of industrial robot in 1954. Now, with advances in mechatronics, materials science, artificial intelligence, and other relevant areas, increasingly capable units, some of them humanoid, are becoming available for use in industry, homes, and hostile environments.

The related area of bionics is potentially even more important. The origins of this technology are scattered and diffuse, harking back centuries, and even millennia, to crude prosthetic replacements for missing limbs, such as wooden legs. Bionic technology is now producing sophisticated artificial limbs and bodily organs, as well as devices that connect directly to the neural system or the brain; for example, cochlear implants that restore hearing to the deaf. Bionic additions such as these promise to address important issues for the injured and ill and perhaps someday to enhance human capabilities.

These and other potential results of bionic and robotic technology, however, such as the displacement of human workers, raise complex ethical issues that require careful consideration. Although these technologies are at the beginning stages of development, it is not too early to survey the state of the art and its implications.

6.4.1 Robots for Industry and the Home

The year 2004 marked the fiftieth anniversary of the patenting of the first industrial robot by George Devol, an engineer. With his partner Joseph Engelberger, Devol began making and selling a one-armed, programmable unit called the UNIMATE. Engelberger envisioned robotic devices as "help[ing] the factory operator in a way that can be compared to business machines as an aid to the office worker." General Motors (GM) bought its first UNIMATE in 1961, but for a variety of economic and societal reasons, the Japanese were the first to widely use robots in automobile factories. In 1978, however, GM installed an assembly line using a PUMA (programmable universal machine for assembly), and robots began to appear in U.S. industry in substantial numbers.

A typical industrial robot is fixed in position and consists principally of a powerful multijointed mechanical arm that is nearly as flexible as a human arm and that can be programmed to carry out intricate manipulations of components large and small. Table 6.4.1 shows that Japan still

Table 6.4.1
World population of industrial robots for selected years (in thousands of units)

	1984	1988	2001	2002	2003[a]	2006[b]
Japan	67	176	360	350	344	333
U.S.	13	33	97	104	111	135
Europe	17	53	231	244	259	316
Asia/Australia[b]	0.7	2	57	60	64	73
Total	100	266	757	772	838	875

[a] Projected
[b] Excluding Japan
Note: Some totals incorporate values for other countries, including USSR/Russia. If estimated uncounted units are included, the total for 2002 and later is thought to exceed 1,000,000 units.
Sources: NRC, 1996; UNECE, 2003.

leads in the use of these robotic workers, that roughly 1 million such units are now operating worldwide, and that global use is increasing. This growth can be expected to continue, if only because of the economic imperative—as the costs of human workers are increasing, those of robotic workers are falling. Robots already form some 10% of the workforce in the Japanese, Italian, and German automobile industries, illustrating the potential for robotic labor to supplant humans, with consequent disruptions, especially for older workers.

Assembly-line robots will continue to play an important industrial role, and indications are that ongoing technical advances will also produce robots suitable for nonindustrial applications in the home and in dangerous and demanding environments. These advances include new artificial physical, sensory, and mental capabilities, as illustrated by three particular units designed and built in the past few years: ASIMO (advanced step in innovative mobility), a child-size humanoid robot from the Honda Corporation; QRIO, a 2-foot-tall humanoid robot created by the Sony Corporation; and KISMET, a robotic humanoid head and face designed and built by robot engineer Cynthia Breazeal at MIT. Together these three units display a range of physical abilities that also draw on artificial sensory capabilities, such as vision, walking, climbing stairs, adjusting gait for different surfaces, avoiding obstacles while walking, recovering from a fall, dancing, carrying objects, and responding to humans by shaking hands, showing facial expressions, and waving goodbye.

These abilities also require a degree of intelligent behavior, defined as behavior that helps an organism survive and thrive by providing effective responses to changing circumstances. According to Harvard psychologist Howard Gardner, this adaptive property in humans encompasses seven different types of intelligence: logical-mathematical, linguistic, musical, bodily-kinesthetic, spatial, interpersonal, and intrapersonal (Gardner, 1999). The latter (intrapersonal) touches on the perplexing issue of whether an artificial brain can be truly conscious of itself as we humans are, a question that is unlikely to be answered in the near future. That question aside, ASIMO, QRIO, and KISMET clearly display the low-level rudiments of intelligent behavior. For instance, they can memorize and recognize human names and faces, and even hold limited conversations.

With advances in mobility and physical versatility, sensing abilities, and intelligence, robots are becoming suitable for home and office use, although many are not humanoid. The most popular examples are robotic pets, such as the artificial dog AIBO made by Sony. Designed purely for entertainment, AIBO was introduced in 1999 and quickly

Table 6.4.2
World population of nonindustrial robots for selected years (in thousands of units)

	2002	2006[b]
Household (vacuum, lawn)	50	500
Professional services[a]	19	49
Entertainment	545	1,500
Total	625	2,200
Value (billions of $)	3.6	6.0

[a] These include surgical devices, units for underwater exploration, surveillance, and hazardous duty, units that assist disabled people, etc.
[b] Projected
Source: UNECE, 2003.

enjoyed brisk sales. The AIBO dogs display sufficient intelligence and manipulative ability to be formed into soccer teams, which roboticists use to study how groups of robots interact. A more practical example is Roomba, a robotic vacuum cleaner that uses intelligent decision making to avoid furniture as it vacuums every square inch of a floor, without human guidance.

The humanoid robots ASIMO and QRIO are not yet on sale for general use, but ASIMO can be rented from Sony as a robot receptionist and guide and has been designed to interact with humans in future applications; and QRIO has clear possibilities for entertainment. Table 6.4.2 shows the recent spectacular growth in these and other nonindustrial applications, with the number of units already far outstripping the number of industrial robots. With other developments on the horizon, such as intelligent manipulation of objects by robot hands and fingers (driven by research at the National Aeronautics and Space Administration and by the development of surgical robots for delicate medical procedures) and high-speed object recognition and obstacle avoidance (exemplified by projects sponsored by the Defense Advanced Research Projects Agency [DARPA] and the Daimler Chrysler Corporation), a multitude of new applications may be expected to develop, including many for the military, such as self-guided, intelligent weapons.

6.4.2 Bionic Humans

As robots are becoming more natural by taking on human physical and mental characteristics, humans are becoming more artificial. At present, bionic technology is less well developed than robotic technology, and there seems to be no compilation of worldwide activity in bionics comparable with the summaries for robotics. However, according to one recent estimate from the National Institutes of Health, 8% to 10% of the U.S. population—that is, about 25 million people—has artificial parts, from breast implants to coronary stents to prosthetic limbs to cochlear units, suggesting a substantial economic impact. A list of recent highlights in the development of bionic additions, including the growing area of neural (or brain–machine) interfaces, indicates some current and future possibilities for this technology:

- More than 30,000 cochlear implants are in use, including some placed in deaf children as young as 1 to 2 years of age.
- Several research laboratories and corporations are pursuing the implantation of electronic devices in the retina or the cortex of the brain to restore sight to the blind.
- DARPA is funding research in brain–machine interfaces for direct mental control of military aircraft, powered exoskeletons that give soldiers increased strength and mobility, and other applications.
- Living monkey brains have been used to operate robotic arms, and a sea-lamprey brain has been used to operate a wheeled robotic body. The aim is to make artificial limbs and other devices for humans that function under direct mental control.
- At least one research group has made it possible for a fully paralyzed person to operate an external device, a computer, by mental control alone.
- Researchers are developing the "neuron on a chip," that is, a living neuron grown on a standard electronic computer chip so that the neuron and the chip can directly exchange electrical signals and hence information.

The potential to relieve a variety of human ills and injuries is clear, and the science-fictionish aspiration of actually improving human physical, mental, and even emotional capabilities by artificial additions may be attainable someday.

6.4.3 Ethical and Societal Issues

Despite these potential benefits, robotic and bionic technologies also have troubling aspects. In robotics, replacement of expensive human workers by cheaper robots may loom large in the automobile industry and other applications, such as using intelligent robots as caregivers for the ill and elderly. The latter application raises another fundamental question: Do we really want a society where human needs are met by machines, not people?

For bionic humans, ethical issues arise from the use of neural connections and brain–machine interfaces, centered around the question of what it means to be human. Certainly, a person who has a natural limb replaced with an artificial one has not become less human or lost a significant degree of "personhood." But suppose a majority of organs in an injured person is replaced by artificial components; or, suppose the artificial additions change mental capacity, memory, or personality. Is such a heavily artificial person somehow less than human? Would the established legal, medical, and ethical meanings of personhood, identity, and so on, have to be altered?

We are only in the early stages of understanding brain–machine interfaces and do not grasp all of the potential side effects. Neural implants have been shown to change the brain through its plasticity, that is, the innate ability of neurons to reform the connections among themselves to record new knowledge as the brain learns. Although this could be beneficial—for instance, by enabling a person to incorporate an artificial limb into his or her overall body image—we do not yet know if all such changes would be desirable.

Another, more subtle issue is suggested by some experiences with cochlear implants, which usually restore only partial hearing. Most implantees welcome even this incomplete restoration, but some find themselves uncomfortably suspended in a gray area between two cultures—that of the fully deaf and that of the fully hearing. Hence, psychological and even spiritual factors may prove to be barriers to the development of bionic technology. Finally, the process of surgical implantation can raise medical issues, such as infection and rejection by the body or poorly understood side effects. For instance, in 2003, U.S. government agencies issued a warning that young children with cochlear implants might be at increased risk for meningitis.

For both robotic and bionic technology, projected uses in warfare raise a host of issues. Would self-guided weapons violate the Geneva Conventions? Would a heavy dependence on robotic or bionic military units lead to the perception that wars can be fought at minimal human cost, a potentially destabilizing factor in international affairs? These and other issues have already been discussed at one major conference, the International Symposium on Roboethics, held in 2004 to address the ethical, social, humanitarian, and ecological questions raised by robotic technology (Roboethics, 2004).

6.4.4 Conclusion

Although it will be years before we understand the ultimate technological limits for robots and bionic implants, we can already draw some conclusions: The population of established types of industrial robots has grown eightfold or more in the past 20 years and will have an increasing impact on manufacturing industries. The latest extensions in robotic capabilities offer new opportunities for household and entertainment robots, with enormous growth projected in the near future, and may offer new manufacturing uses. A potentially huge impact will come from medical uses of robots and human–machine hybrids. These include surgical and caretaking units and the replacement, and even enhancement, of human physical and mental abilities. We must begin considering the many ethical challenges and societal changes that would accompany the widespread introduction of robotic and bionic technology.

Any new technology can have both positive and negative outcomes for society. Because robotics and bionics involve simulating, altering, and perhaps even changing the essential nature of humans, they have a special significance. Only the best efforts of everyone involved, from technological and medical experts to political decision makers and ordinary citizens, can ensure that these rapidly evolving areas will bring more good than harm.

Notes

1. Editor's note: This reading originally appeared in 2005 in *Bridge*, volume 35, pages 21–25, and is used with permission.

References

Gardner, H. (1999). *Intelligence reframed*. New York: Basic Books.

NRC (National Research Council). (1996). Approaches to robotics in the United States and Japan: Report of a bilateral exchange. Washington, DC: National Academy Press. Available at http://books.nap.edu/cataloq/9511.html.

Roboethics. (2004). The ethics, social, humanitarian, and ecological aspects of robotics. First International Symposium on Roboethics, Sanremo, Italy, January 30–31, 2004. Available at http://www.scuoladirobotica.it/roboethics/.

UNECE (United Nations Economic Commission for Europe). World robotics 2003: Statistics, market analysis, forecasts, case studies and profitability of robot investment. Geneva, Switzerland: United Nations Publications. Available at http://www.unece.org/press/pr2003/03stat p01e.pdf.

Reading 6.5

From Neurons to Politics—Without a Soul[1]

Nancey Murphy

My lecture this afternoon is an attempt to take the topic that is near and dear to my heart these days—neuroscience and philosophy of mind—and apply it to neuroethics. But rather than go directly from brain science to ethics, I plan a roundabout route from neuroscience to some reflections on Christians' political involvements.

I'll begin with the thesis that recent developments in neuroscience are making it more and more difficult to be an intellectually fulfilled anthropological dualist. In fact, I join a large majority of current philosophers in adopting a "nonreductive physicalist" account of the person. The prevalence of physicalism in the academic world calls on Christians to reevaluate centuries of biblical interpretation and theology. Despite being in conflict with much of the tradition, I claim that Christian theology can and should incorporate a physicalist anthropology. Then I shall be a bit speculative regarding the differences that a physicalist theology would have made in Christian attitudes toward politics throughout Western history. In brief, I speculate that if there had been no such thing as souls to save, Christians would have had to find something else to worry about, and maybe, just maybe, they would have concerned themselves more with Jesus' teaching about the real, and present, and realizable kingdom of God on Earth.

6.5.1 Human Nature in the Bible

It is certainly the case that most Christians throughout most of Christian history have been dualists of some sort, and have seen dualism (or a more elaborate tripartite account involving body, soul, and spirit) as the teaching of the Bible. One could say that contemporary neuroscience, with its discoveries of neural mechanisms underlying human psychology, calls Christians to reevaluate their thinking. It has certainly brought this issue into public view. What most Christians in the pews do not know, though, is that dualism has been questioned by Christian scholars for over a century. Beginning over a hundred years ago, biblical scholars came to recognize that the Hebrew Scriptures have been badly translated. The Septuagint is a Greek translation of the Hebrew scriptures, probably dating from around 250 BCE. This text translated Hebrew anthropological terms into Greek, and it then contained words that could be understood in the way those terms were defined in Greek philosophy. The clearest instance of this is the Hebrew word *nephesh,* which was translated as *psyche* in the Septuagint and later translated into English as "soul." It is now widely agreed that *nephesh* did not mean what later Christians have meant by "soul." In most cases, it is simply a way of referring to the whole living person. For example, a passage in Genesis that dualists have often relied on read in the old King James translation: "And the Lord God formed man of the dust of the ground, and breathed into his nostrils the breath of life; and man became a living soul." The updated Revised Standard Version reads: "the Lord God formed man of the dust from the ground, and breathed into his nostrils the breath of life; and man became a living *being*."

In the half of the Christian scholarly world that we might designate as liberal, there was a wide consensus by the middle of the twentieth century that interpretations of New Testament teaching had also been distorted by reading Greek philosophical conceptions back into them. However, this is still being debated among more conservative scholars. And here it is puzzling why the disputes cannot be easily settled. New Testament scholar Joel Green (e.g., 2008) points to differences of interpretation being due to different readings of non-Canonical books from the intertestamental period—particularly regarding the question of the "intermediate state": does the New Testament teach that there will be a period of conscious existence between death and bodily resurrection? If so, this would seem to require that we have souls to fill in that bodily gap. This leads me to ask: do Christians really need to work through a long list of non-Canonical books in order to determine what the Bible teaches on this issue? The unlikelihood of an affirmative answer leads me to this conclusion: The New Testament authors are not intending to teach *anything* about humans' metaphysical composition. If they were, surely they could have done so much more clearly!

Helpful support for this conclusion comes from New Testament scholar James Dunn. Dunn distinguishes between what he calls "aspective" and "partitive" accounts of human nature. Dunn writes:

... in simplified terms, while Greek thought tended to regard the human being as made up of distinct parts, Hebraic thought saw the human being more as a whole person existing on different dimensions. As we might say, it was more characteristically Greek to conceive of the human person "partitively," whereas it was more characteristically Hebrew to conceive of the human person "aspectively." That is to say, we speak of a school *having* a gym (the gym is part of the school); but we say I *am* a Scot (my Scottishness is an aspect of my whole being) (Dunn, 1998).

So the Greek philosophers were interested in the question: What are the essential parts that make up a human being? In contrast, for the biblical authors each "part" ("part" in scare quotes) stands for the whole person thought of from a certain angle. For example, "spirit" stands for the whole person in relation to God. What the New Testament authors are concerned with, then, is human beings in relationship to the natural world, to the community, and to God. Paul's distinction between spirit and flesh is not our later distinction between soul and body. Paul is concerned with two ways of living: one in conformity with the Spirit of God, and the other in conformity to the old aeon before Christ.

So I conclude that there is no such thing as *the* biblical view of human nature *insofar as we are interested in a partitive account.* The biblical authors, especially the New Testament authors, wrote within the context of a wide variety of views, probably as diverse as in our own day, but did not take a clear stand on one theory or another. What the New Testament authors *do* attest is, first, that humans are psychophysical unities; second, that Christian hope for eternal life is staked on bodily resurrection rather than an immortal soul; and, third, that humans are to be understood in terms of their relationships—relationships to the community and especially to God.

I believe we can conclude, further, that this leaves contemporary Christians free to choose among several options. It would be very bold of me to say that dualism *per se* is ruled out, given that it has been so prominent in the tradition. However, the radical dualisms of Plato and Descartes, which take the body to be unnecessary for, or even a hindrance to, full human life, are clearly out of bounds. Equally unacceptable is any physicalist account that denies human ability to be in relationship with God. Thus, reductionist forms of physicalism are also out of bounds.

6.5.2 Physicalism and Theology

I turn now to the question of what difference a physicalist anthropology might make to theology. When I first got interested in this topic, it seemed to me that all a physicalist anthropology strictly requires is one or two adjustments: Most important, one needs to understand resurrection differently: not re-clothing of a "naked" soul with a (new) body, but rather restoring the whole person to life—a new transformed kind of life.

But the more I think about it, the more I think that the whole of Christian thought and practice might have gone differently if dualism had not crept into Christian teaching. There is no way to know for sure, but I want to present a "just so" story about the influences of dualism, and some speculations about how it might have been different, and how the world today, in consequence, might be a radically different place. Some of you who are Christians are going to feel that I'm caricaturing your faith, and I don't mean to do that. Think of me as developing so-called ideal types, radical versions of two possible forms of Christianity, that in actuality are usually mixed rather than pure. These types are derived from two different conceptions of human nature, the Hebraic, which has been lost, and the Hellenistic, inherited particularly from Augustine, writing in the early 400s.

Old Testament Scholar Aubrey Johnson (1961) emphasizes one important aspect of the Hebraic conception of personhood, which may be contrasted with modern individualism. For moderns, individuals are thought to be "self-contained" in two senses: The first is that they are what they are apart from their relationships. The second is the idea that the real self—the soul or mind or ego—is somehow contained within the body. In contrast, Johnson argues, the Hebraic personality was thought to be extended in subtle ways throughout the community by means of speech and other forms of communication. This extension of personality is so strong within a household that in its entirety it is regarded as a "psychical whole." "Accordingly, an Israelite thought the individual, as a [*nephesh*] or centre of power capable of indefinite extension, is never a mere isolated unit. . . ." (Johnson, 1961).

In contrast, Augustine made a conceptual innovation that has led to the most extreme form of individualism. From Augustine to the present we have had a conception of the self that distinguishes the inner life from the outer, and spirituality has been associated largely with the inner (Cary, 2000). The combination of the Neoplatonic emphasis on the care of the soul with Augustine's metaphor of entering into one's own self or soul in order to find God constituted a complex of ideas that has shaped the whole of Western spirituality from that point onward.

For example, when the human person is identified with a solitary mind, God tends to be conceived as a *disembodied* mind, as in the case of so-called classical theism. The most common criticism of this conception of God, influenced by Greek metaphysics, is that God is taken to be impassible, that is, not affected by relations with anything outside of himself. Although all traditional accounts of God emphasize that God is personal, it is difficult to imagine what it could mean to have a personal relationship with such a God.

In addition, with the development of the Newtonian worldview as a closed causal system, it became difficult to imagine how God could have an effect on the created order without violating the laws he himself had ordained. The typical strategy of liberal Christian theologians was to deny that God engages in any specific acts in nature or history. Rather, as Gordon Kaufman says, if we are to understand the phrase "act of God" we should use it only to designate the master act in which God gives the whole world its structure and history its direction. So there are two contrasting conceptions of God, associated with physicalist and dualist anthropologies respectively: Christians have left behind a concept of God as present and active in the particularities of history in favor of one neither affected by nor affecting particular events.

The influence of these contrasting anthropologies reaches even further. For the dualist type of Christian, what Christianity is basically all about is repenting of sin, believing in Christ, being forgiven, so that at death one's soul will depart the body and ascend to Heaven to enjoy the presence of God. At the end of time, dead bodies will be raised and reunited with their souls, all will face judgment. The saved will remain in the company of God and the damned will be consigned to permanent suffering in Hell. There are many subtle variants on this story, especially regarding how things will go at the end, but the central core is the idea that souls will be saved for heaven if their sins are forgiven.

There is a competing theme regarding the end of history that actually predominates in Christian Scripture. This is the theme of the Kingdom of God. This was central to Jesus' preaching. He claimed that the Kingdom had come near in him. It has already arrived to the extent that people live according to his teachings. Its fulfillment will be a complete transformation of heaven *and earth*. The transformation involved can be glimpsed only from the hazy and somewhat contradictory stories of Jesus after his resurrection. We await resurrection, and the whole of creation awaits transformation.

Ted Peters whimsically describes the dualist account of salvation as "soul-ectomy." If souls are saved *out of* this world, then nothing here matters ultimately. If instead it is our bodily selves that are saved and transformed, then bodies and all that go with them matter—families, history, and all of nature.

Jewish scholar Neil Gillman lends weight to my suggestion. His book, titled *The Death of Death*, argues that resurrection of the body, rather than immortality of the soul, is the only authentically Jewish conception of life after death. Why are physicalism and resurrection important to Jews? For many reasons, Gillman replies:

> Because the notion of immortality tends to deny the reality of death, of God's power to take my life and to restore it; because the doctrine of immortality implies that my body is less precious, important, even "pure," while resurrection affirms that my body is no less God's creation and is both necessary and good; because the notion of a bodiless soul runs counter to my experience of myself and others.... (Gillman, 1997, p. 238.)
>
> It is indispensable for another reason. If my body inserts me into history and society, then the affirmation of bodily resurrection is also an affirmation of history and society. If my bodily existence is insignificant, then so are history and society. To affirm that God has the power to reconstitute me in my bodily existence is to affirm that God also cares deeply about history and society. (Gillman, 1997, p. 262.)

6.5.3 Physicalism, Dualism, and History

I am suggesting that original Christianity is better understood in sociopolitical terms than in terms of what is currently thought of as religious or metaphysical. The adoption of a dualist anthropology provided something different—different from sociopolitical and ethical concerns—with which Christians became primarily preoccupied.

I am suggesting, further, that adoption of a physicalist understanding of human nature provides a critical opportunity to evaluate and reject many of the ways Christians have sought to encounter God in the privacy of their closets, rather than in the messy world of poverty-fighting, healing, justice-seeking, and peacemaking. Even more critically, I'm asking how different the world might be now if they (we) had been doing so all along. If Christians had been focusing more, throughout all of these centuries, on following Jesus' teachings about sharing, and about loving our enemies at least enough so as not to kill them, how different might world politics be today? What *would* Christians have been doing these past 2000 years if there were no such things as souls to save?

Notes

1. Editor's note: This reading is based on a talk delivered at the University of Pennsylvania in December 2006 and was abridged with the approval of the author. A fuller treatment of the issues discussed here can be found in the author's

2006 book, *Bodies and Souls, or Spirited Bodies?*, published by Cambridge University Press.

References

Cary, P. (2000). *Augustine's invention of the inner self: The legacy of a Christian Platonist.* Oxford: Oxford University Press.

Dunn, J. D. G. (1998). *The theology of Paul the Apostle.* Grand Rapids, MI: Eerdmans.

Gillman, N. (1997). *The death of death: Resurrection and immortality in Jewish thought.* Woodstock, VT: Jewish Lights Publishing.

Green, J. B. (2008). *Body, soul, and human life: The nature of humanity in the Bible.* Grand Rapids, MI: Baker.

Johnson, A. R. (1961). *The one and the many in the Israelite conception of God.* Cardiff: University of Wales Press.

Contributors

Zenab Amin Senior Medical Writer, MedKnowledge Group, Portland, OR

Ofek Bar-Ilan Stanford Center for Biomedical Ethics, Palo Alto, CA

Richard G. Boire Richard Glen Boire Law Firm, Davis, CA

Philip Campbell Editor-in-Chief, *Nature*, London, UK

Turhan Canli Department of Psychology, Stony Brook University, Stony Brook, NY

Jonathan Cohen Department of Psychology, Center for the Study of Brain, Mind, and Behavior, Princeton University, Princeton, NJ

Robert Cook-Deegan Center for Genome Ethics Law and Policy, Department of Public Policy Studies, Duke University, Durham, NC

Lawrence H. Diller Department of Pediatrics, University of California San Francisco Medical School, San Francisco, CA

Carl Elliott Center for Bioethics, Department of Philosophy, University of Minnesota, Minneapolis, MN

Martha J. Farah Center for Neuroscience & Society and Center for Cognitive Neuroscience, University of Pennsylvania, Philadelphia, PA

Rod Flower Department of Biochemical Pharmacology, The William Harvey Research Institute, London, UK

Kenneth R. Foster Center for Neuroscience & Society, Department of Bioengineering, University of Pennsylvania, Philadelphia, PA

Howard Gardner Graduate School of Education, Harvard University, Cambridge, MA

Michael Gazzaniga Sage Center for the Study of Mind, University of California, Santa Barbara, Santa Barbara, CA

Jeremy R. Gray Psychology Department, Yale University, New Haven, CT

Henry Greely Stanford Law School, Stanford, CA

Joshua Greene Department of Psychology, Harvard University, Cambridge, MA

John Harris Institute for Science Ethics and Innovation, Wellcome Strategic Programme in the Human Body, Its Scope, Limits and Future, University of Manchester, Manchester, UK

Andrea S. Heberlein Department of Psychology, Harvard University, Cambridge, MA

Steven E. Hyman Office of the Provost, Harvard University, Cambridge, MA

Judy Illes Department of Neurology, National Core for Neuroethics, University of British Columbia, Vancouver, BC, Canada

Eric Kandel Center for Neurobiology and Behavior, Columbia University, New York, NY

Ronald C. Kessler Harvard Medical School, Department of Health Care Policy, Boston, MA

Patricia King Georgetown University Law Center, Washington, DC

Adam J. Kolber University of San Diego School of Law, San Diego, CA

Peter D. Kramer Brown Medical School, Providence, RI

Daniel D. Langleben Department of Psychiatry, Center for Neuroscience & Society, University of Pennsylvania, Philadelphia, PA

Steven Laureys Cyclotron Research Centre and Neurology Department, Université de Liège, Liège, Belgium

Stephen J. Morse Center for Neuroscience & Society, University of Pennsylvania School of Law, Philadelphia, PA

Nancey Murphy Fuller Theological Seminary, Pasadena, CA

Eric Parens The Hastings Center, Garrison, NY

Sidney Perkowitz Department of Physics, Emory University, Atlanta, GA

Elizabeth A. Phelps Department of Psychology, New York University, New York, NY

Eric Racine Neuroethics Research Unit, Institut de recherches cliniques de Montréal, Department of Medicine and Department of Social and Preventive Medicine, University of Montreal, Montreal, Quebec, Canada

Barbara Sahakian Department of Psychiatry, University of Cambridge, Cambridge, UK

Laura A. Thomas Center for Cognitive Neuroscience, Duke University, Durham, NC

Paul M. Thompson Laboratory of Neuroimaging, Department of Neurology, University of California, Los Angeles, Los Angeles, CA

Stacey A. Tovino Drake University Law School, Des Moines, IA

Paul Root Wolpe Center for Ethics, Emory University, Atlanta, GA

Name Index

Subject Index

Basic Bioethics

Arthur Caplan, editor

Peter A. Ubel, *Pricing Life: Why It's Time for Health Care Rationing*

Mark G. Kuczewski and Ronald Polansky, eds., *Bioethics: Ancient Themes in Contemporary Issues*

Suzanne Holland, Karen Lebacqz, and Laurie Zoloth, eds., *The Human Embryonic Stem Cell Debate: Science, Ethics, and Public Policy*

Gita Sen, Asha George, and Piroska Östlin, eds., *Engendering International Health: The Challenge of Equity*

Carolyn McLeod, *Self-Trust and Reproductive Autonomy*

Lenny Moss, *What Genes* Can't *Do*

Jonathan D. Moreno, ed., *In the Wake of Terror: Medicine and Morality in a Time of Crisis*

Glenn McGee, ed., *Pragmatic Bioethics,* 2nd edition

Timothy F. Murphy, *Case Studies in Biomedical Research Ethics*

Mark A. Rothstein, ed., *Genetics and Life Insurance: Medical Underwriting and Social Policy*

Kenneth A. Richman, *Ethics and the Metaphysics of Medicine: Reflections on Health and Beneficence*

David Lazer, ed., *DNA and the Criminal Justice System: The Technology of Justice*

Harold W. Baillie and Timothy K. Casey, eds., *Is Human Nature Obsolete? Genetics, Bioengineering, and the Future of the Human Condition*

Robert H. Blank and Janna C. Merrick, eds., *End-of-Life Decision Making: A Cross-National Study*

Norman L. Cantor, *Making Medical Decisions for the Profoundly Mentally Disabled*

Margrit Shildrick and Roxanne Mykitiuk, eds., *Ethics of the Body: Post-Conventional Challenges*

Alfred I. Tauber, *Patient Autonomy and the Ethics of Responsibility*

David H. Brendel, *Healing Psychiatry: Bridging the Science/Humanism Divide*

Jonathan Baron, *Against Bioethics*

Michael L. Gross, *Bioethics and Armed Conflict: Moral Dilemmas of Medicine and War*

Karen F. Greif and Jon F. Merz, *Current Controversies in the Biological Sciences: Case Studies of Policy Challenges from New Technologies*

Deborah Blizzard, *Looking Within: A Sociocultural Examination of Fetoscopy*

Ronald Cole-Turner, ed., *Design and Destiny: Jewish and Christian Perspectives on Human Germline Modification*

Holly Fernandez Lynch, *Conflicts of Conscience in Health Care: An Institutional Compromise*

Mark A. Bedau and Emily C. Parke, eds., *The Ethics of Protocells: Moral and Social Implications of Creating Life in the Laboratory*

Jonathan D. Moreno and Sam Berger, eds., *Progress in Bioethics: Science, Policy, and Politics*

Eric Racine, *Pragmatic Neuroethics: Improving Understanding and Treatment of the Mind-Brain*

Martha J. Farah, ed., *Neuroethics: An Introduction with Readings*